Heinz Hampel
Automatic Wristwatches
from Switzerland

Heinz Hampel

AUTOMATIC Wristwatches From Switzerland

Self-winding
Wristwatches

Schiffer Publishing Ltd
77 Lower Valley Road, Atglen, PA 19310

Library of Congress Catalog Number: 94-65607

Originally published by Verlag Georg D.W. Callwey GmbH & Co., München, 1992

Translated by
Dr. Edward Force, Central Connecticut State University.

Jacket Design: Baur + Belli Design, München.

Printed in the United States of America
ISBN: 0-88740-609-2

We are interested in hearing from authors
with book ideas on related topics.

Published by Schiffer Publishing Ltd.
77 Lower Valley Road
Atglen, PA 19310
Please write for a free catalog.
This book may be purchased from the publisher.
Please include $2.95 postage.
Try your bookstore first.

Contents

Foreword

Since the beginning of the 1980s, the mechanical wristwatch has displaced the previously dominant pocket watch in popularity among watch collectors. The collectability of these utilitarian articles, which are being worn again in growing numbers, is directly connected to the flooding of the market with quartz watches. This also includes a greater appreciation of the superb technology, widely unappreciated until then, of their micromechanical features which are reflected very impressively in the development of the automatic watch.

While the hand-wound wristwatch basically represented merely a smaller version of the pocket watch, the automatic watch can be seen as an independent, new creation. Its development was managed almost exclusively by the Swiss watch industry, which devoted a period of more than fifty years to the perfection of this successful export. Limiting this study to Swiss automatic-watch production thus does not involve any essential limitation of the thematic variety. Even today it is difficult to discern if a watch is automatic, since the automatic watch differs externally from the wind-up watches only in the lettering on the dial. Aside from the dial, the automatic wristwatch can be identified in only two ways: there is no need to wind it daily, and the self-winding mechanism is the dominant component of the movement when the watch is opened.

Judging by the vast variety offered at watch sales and auctions, collecting automatic watches ranks along with that of the ever-popular chronographs as another area that can be fascinating to the technically interested aficionado, and also offers the advantage of a greater wealth of variety.

In this book, some 200 automatic watches will be portrayed in three photos each. This selection represents nearly 85% of all the different movement designs produced in Switzerland, and the remaining types are included in the tables.

Each of the manufacturers' chapters begins with an inclusive summation of the firm's caliber development. A special feature here is the background information, on such topics as the cooperation of various watchmaking firms, etc. In addition, all firm calibers are listed in tables, and the illustrated types are described in detail.

The photographs offers not only a look at the earlier automatic watches, but also an overview of developmental history in the years between 1929 and 1978. In the appendix the technology of the self-winding mechanism and that of the actual movement will be explained, as will important technical concepts and terms.

Thus the collector will be given the opportunity to classify his own watch in terms of time and technology, either through visual comparison with the illustration or by locating the applicable caliber designation in the tables.

Since most of the watch manufacturers portrayed here no longer exist, it was not always possible to obtain representative information. Therefore technical data is often missing or incomplete, which, unfortunately, applies to the dates on which numerous calibers were put on the market. Even determining the manufacturers of foreign movements that have been used by various manufacturers with their own identifying marks could not always be achieved.

At this point I would like to thank all those who have generously provided watches, photos or other materials, and to assure them that this book would have remained incomplete without their support.

They are, in alphabetical order:
Audemars Piguet, Le Brassus; Dr. A. L. Benz, Bern; Volkhard Cremer, Berlin; Wolfgang Hirschfeld, Berlin; Jaeger LeCoultre, Le Sentier; Hans Kocher, Biel; Prince Kretzulesco, Berlin; Joachim Ortel, Berlin; Piaget, Co^*te-aux-Fées; Frederic Piguet, Le Brassus; Edmund Pusl, Munich; Axel Schneider, Munich; Vacheron Constantin, Geneva; Peter Winkler, Berlin; and Christian Pfeiffer-Belli, the editor-in-chief at Callwey-Verlag. And I must not forget my wife Norma and my children, whose patience while I was writing this book was often overly strained—therefore my special thanks go to them as well.

History of the Automatic Wristwatch

Since the wristwatch was commonly used at the beginning of World War I, the time was also right at the beginning of the 1920s for the idea of a self-winding watch.

The concept of attaining "perpetual motion" by building an automatic winding apparatus into a pocket watch was first achieved by Abraham-Louis Perrelet, Sr. (1729-1826) in 1770. But subsequent efforts, some of them by renowned watchmakers, to improve the running reliability of the automatic winding came to grief. This was mainly because the watch, kept in a pocket close to the body, was moved only to a minimal extent. But since shocks had a negative influence on a watch's running precision, even the watchmakers themselves sought to limit the popularization of the wristwatch. They were aided by early problems in manufacturing the small movements, which often had to be improved by them later.

But the very position of the watch on the wrist, which was chosen in the first place only for ease of reading, offered an exceptional advantage for self-winding, namely that of guaranteed motion.

In 1924 the English watchmaker John Harwood applied in Switzerland to patent a wristwatch with automatic hammer winding. This was a winding principle that was developed further and made ready for series production in collaboration with the Swiss firms of Anton Schild S.A. and Fortis S.A. in Grenchen, and first put on the market in 1929.

Although Léon Leroy of Paris had already built a small series of self-winding watches for Sir David Salomons in 1922, Harwood is regarded as the actual inventor of the "automatic." He provided a concept that was aimed at the start for a later series production and so his efforts became the starting point of the further development of the watch as a consumer product.

While the subsequent designs made at that time, with their self-winding mechanisms that look primitive from a present-day standpoint were characterized by the rectangular form that was stylish in those days, the Rolex "Perpetual" of 1931, developed by Hans Wilsdorf, stands out as an exception. Wilsdorf may well have had Perrelet's invention in mind when he created the first rotor automatic to go into series production and set it in a round case.

A.-L. Perrelet (1729-1826) was the inventor of automatic winding in the portable clock.

John Harwood (1881-1960) initiated the industrially produced automatic wristwatch.

Hans Wilsdorf (1881-1960), director of the Rolex firm, developed the first automatic wristwatch with rotor winding.

The round case offers very definite advantages over the other shapes, including better sealing against water and dust, optimal space utilization with the shape of the movement determined by the rotor, simpler manufacturing for mass production, and also a more easily readable dial.

The combination of effective self-winding, a precision movement and a waterproof case in the Rolex "Oyster Perpetual" can be seen from a present-day standpoint as the design that set the standard by which all subsequent developments were measured. At this point, the automatic wristwatch was recognized as the optimal portable clock.

The precision of the movement is improved by the constant delivery of energy from the mainspring, which is constantly kept wound with the help of the effective winding rotor to an extent that no comparable hand-wound watch

The course of driving power in hand-wound and automatic watches: while in hand-wound watches the power decreases strongly during the night (left), the power of the automatic watch, constantly wound during the day is clearly more even, which has a favorable effect on running regularity, especially at night (right).

can attain. Exposed to constant motion by being worn on the wrist all day, the watch needs no special handling; it does not need to be wound by hand, because its up-down indicator extends far beyond its nightly resting phase and, enclosed in a case impervious to water and dust, guarantees trouble-free performance for years.

Naturally, all the qualities of its day embodied in the Oyster Perpetual" could not yet be housed in an elegant case. But many of the other automatics developed parallel to it, with their relatively flat rectangular cases, even appealed to the ladies.

The many new designs of the years around 1930 can be classified according to their type of winding for the sake of simplicity. In addition to the hammer winding that first appeared in the "Harwood" and the rotor used in the Rolex "Perpetual", the so-called "rolling" winding of the "Rolls" and the "Wig-Wag", the lever winding of the "Autorist" and the Wyler-Automatic, and the pendulum winding of the "Perpetual" of the Frey & Co. firm and the Bulova Automatic were also used.

The last three inventions led to a dead end in technical development, even though the case and movement designs of these early automatics are fascinating and technically interesting.

Thus it is all the more astonishing that after the Rolex development the pendulum winding was still used in the 1940s by the Buren firm and perfected with a swinging mass sunk in the movement and working in both directions. Something similar is true of the rolling apparatus of the Pierce Automatic, with a sliding weight mounted on rails.

Although the Rolex firm seemed to have pointed the way toward the further development of the automatic watch with its "Perpetual", no other manufacturer used the rotor in the 1930s, rather favoring the hammer winding introduced by Harwood. The reasons for this decision included the more stable mounting of the hammer's swinging mass and the flatter and simpler construction of the apparatus. In addition, in 1931 the firm of E. Meylan of La Chaux-de-Fonds introduced a movement design in which the self-winding component was a separate entity that could be removed from the actual movement without taking out screws, thus making this type of winding especially easy to repair.

There were many reasons that led to more and more complaints from automatic watch owners. What with the dearth of movement in modern everyday activity, complaints piled up from automatic watch owners whose watches had stopped for reasons they could not understand. Some designers tried to improve winding performance by using the other direction of the hammer's movement, as for example in the Revue Automatic made by the Thommen firm of Waldenburg in 1948; others tried to make the functioning condition of their watches clearer by putting a running reserve, or up-down indicator, on the dial, first done in 1948 with the "Powermatic." This was used in great quantities by the firm of Jaeger LeCoultre, and the "Autographic" of the Zodiac firm. It must be noted, though, that the Breguet firm had already sold a platinum watch to Sir Percival David in 1933, a one-off creation that was equipped with a up-down indicator such as was well known from marine chronometers.

Under pressure from consumers and watchmakers in the 1940s, all the automatic watch manufacturers were ready to reintroduce hand-winding, which Harwood had viewed as superfluous. For until then, without exception, only hand-setting could be done by using the crown, which made it necessary to spend time shaking the watch to start it after a long illness or lengthy repairs. But in 1955 Jaeger LeCoultre was still praising its "Futurematic" as the "first fully automatic" watch, calling attention to its lack of hand-winding. In this case, though, the hammer automatic system worked very effectively in both directions, and for the first time since Perrelet (1770), the swinging mass was blocked to protect the mainspring from overwinding. The process was regulated via the up-down indicator.

After 1931, the year the Rolex Perpetual was introduced, only a few developmental departments did any work on the further development of this type of winding. The leader among them was the Felsa raw movement plant of Ebauches S.A. in Grenchen, which first utilized both directions of the rotor for winding in its "Bidynator" caliber, developed for production in 1942. This movement, built to be produced by the most modern technology, made it possible for many entrepreneurs to offer reasonably priced and reliably operating automatic watches and thus compete with the established hammer automatics.

The essential further development of the automatic now was no longer carried on by individual inventors or small firms, but consisted of intensive, systematic developmental work in the development departments, particularly those of the big raw-movement producers of Ebauches S.A., such as AS, Eta and Felsa.

Their efforts were now directed at important individual improvements. The EtáEterna firm, for example, was developing a miniature ball bearing for the rotor mounting of the 1948 ladies' caliber and a low-friction switching gear to utilize both turning directions. With the help of new transmission routes in the actual movement, first offered by the Zenith firm in their hammer automatic in 1948 and known in the trade as "direct" center seconds and minute wheel mounted out of the center, and at the same time by the Eta firm, the use of especially thin geared wheels made of beryllium (1956), not only increased the transmission capability of the gears but also decreased their construction height considerably.

The eccentric winding first offered by the Longines firm in 1945 offers an alternative to the "bidynator" winding. Here too, the rotor winds in both directions, and the automatic apparatus can be made simple and robust. The automatic design of the IWC firm, introduced in 1950, likewise functions with this type of winding gear, though in improved form.

By the beginning of the 1950s, the price range of the automatic watch was about double that of a hand-wound watch of comparable quality. The Swiss manufacturers, of which Jaeger LeCoultre, Longines and Rolex merit particular mention, exported chiefly to the American market, since there was not sufficient buying power in postwar Europe.

Practical production methods, such as the assembly line and automatic assembly, brought the prices down significantly in the mid-1950s and, along with the onset of economic recovery, resulted in the automatic watch being bought by greater numbers as an everyday watch.

In its essential details, the automatic can be regarded as fully developed by the end of the 1950s. Even in terms of details, almost all the alternative solutions had been worked out, as shown by the many patents for click wheels and releases. But progress in the realm of the electric wristwatch, as the "Accutron" of 1959 emphatically attests, put a certain pressure on the automatic manufacturers, forcing them on to further efforts that essentially extend into three areas of development. They concern the improvement of running regularity, the reduction of movement height and the combination of the automatic with additional functions such as the alarm or chronograph.

Thus by doubling the swinging rate, or beats per second, of the balance, which was increased in stages from 18,000 to 36,000 BPH over the years, the running regularity of these fast-beat watches was improved to the extent that even medium-priced watches were sold with chronometer certificates. An example is the "Gyromatic-HF-Chronometer" introduced by the Girard Perregaux firm in 1966. Successful participation in the official chronometer competitions at the observatories of Geneva and Neuenburg thus no longer indicated any proof of outstanding performance that a manufacturer could utilize in advertising.

The automatic manufacturers also made extraordinary efforts to reduce the height of their calibers.

Only the most noteworthy designs will be noted chronologically below, with the Zenith hammer automatic, measuring 4.9 mm, cited as the flattest watch of the 1940s as a basis of comparison.

In 1957 the Buren firm introduced its "Super-Slender" watch with a small "planetary" rotor set into the movement, at the same time as an almost identical development, the "Microrotor" caliber of the Universal firm. While these movements, with center seconds, measured 4.2 and 4.1 mm respectively, Piaget introduced its flat Caliber 12 P in 1959, measuring 2.3 mm and likewise equipped with a small rotor, but now made of gold. It held the record for two decades.

In the same year, the raw movement producer FHF of Fontainemelon took a completely different course with its "Fontomatic" caliber, putting an axle-less rotor running on a ring of ball bearings around the edge and using the available interior space to house the automatic drive.

The firm of Eterna also spoke of a record in 1962, when their "Eterna-Matic 3000", equipped with a strikingly large rotor bearing and including date indication and center seconds, measured only 3.6 mm in height.

A caliber with the same additional features, but reserved for Audemars Piguet and Vacheron & Constantin, was made by Jaeger LeCoultre in 1967 and measured only 3.4 mm—and only 2.45 mm without the special features. Its characteristic component is a gold roller with peripheral ruby rollers.

The limits of what could actually be built were reached by a development of Bouchet-Lassale S.A. in 1978 with its Caliber 2000, whose height measured only 2.08 mm (without second indication!). Only with the development of special miniature bearings was it at all possible to dispense without the hitherto customary second plate to hold the wheel arbors securely and mount all the wheels on only one side, in "flying" fashion, with only 18 ball bearings needed per movement. This superior achievement of micromechanical high technology could be achieved only through intensive cooperation by a variety of Swiss specialist firms, and can be regarded as the last great achievement of the Swiss watch industry in view of quartz-watch technology.

Thus after the thickness of the caliber had been reduced by half in the greatest variety of drive types, the next step was to install automatic winding in hitherto massive-looking alarm watches and wrist chronographs. It was not by chance that three such watches were introduced in 1969.

For the first time, the Zenith firm was able to combine the normal chronograph mechanism with a fast beat escapement and automatic winding in "El Primero" without letting the height, including date indication, go beyond 6.5 mm.

If this height could only be attained with a complex movement construction, the "Chronomatic", a competing development by the cooperating firms of Breitling, Buren and Hamilton as well as Heuer-Leonidas, indicated the course of technology with its so-called sandwich construction. This consisted of a stack of two or more complete components, also known as modules. The "Chronomatic" consists of a further development of the Buren firm's planetary rotor module and a newly developed chronograph module of the Dubois-Depraz firm. The movement, measuring 7.7 mm, obviously looked rather thick.

The third newly introduced model was the "Memomatic" of the Omega firm, an alarm wristwatch in which, for the first time, the striking apparatus was wound automatically along with the movement and the waking time could be set precisely to the minute. In the first automatic alarm watch, the "Memovox" by Jaeger LeCoultre, made in 1956, the alarm still had to be wound separately by hand.

Also worthy of mention are the developments of individual Swiss manufacturers who, in this era already affected by competition from the quartz watch, tested new types of apparatus with the goal of simplifying the automatic mechanism, making it more resistant to wear and friction, or improving its transmission characteristics in particular.

The "Zenitissimo", an automatic caliber for ladies' watches built by the Zenith firm in 1967, is an interesting example. Here the rotor, moving around the barrel, gives its energy directly to the mainspring via a tiny planetary equalization drive. The drive spring moves only when the rotor stops or delivers too little energy.

The Longines firm took a different path with its double-barrel caliber, introduced in 1975. The two barrels are made to function one after the other, and what with the increase in their number of turns, can deliver driving energy to the movement more evenly.

On the other hand, the two barrels of the "Duomatic", introduced by Favre-Leuba in 1968, work parallel, or simultaneously, driving an intermediate wheel.

From 1975 on, the automatic scarcely had a chance of surviving against the overwhelming competition of the reasonably priced and also more precise quartz watch. Many watch companies had to close, and their workers had to look for new lines of work.

When the demand for the mechanical wristwatch, linked with an individual image, increased again in the 1980s, only a few firms were in any condition to reissue old factory stocks.

New developments concern almost exclusively the realm of added complications, such as the so-called perpetual calendar, the chronograph, and the striking movement. The power source of these watches, made as they are by the "sandwich" method, is usually a purchased automatic movement.

An exception is the raw-movement producing firm of Frederic Piguet S.A. in Le Brassus, which has recently developed extraordinarily innovative high-quality automatic calibers.

Early Automatic Watches

The early automatic watches include all the developments of the first experimental phase, from 1929 to 1935, with numerous designs that can be classified by their types of winding:
Hammer winding: Harwood, Glycine
Roller winding: Rolls, Bucherer-Bulova, Wig-Wag
Pendulum winding: Perpetual, Mars-Bulova
Lever winding: Wyler, Autorist/Universe
Rotor winding: Rolex (see special chapter on firm)

The Harwood Automatic

The idea for this first automatic watch to go into series production came to John Harwood, the son of a British watchmaker, while serving at the front as a soldier in World War I. Under the extreme conditions of position warfare in trenches, his watch stopped frequently. If the cause was not penetrating dust or water, as it so often was, then it was forgetting to wind the watch by hand. Harwood became inspired—he had to develop a self-winding wristwatch, and one which would be better protected from dust and moisture by eliminating the winding crown. A turning bezel would be used for setting the hands. The problem of sealing the watch could not be solved any better.

The winding mass was a hammer that was mounted on both sides of the movement but wound the watch only in one direction as it would swing. Its impact was dampened by spring buffers. The friction clutch on the hammer, provided protection against the overwinding of the mainspring.

With his partner Harry Cutts, Harwood worked until 1923 to perfect this design and applied for a Swiss patent in the same year. The patent, numbered 106583, was issued in 1924. On his second visit to Switzerland, he showed the finished prototype of his automatic watch to, among others, Ernst Schild. As the director of the raw-movement factory of Anton Schild S.A. (AS), Schild was in charge of the preparations for production, which lasted until 1929.

The raw movement produced by AS was finished by the Fortis firm, also located in Grenchen. Contact with Fortis director Walter Vogt was made easily, for Ernst Schild was also an active partner of Fortis S.A. At the Basel Fair, the Harwood Automatic was first displayed at the Fortis stand in 1929.

By the terms of the contract, Harwood had retained the right to issue production licenses to other interested parties as well. Thus the Swiss firm of Blancpain produced watches with this caliber for the French market, and so did the Selza Watch firm at a later date. In the USA it was also sold by the Perpetual Selfwinding Watch Company.

In the end, there were several reasons for the economic failure of this first automatic watch. They included the relatively complicated design, which was expensive to produce; the meager running reserve of only 12 hours without the possibility of being wound by a winding stem; and the worldwide economic crisis that peaked in 1931.

The Rolls Automatic

The firm of Léon Hatot S.A. of Paris licensed the Swiss manufacturer Blancpain of Villeret to produce its roller-automatic watch from 1930 to 1934. In this design the entire movement glided back and forth over balls in a frame, whereby winding took place in only one direction. Blows to the case were absorbed by leaf springs. The entire mechanism was housed in a smooth rectangular hinged case that had to be opened to reach the hidden panel on the edge by which the hands were set.

Roller automatics according to the Hatot patent were built in three calibers. A disadvantage was the limited degree of effectiveness of the self-winding mechanism with no possibility of being wound by a winding stem, and the irritating movement of the dial (3 mm).

The Glycine Automatic

The firm of E. Meylan S.A. (E.M.S.A.) in La Chaux-de-Fonds obtained patents on a hammer automatic design in 1930 and 1931. This was a complete automatic unit, designated Caliber 35, which was formed as a ring with a U-shaped profile and could be installed in a previously developed basic movement measuring 8.75 lignes. For this purpose, EMSA usually utilized its Caliber 20 with eccentric or Caliber 43 with center seconds. Although both basic movements were prepared for the installation of hand-winding, they were usually delivered without it; the crown thus served only for hand-setting. The two components could be attached to each other without screws and were thus extremely well suited for servicing.

The movement of the hammer, which wound in one direction, was limited by spring buffers, such as were used in the "Harwood." When the watch was fully wound, damage to the balance was prevented by the barrel being set up, which simultaneously kept the mainspring wound to some extent so the watch would start after just a few arm movements.

Since the hammer movement was not limited when the watch was fully wound, a wheel formed as a spring coupling served to prevent excessive stress on the movement.

The Wig-Wag Automatic

Under the two brand names of "Mars" and "Aster", the Biel-based firm of La Champagne S.A. Louis Müller & Cie. introduced a roller automatic known by the eye-catching name of "Wig-Wag". The Swiss patent for it was granted on March 11, 1931 and numbered 153875.

The entire movement served as a swinging mass, mounted on two levers that swung out to the sides, and swung back and forth in a case frame, using only one direction for winding.

The basic movement was a complete round hand-wound caliber, which was installed from the front in the rolling weight mass that also held the automatic drive. The winding and setting crown lay hidden beneath the rear couvette that was hinged upward.

The Wyler Automatic

The firm of Wyler S.A. in La Chaux-de-Fonds patented its self-winding design in Switzerland in 1931, under number 158881. This was a lever-winding system in which a pull on the band moved the projecting, spring-mounted double bottom of the case and thus wound the mainspring via a pin in the bottom and an apparatus inside the case. The winding took place not so much from swinging arm movements as from changes in the shape of the wrist. To maintain a sufficient running reserve, the watch had to be pressed against the wrist several times. The rim panel on the bottom of the case served only for hand-setting.

The Autorist Automatic

Like the first hammer automatic, this automatic watch with lever winding was also inspired by John Harwood, developed in cooperation with the raw-movement factory of A. Schild S.A. and the firm of Fortis S.A., and introduced in 1931.

Here the motive power coming from the band was transmitted into the watch by a moveable lower attachment and wound the mainspring with the help of an intricate system of levers. A sprung lever served to prevent over-winding. To be able to utilize the slightest lever movement, the clutch device was built as a smooth, toothless disc that was moved forward with the help of wedge clicks.

A prerequisite for safe winding was a non-elastic and firmly attached band, but such a band offers the wearer no comfort. The crown at the side was used for only hand-setting.

This lever automatic was produced in two sizes: under caliber number 796 as "Autorist" or "Universe" on the larger AS Caliber 624 formed movement, with the hand-winding mechanism omitted, and as a ladies' model of the "Universe" on the AS Caliber 475 formed movement.

The Bucherer Automatic

In 1933 the Swiss Patent Office registered patent no. 169089, another type of roller winding, for the Bulova Watch Company. The Bucherer automatic watch was built according to this principle, but it has never been determined who produced it (Bucherer is a Swiss watch and jewelry business)

In this design, the movement worked together with the case as a swinging mass and moved back and forth on parallelogram levers over a metal frame to which the band was also attached. Via the upper of the two lever axles, the rolling movement was transmitted into the inside of the case and utilized for winding via switching click and wheel.

A particular disadvantage of this automatic was that when it was worn, its movement was hindered by the wearer's sleeve cuff, so that a sufficient running reserve often did not build up. Since hand-winding via the crown was not possible, the watch often stopped. To set the hands, the upper part of the case had to be folded up to reach the hidden control panel.

The Perpetual Automatic

This automatic watch, sold by the Perpetual Selfwinding Watch Co. of New York, was produced by the firm of Frey & Co. in Biel around 1935 and functioned with a pendulum winding working in one direction. The basic movement was a small formed movement without a hand-winding mechanism, surrounded by the automatic component in the form of a frame. The winding click, mounted on the pendulum on a spring, provided protection against over-winding by a clutch device. The pendulum's swing was not limited by buffers, but by pins, and a loud clicking was heard while winding. In some models, the hand-setting crown was located on the left side by the numeral 9.

The Aster Automatic

This was a pendulum automatic watch developed by Bulova and was patented in Switzerland in 1933 under number 166843. The firm of La Champagne Ls. Muller & Cie., likewise located in Biel, produced the watch under its "Aster" brand name as Caliber 28-20 and for Bulova as Caliber 13 AP.

A plate-like swinging mass swung over a specially developed formed movement into which the automatic mechanism was integrated, and wound in one direction with the help of an eccentric click. The hand-winding mechanism was naturally disconnected and thus limited to some extent the friction losses in automatic operation.

Characteristic of this watch was the extra dial for the eccentric seconds hand under the 12.

Harwood

The first worldwide production automatic watch, based on Harwood's work, made circa 1930 by AS and Fortis and supplied to the Selfwinding Watch Co. of New York.

Chromed metal case with stainless steel snap-on bottom. Ribbed turning bezel for hand-setting. The red dot in the dial aperture above the 6 shows that the external setting mechanism is uncoupled from the hands and the movement is not braked. Hand winding is not available. Hammer swinging mass mounted on both sides of the movement, with buffer springs, winding in one direction via a hammer click on the centrally mounted winding wheel, overwinding protection in the form of a sliding coupling on the hammer.

Cutout bimetallic screw balance, flat hairspring, no shock resistance.

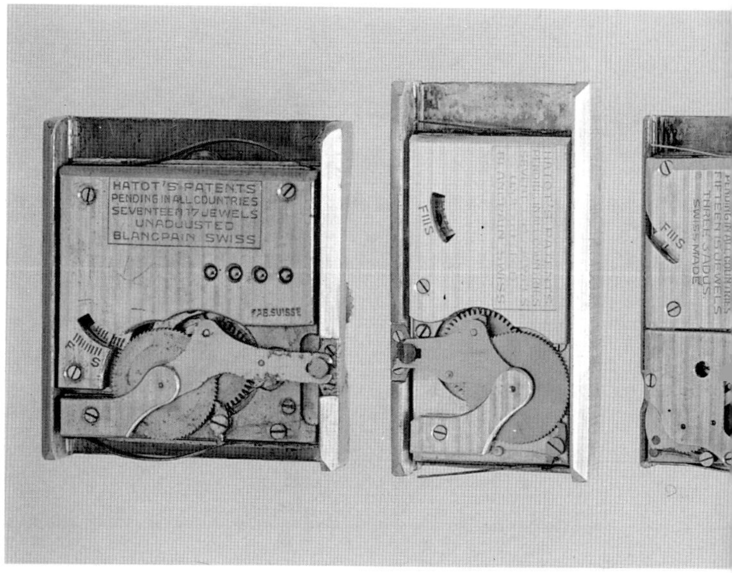

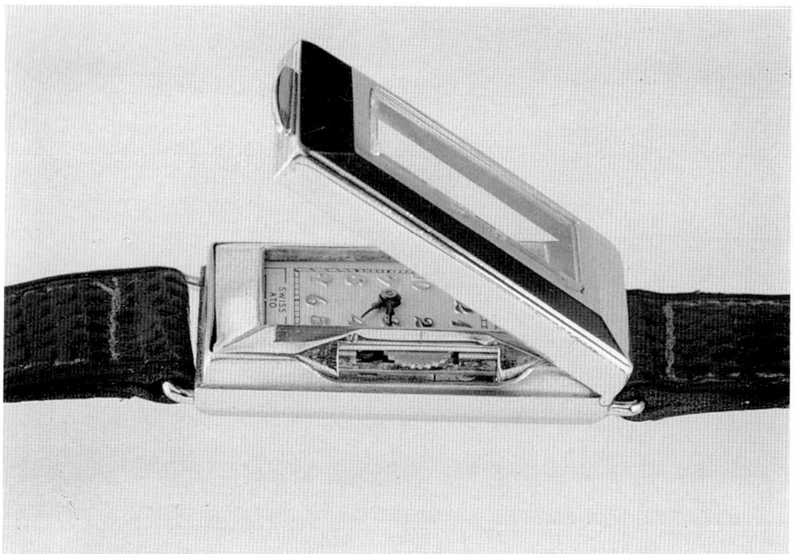

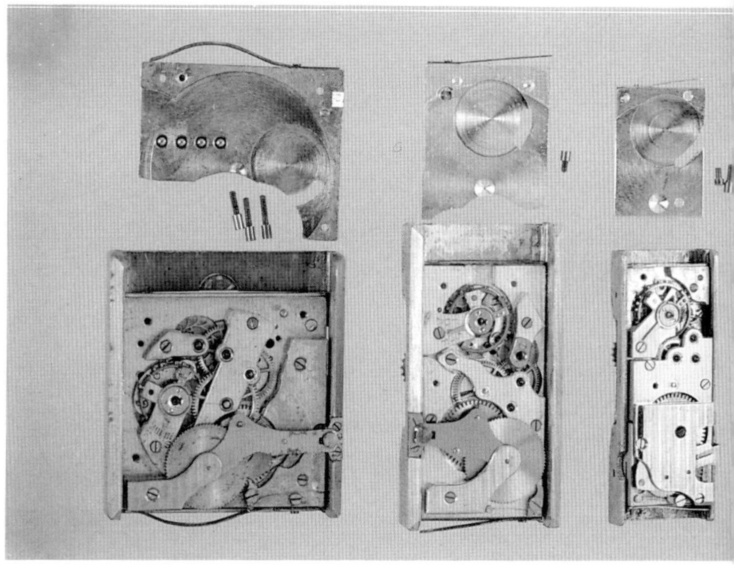

Rolls

The "Rolls" roller automatic made by Blancpain between 1930 and 1934, after a Hatot patent, in men's, medium and ladies' models. Chromed metal case with hinged bezel and hidden rim disc (only for hand-setting).
The whole movement including added weight serves as a swinging mass and slides straight back and forth on four balls in a U-shaped frame (4 extra balls later pressed into the added weight of the men's Rolls). Winding in one direction.
Cutout bimetallic screw balance, flat hairspring, no shock resistance.

Opposite Page: Glycine

Medium-sized Glycine automatic watch, circa 1935.
Independent automatic unit in the form of a ring-shaped module (Caliber 35).
Stainless steel case with screw bottom. Rosédial. Crown only for hand-setting. Hammer with buffer springs (in U-shaped outer ring), winding in one direction by a drive of the Harwood type.
Nickel screw balance, flat hairspring, Shock-Resist shock resistance.

GLYCINE
BIENNE GENÈVE

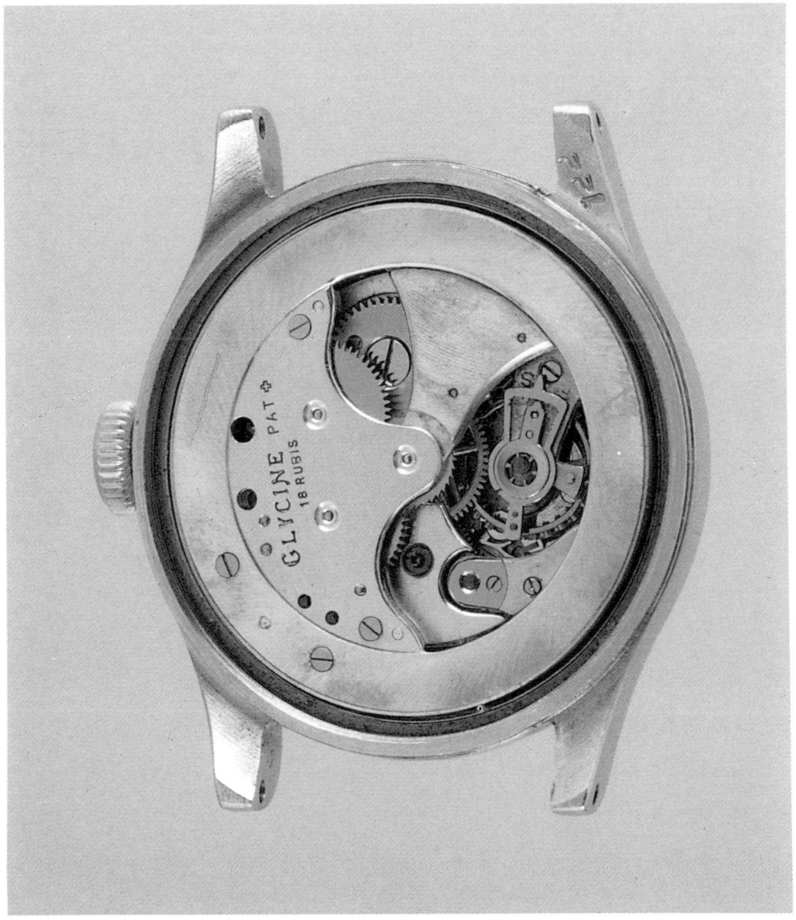
GLYCINE PAT
18 RUBIS

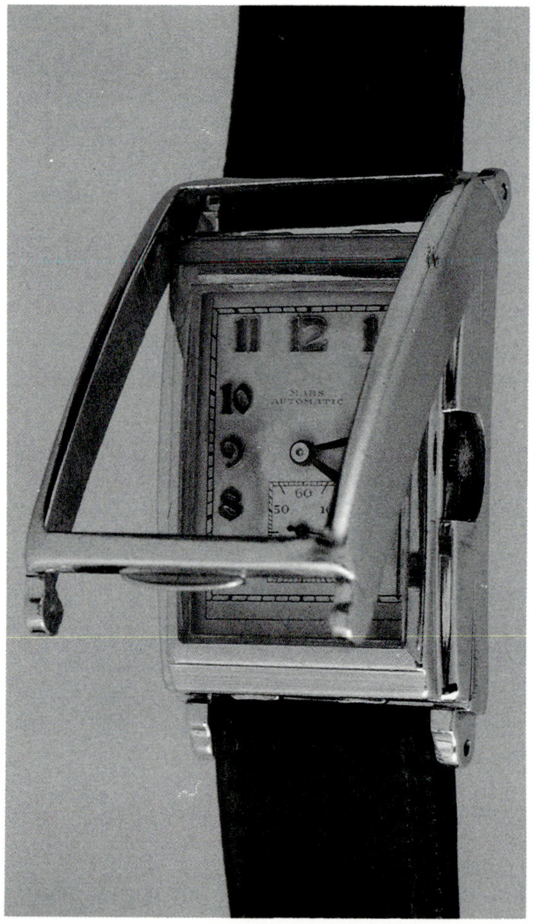

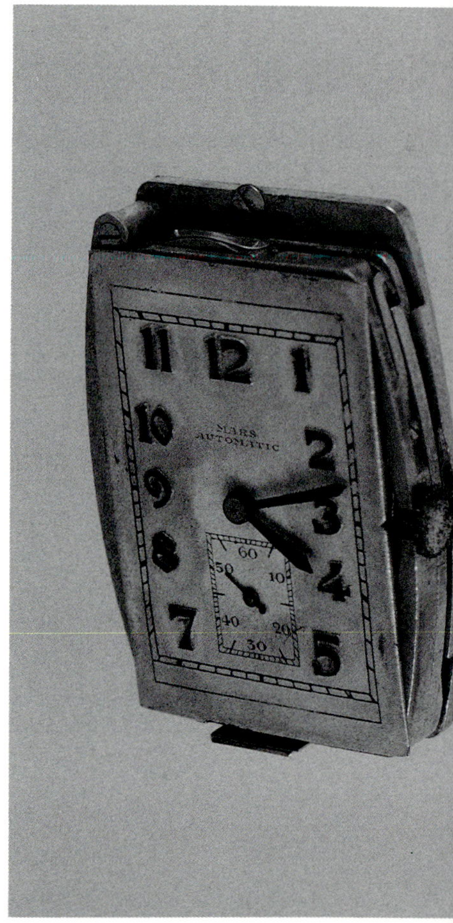

Mars / Wig-Wag

Roller automatic of the "Mars" brand of 1931, also known by the name "Wig-Wag", made by La Champagne S.A. Louis Müller & Cie. Chromed rectangular case with hinged bezel and concealed winding crown.

The complete movement rolls back and forth in a frame, guided by two lateral parallelogram levers (visible at 7 and 11:00). The automatic drive is housed in the barrel-shaped swinging-mass weight (at 1:00). Cutout bimetallic screw balance, flat hairspring, no shock resistance.

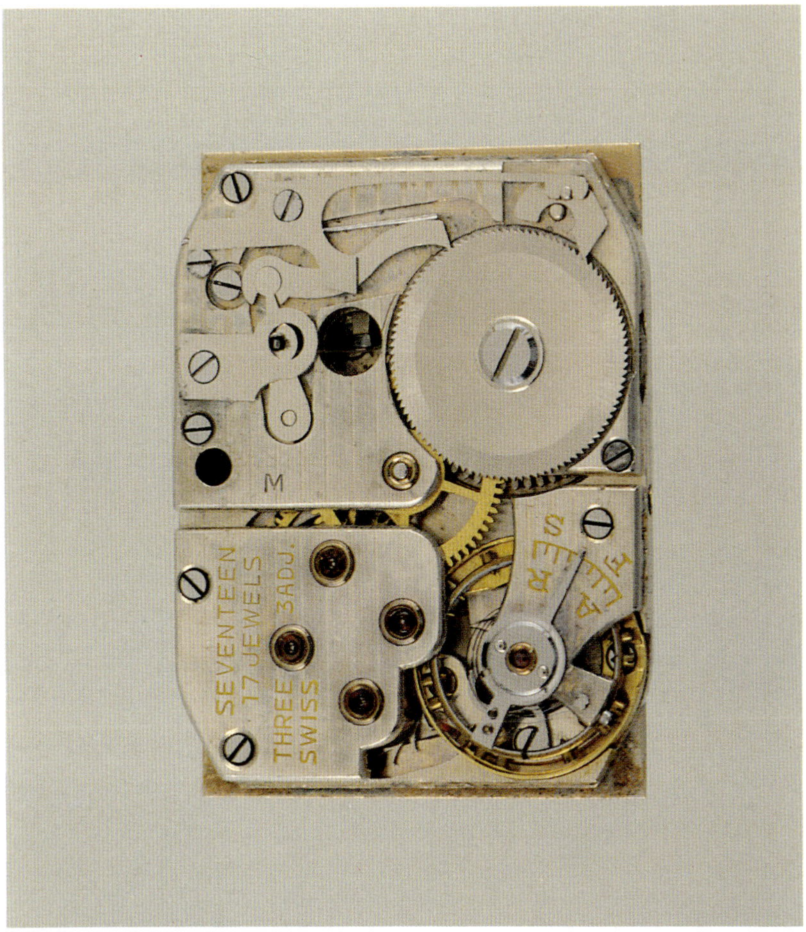

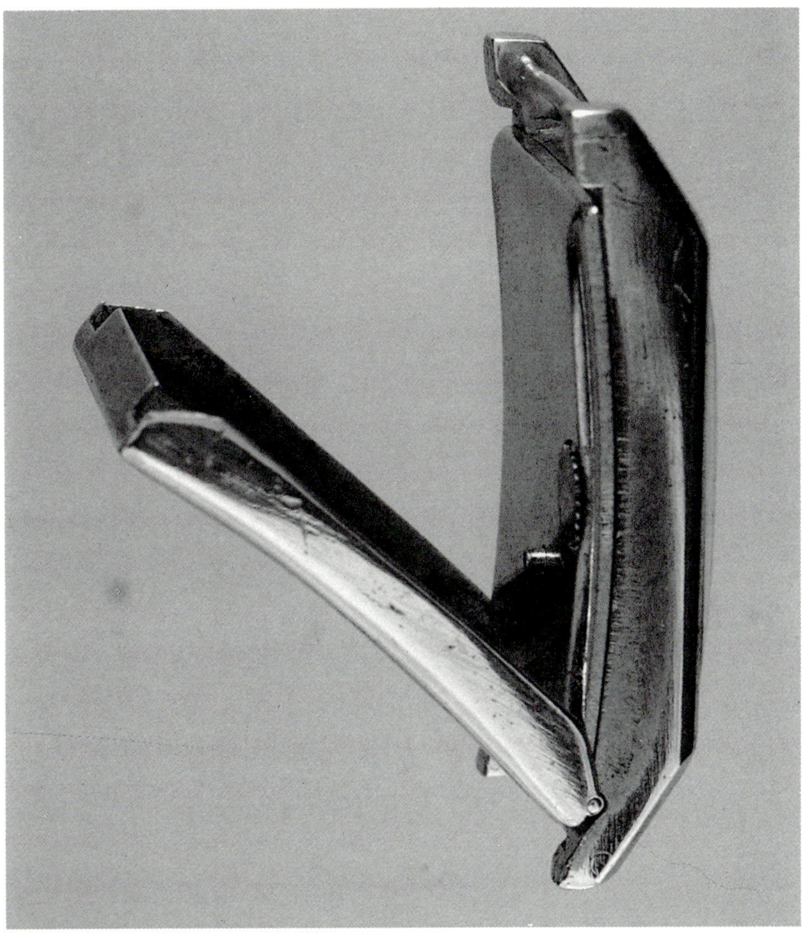

Wyler

Early automatic with lever winding, made in 1931. Chromed rectangular metal case. The disc at the side of the base only sets the hands; the possibility of hand-winding exists only through repeated folding of the hinged bottom.

Winding takes place indirectly from changes in the shape of the wrist, pushing the double bottom against the case. Its movement presses a pin through the inner case bottom onto an angled lever in the movement (in the large hole by the winding wheel), the point of which (visible at left) pushes and turns the winding wheel via a lever system. Cutout bimetallic screw balance, flat hairspring, patented "Wyler" shock resistance system with elastic balance shanks and cage.

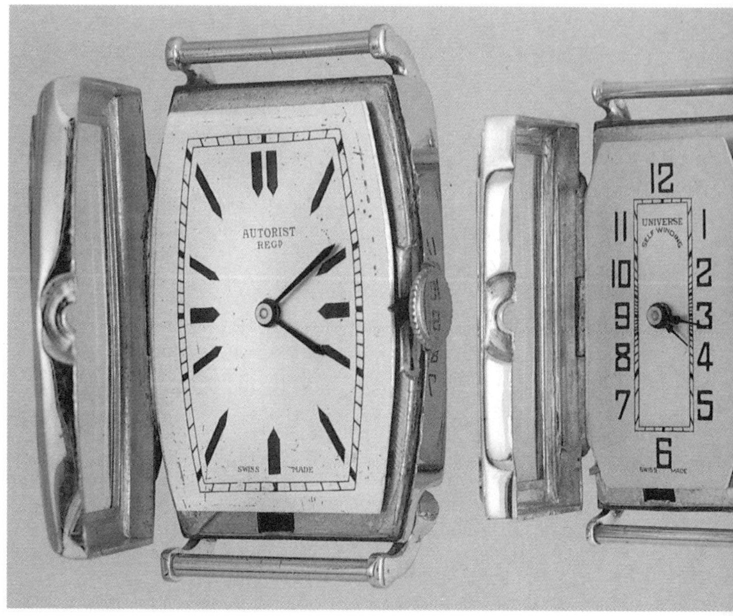

Autorist / Universe
Lever automatics of the "Autorist" and "Universe" brands, circa 1932 (made by Fortis, raw movement by AS).
Barrel-shaped chromed or rectangular silver case with side hinge. Flat crown only for hand-setting. Pulls on the band are transmitted to the movement by the flexible lower band attachment and used for winding.
Closed screw balance, flat hairspring, no shock resistance.

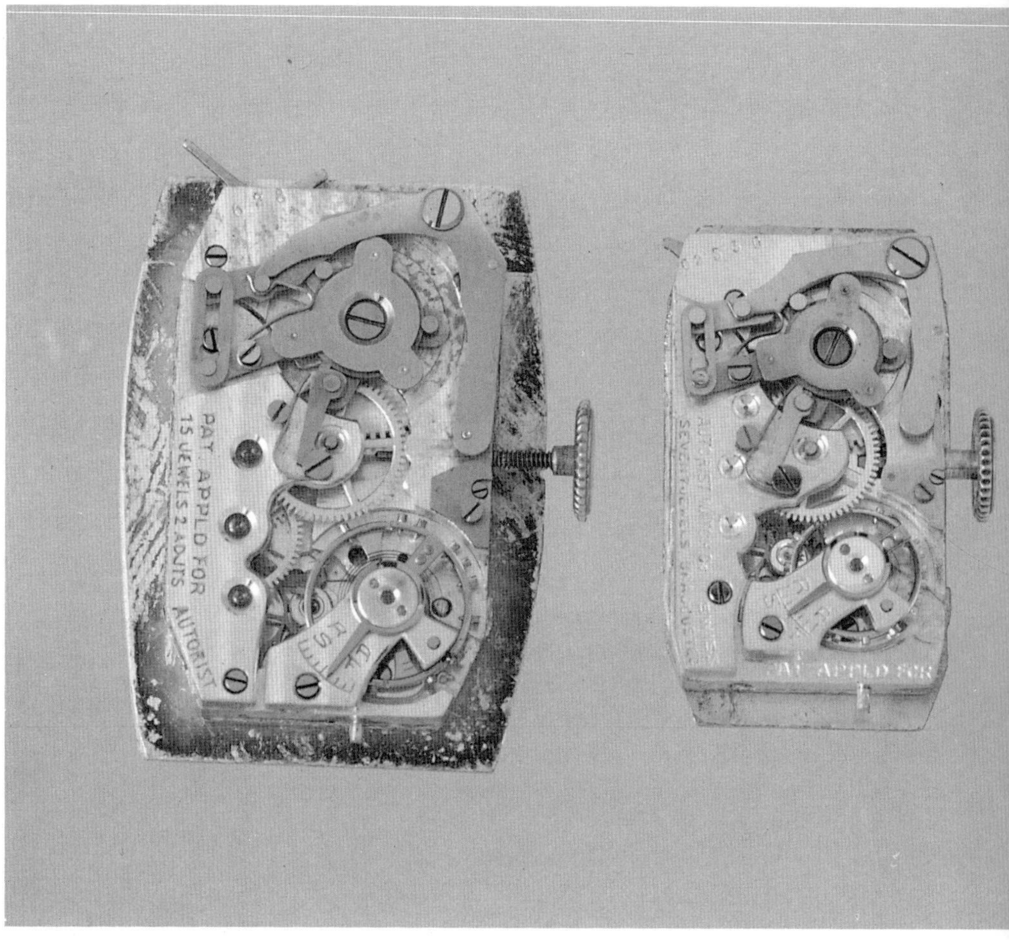

Caliber Notes

Harwood Automatic
D = 24.5 mm (10.5 lignes), H = 6.05 mm, 15 jewels
Basic movement: D = 19.5 mm (9.75 lignes), H = 3.3 mm
AUTOMATIC MODULE screwed onto basic movement. Hammer mounted on both sides of the movement, turning angle of 130 degrees, striking damped by spring buffers, winding in one direction via click on the hammer, centrally mounted winding wheel, two reduction wheels, the first with a locking click, on the locking wheel. Overwinding protection by sliding coupling between the friction spring attached to the hammer and the centrally mounted plate with the winding click.
BASIC MOVEMENT: two-shank cutout bimetallic screw balance, flat hairspring, 18,000 BPH, without shock resistance. Hand-setting by turning bezel with gearing on the setting gear, without hand-winding. Running reserve of 12 hours.

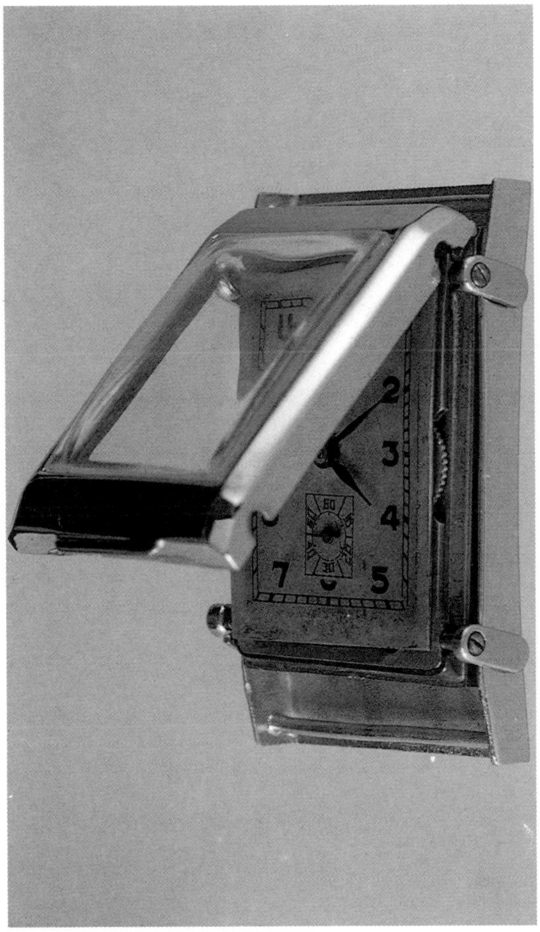

Bucherer - Bulova Patent

Roller automatic of the "Bucherer" brand to a Bulova patent of 1933 (manufacturer unknown). Rectangular chrome plated case with hinged glass rim and hidden disc only for hand-setting.
The entire watch (movement and case) rolls, as a swinging mass, on a band plate, moving over parallelogram levers. Winding takes place via the upper lever arbor (in the front view).
Closed screw balance, flat hairspring, no shock resistance (case chromed later).

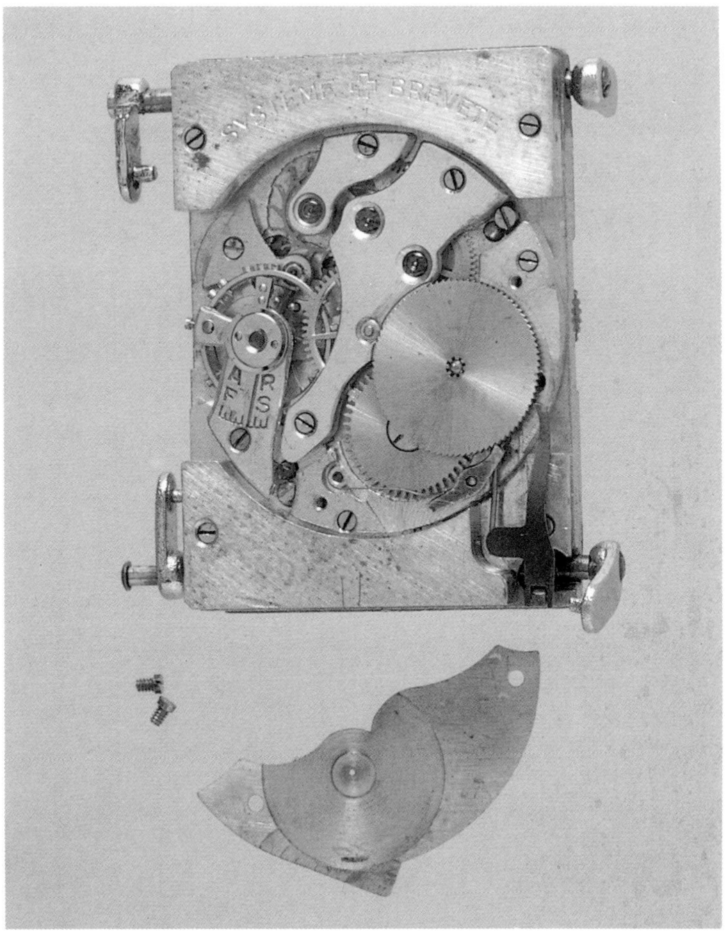

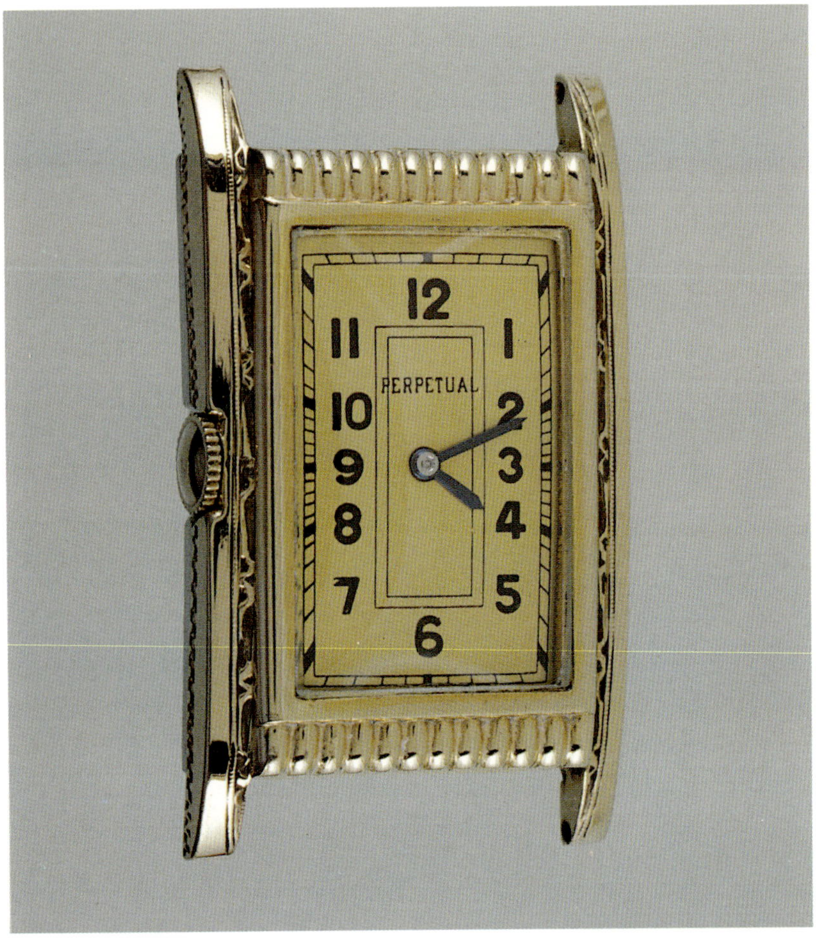

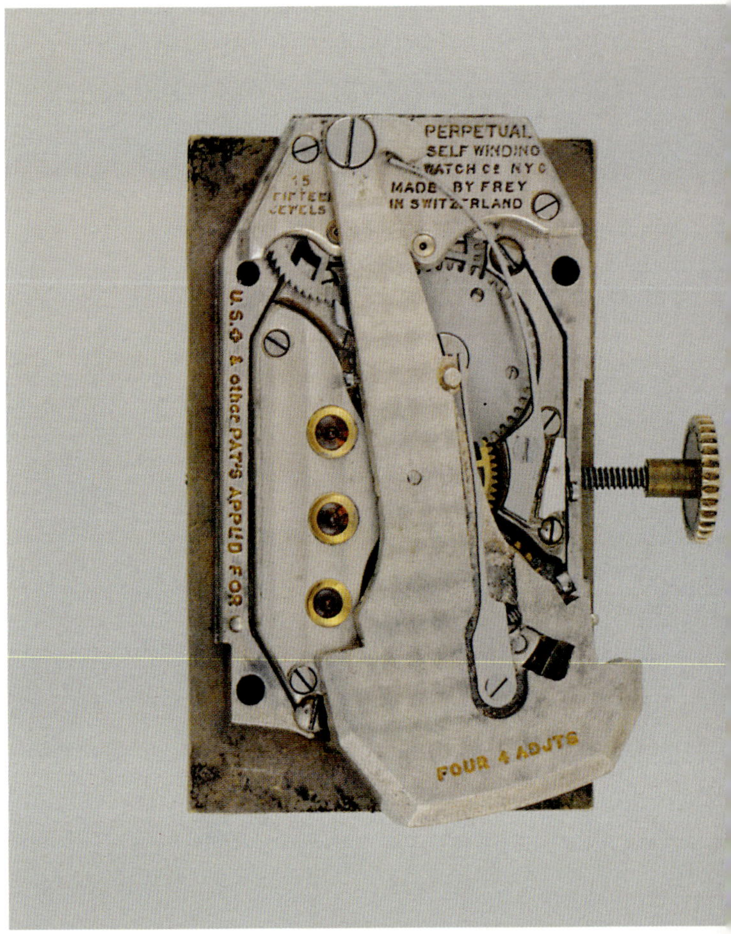

Perpetual

Pendulum automatic of the "Perpetual" brand, made circa 1935 with a movement by the Frey & Co. firm of Biel, reserved for the Perpetual Selfwinding Watch Co. of New York. Rectangular gilded case with crown (only for hand-setting) by the 9.

The automatic module is U-shaped and fits around the basic formed movement, the pendulum is weighted by a lead inlay and winds in one direction via a Harwood-type drive.

Closed screw balance, flat hairspring, no shock resistance.

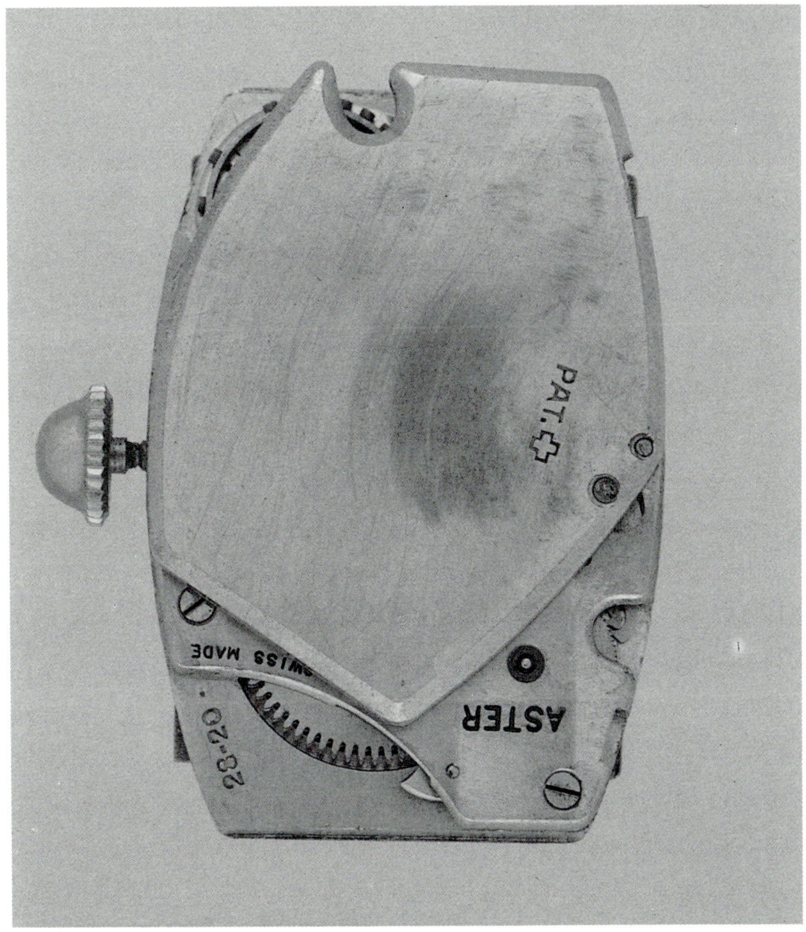

Aster - Bulova Patent

Pendulum automatic of the "Aster" brand to a Bulova patent of 1933 (also made by Bulova itself), produced by La Champagne Louis Müller & Cie. of Biel. Rectangular chromed Staybrite case, small second by the 12.

Flat pendulum with lead inlay, the first watch with the automatic drive integrated into the movement, hand-winding mechanism can be uncoupled. Closed screw balance, flat hairspring, no shock resistance, side anchor.

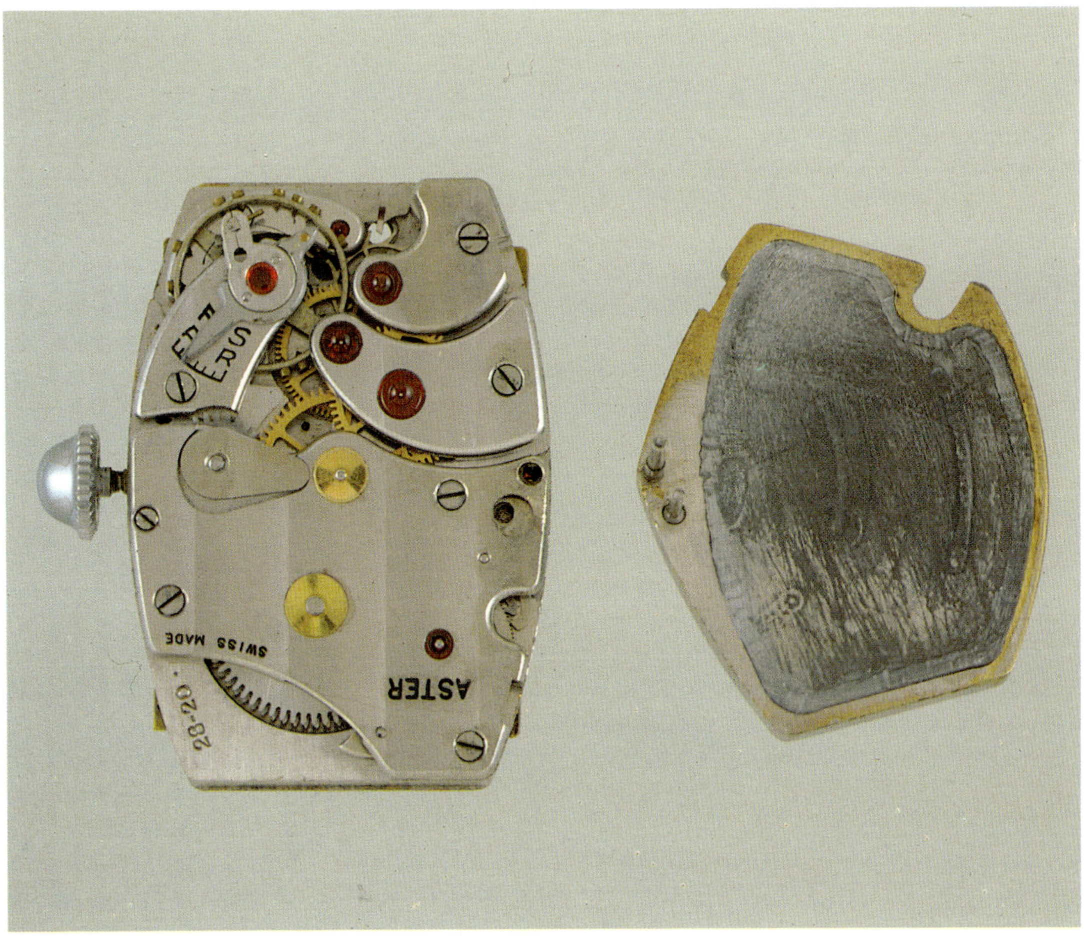

Rolls-Automatic
Large Model:
AUTOMATIC MODULE screwed to basic movement. Winding in one gliding direction via a click lever mounted on the frame, free to turn, with a combined winding and reduction wheel on the locking wheel, gliding guide on ball bearings for the watch movement which functions as a swinging mass, with additional weight, buffered by leaf springs (extent of movement 3 mm).
BASIC MOVEMENT: two-shank cutout bimetallic screw balance, flat hairspring, 18,000 BPH, without shock resistance, eccentric seconds hand, without hand-winding, hand-setting via rim screw.
Medium version: dimensions 30 x 17 mm, H = 6.8 mm, 17 jewels.
(without frame: 24.5 x 12.5 mm, H = 5.5 mm)
See large version of the Rolls, but without second.
Small model:
See large model of the Rolls, but with a second reduction wheel, an anchor version with side fork, without second.

Glycine Automatic
D = 26.0 mm (11.5 lignes), H = 5.8 mm, 18 jewels
(Basic movement Caliber 20 with D = 8.75 lignes)
AUTOMATIC MODULE as independent ring-shaped unit (Cal. 35) linked with basic movement without screw attachment. Centrally mounted hammer on two pressed-in balls above and two below, gliding on the inside of the U-shaped outer ring and limited by sprung buffers, winding in one direction via click on the hammer, centrally mounted winding wheel, two reduction wheels on the locking wheel, spring coupling in first reduction wheel to protect mechanism when fully wound, barrel setting for protection against damage to balance wheel and starting assistance through pre-winding of mainspring.
BASIC MOVEMENT: two-shank nickel screw balance, flat hairspring, 18,000 BPH, Shock-Resist shock resistance, gold anchor, escape wheel between cap jewels, eccentric seconds hand, without hand-winding, hand-setting crown, running reserve of 30 hours.

Wig-Wag Automatic (Mars)
15 jewels
AUTOMATIC MODULE: One-sided parallelogram guidance of the movement as swinging mass on two levers in a roller frame. Winding in one direction from a click spring on the upper lever to the switching wheel, one reduction wheel on the locking wheel.
BASIC MOVEMENT (Mars Caliber 19-4): two-shank cutout bimetallic screw balance, flat hairspring, 18,000 BPH, no shock resistance, eccentric seconds hand, crown for manual winding and hand-setting.

Wyler Automatic
Dimensions: 8.75 x 11 lignes, 17 jewels
AUTOMATIC MODULE mounted as a rail on a formed movement. External winding via outside hinged bottom, vertical pin extending through the fixed case bottom, spring-mounted angled lever, rail with plate joint, pushing switching click as locking wheel.
BASIC MOVEMENT: two-shank cutout bimetallic screw balance, flat hairspring, 18,000 BPH, shock-resistance system after Wyler (sprung curved balance shank, concentric catch ring around the balance ring and top buffer), eccentric seconds hand, without hand-winding, hand-setting via rim panel on case bottom.

Autorist Automatic (AS Caliber 796)
Dimensions: 15.2 x 24.8 mm (1.75 x 11 lignes), 15 jewels
AUTOMATIC MODULE as lever rail mounted on AS Caliber 624 formed movement. External lever winding via moveable band attachments and lever mechanism, clamping winding click on the locking wheel, which is built as a toothless smooth disc. Overwinding protection by sprung double lever.
BASIC MOVEMENT: two-shank screw balance, flat hairspring, 18,000 BPH, without hand-winding, crown on side only for hand-setting.

Universe Automatic
Dimensions: 11.8 x 19 mm (5.25 x 8.5 lignes), 7 jewels
See Autorist for details, but the BASIC MOVEMENT used is the AS Caliber 475 formed movement.

Bucherer Automatic (Bulova Patent)
Dimensions: 19.6 x 27 mm
AUTOMATIC MODULE: Movement including case forms swinging mass that is hung on parallelogram levers. Winding in one direction of rolling movement via the upper lever arbor, switching click, switching wheel on locking wheel.
BASIC MOVEMENT: two-shank nickel screw balance, flat hairspring, 18,000 BPH, without shock resistance, eccentric seconds hand, rim panel only for hand-setting.

Perpetual Automatic
Dimensions: 15.6 x 28 mm, 15 jewels (basic movement 13 x 18.5 mm)
AUTOMATIC MODULE as complete unit set onto the basic movement like a frame. Pendulum with lead inlay and return spring against contact pin, winding in one direction via click on pendulum, winding wheel, one reduction wheel on the locking wheel. Overwinding protection via winding click spring-mounted on pendulum.
BASIC MOVEMENT: two-shank screw balance, flat hairspring, 18,000 BPH, without shock resistance, crown only for hand-setting.

Aster Automatic (Bulova Patent)
Dimensions: 8.75 x 12 lignes, 18 jewels
AUTOMATIC MODULE integrated into barrel-shaped basic movement. Longitudinally mounted flat pendulum with lead inlay, winding in one direction via winding click through eccentrically attached pendulum pin, combined winding and reduction wheel on locking wheel. Separation of hand-winding mechanism by crown-wheel rocker.
BASIC MOVEMENT: two-shank nickel screw balance, flat hairspring, 18,000 BPH, without shock resistance, anchor with side fork, eccentric seconds hand by the 12.

Automatic Catalog (by manufacturers)

Only those Swiss watchmaking firms that put their own automatic movements on the market are presented here in alphabetical order. Many well-known firms that turned to movements from outside their firms for use in automatic wristwatches will therefore not be considered here, but may be noted under the name of the producer of their raw movements.

The manufacturers who had movements developed especially for them by outside firms (such as Vacheron Constantin), or who made a particular contribution to the development of automatic watches, such as Mido or Zodiac, will be treated as exceptions.

Every firm's section will be organized according to the same principles. It begins with the firm's own caliber development, including a brief history of the firm, followed in chronological order by the illustrations of the firm's automatic watches, plus a caliber list that will include every caliber made by that firm, in numerical order. The conclusion will consist of the caliber data with thorough descriptions of the movements.

Terms and Abbreviations (Caliber List)

Caliber: Firm's own movement number, usually found under or near the balance wheel

Diameter: Overall dimensions of the movement, given in lignes and millimeters (1 ligne = 2.26 mm)

Height: Distance from the dial plane of the bottom of the movement to the highest point of the bridges or the top of the movement, thus the top of the rotor in an automatic caliber.

Other Abbreviations as used by the Fédération Horlogère Suisse and Ebauches S.A.:

CHR	Chronograph
CLD	Calendar
GMT	24-hour indication (Greenwich Mean Time)
JG	Weekday indication in dial aperture
M	Anti-magnetic
MG	Month indication in dial aperture
PH	Moon-phase indication
QA	Date indication with hand
RDR	Quick set for date
REV	Alarm
RMA	Up-down indication with hand
RMG	Up-down indication in dial aperture
s	Eccentric seconds hand
SC	Center seconds hand
SCD	Direct center seconds hand
SCI	Indirect center seconds hand

Beats per hour (BPH) will convert to Hz as follows:

BPH	Hz
18,000	2.5
19,800	2.75
21,600	3
25,200	3.5
28,800	4
36,000	5

Introduced Year: When production began or caliber's first year on the market

Jewels: Number of stones (rubies) used in the movement, including those without a function

Notes: References to the basic movement, raw-movement or ebauche producer, name of movement or watch model

Photo: Reference to illustration of caliber is xx in this chapter (xx) in another chapter (as indicated)

Running Reserve: Indication of maximum running reserve in hours, which equals the running time of a fully wound resting watch

Further Information: Most of the illustrated watches are equipped with a Swiss ruby anchor escapement, to which no specific reference will be made. In order to make it easier to locate details in the illustrations, a sectored orientation scheme will be used, with a transparent dial added over the illustration. For example, along with the note "Switcher at 2:00", it will be found along the 2:00 radius.

Caliber Development

The Swiss Watchmaking Cooperation was instituted in 1883 by the Swiss watchmaking firms of Kurth (Grenchen), Straub and Gruen (Biel) and other firms, as well as watch dealers, and in 1908 it established its own production facility in Biel under the name of "Alpina union Horlogère".

The organization's only self-developed automatic caliber was introduced at the end of the 1940s as Number 282.

It was a solid movement with a rotor in typical form, winding in one direction.

All additional automatic watches of the "Alpina" brand were fit with dependable movements made by others, such as Caliber WH 245, made by Eta and equipped with the special "Wyler" shock-resistance system.

Caliber Notes

Caliber 582/P82
D = 28.5 mm (12.5 lignes), H = 5.5 mm, 17 jewels.
AUTOMATIC MODULE screwed to basic movement. Rotor movement limited by open buffer springs, winding in one direction via ratchet click, combined winding and reduction wheel on the crown wheel.
BASIC MOVEMENT: double spoked Glucydur screw balance, Nivarox hairspring, 18,000 BPH, Incabloc shock resistance, seconds bit.

Magazine advertisement (from SUZ, 1949)

Caliber List: Alpina

Caliber	Diameter Lignes	mm	Height mm	Other	BPH	Jewels	Running Reserve Hours	Notes	Year	Photo
362	7 3/4 x 6 3/4									
WH 245	11 1/2	25,6		SCD, QG, JG, RDR	21.600	25		Eta-Ebauche (2789)	ca. 74	
571	11 1/2			SC	18.000	17		Hammer-Fremdwerk	1946	
571 C	"			SC, QG	"			"		
582 /P82	12 1/2	28,5		s	18.000	17		Rotor movements not by Alpina	ca. 49	xx
582 -3	"	"	5,5	s	"	"				
584 /P84	"	"	6,1	SCI	"					

Alpina Caliber 582
Men's automatic made by Alpina circa 1950, using Caliber 582. Case in gold and stainless steel screw bottom. Glucydur screw balance, Nivarox hairspring, Incabloc shock resistance.

Caliber Development

The watchmaking firm of Amida S.A. had its headquarters in Grenchen.

It produced watches with both pin-lever and jewel-anchor escapements and marketed them under the brand name of Castell.

In the mid-1950s the firm developed the "Adamatic", its first automatic watch. This Caliber 550 was built on the hand-wound Caliber 530 (10.5 lignes) and equipped with a rotor winding in both directions via ratchet clicks.

The automatic and hand-winding mechanisms are mutually independent of each other.

The subsequent Calibers 552 and 553 also had the same design but were improved in details such as the rotor mounting and the ratchet wheels.

Development of the AS firm within Ebauches S.A.

Adolph Schild, the brother of Eta founder Urs Schild, opened his own raw-movement factory in Grenchen in 1896 under the name of "A. Schild S.A." (ASSA or AS). His company, along with the two firms of FHF and A. Michel (AM), united in 1926 to form the basis of the later so dominant collective of Swiss raw-movement producers, the "Ebauches S.A.", which absorbed other significant firms such as Felsa in 1928 and Eta in 1932.

As early as 1931, a large Swiss company, the ASUAG (Allgemeine Schweizer Uhrenindustrie AG), was founded and absorbed all the special firms that had been united in four companies. They were:

1. Ebauches S.A.,
2. The union of all producers of escapement parts,
3. The union of all balance-wheel producers, and
4. The union of all hairspring producers.

Since 1941 an original spare-parts service for all calibers produced by Ebauches S.A. has been operating in Neuenburg; in 1965 it carried parts for some 800 different calibers.

The individual spare parts are still delivered today in a protective special package with the brand name, caliber number and size, numbers and names of the parts in five languages, as well as information on the most important dimensions and replacement possibilities. For this purpose, Ebauches S.A. published a five-language "Technological Dictionary" in 1948, including the universal numbering of all components of a watch in the form of an illustrated catalog.

To define the concept of "raw movement" (*ebauche* in French), it should be noted that it refers to an assembled movement that is delivered without a mainspring, balance, hairspring and jewelling. Unlike the manufacturers who develop and produce their watches themselves, these "Etablisseure" obtain all the components from other firms. Since 1931 the individual firms that make up the ASUAG assemble them into complete movements in their factories, and then sell them under one of many brand names. This provides an explanation for the often-heard comment that the same movements are often found in watches with a variety of brand names.

The sixteen raw-movement producers that make up Ebauches S.A. produced, for example, 33 million movements, representing two-thirds of the year's Swiss production; the rest was divided among the producers of the pin-lever or Roskopf watches and the manufacturers.

Within Ebauches S.A., production was soon divided among various production facilities. Thus there are factories that produce chiefly automatic movements (AS, Eta, Felsa), or chronograph movements (Landeron, Venus, Valjoux), or hand-wound movements (such as Peseux).
Caliber List: Amida

Caliber Development

AS was the first raw-movement factory that was able, in 1931, to offer in the "Autorist" its own development based on an idea of Harwood.

While the first Harwood automatic watch had hammer winding, the Autorist functions with lever winding. Here the motive power coming from the band is conducted into the interior of the watch via a moveable band attachment and winds the watch with the help of a complicated lever system.

Caliber List: Amida

Caliber	Diameter		Height	Other	BPH	Jewels	Running	Notes	Year	Photo
	Lignes	mm	mm				Reserve			
550	12	26,7	6,15	SCD	18.000	17		Adamatic	ca. 55	
551	"	"		SCD, QG	"	"		"		
552	12 1/2	27,8	6,0	SCD	"	21				
553	"	"		SCD, QG	"	"				
911	12 1/2	27,8		SCD						

For reasons already listed (see "Early Automatic Watches"), the Autorist, made in men's size as Caliber 796, was just as unappealing as the smaller size for ladies, which was sold under the name of "Universe" and based on the AS 475 formed movement.

The first AS hammer automatic, Caliber 913, dates from 1935 and was used by Mido and others. It works with a centrally mounted winding wheel that is moved in one direction by a winding click attached to the rotor. This type of construction (Harwood type) is also found in the various AS calibers made until 1946. Its easy serviceability and characteristic handling of details, including the mounting of the winding and locking clicks under separate plates, is especially interesting.

The slimmer rotor-wound Caliber 1298 was first produced in 1946, and the subsequent Caliber 1302 was produced two years later They are equipped with a partly geared click-carrier disc that fulfills the same function as the frequently used click-carrier rocker.

In the movement with rotor winding that was issued in 1950 (Rotomatic, Caliber 1336), a rocker switcher was used for the first time, surviving to this day as a device to wind the spring regardless of which direction the rotor turned. Unfortunately, the two notches mounted on the rocker, because of their not very well-defined contact with the switched wheels, often caused premature wear, which could be reduced later through the use of more favorable wheel forms and materials. For hand-winding, this was separated by a Breguet coupling to protect the automatic drive.

This caliber was also equipped with a ball-wheel differential drive to show the running reserve in an aperture on the dial or by a small hand on an auxiliary dial. It could also be had with a simple calendar (Caliber 1493).

For Caliber 1477, the smallest automatic caliber of the time around 1954, the AS engineers designed a new switching guide of patented twin click wheels, each functioning with a separate springless anchor click.

The "Maximatic" (Caliber 1580), fully developed in 1958, was likewise equipped with these click wheels, and could also be ordered with the alternative of a modern screwless ring balance.

The first AS automatic with a rotor mounted in ball bearings appeared about 1964 as Caliber 1716. Here the automatic drive was lowered into a small oval cutout in the basic movement and allowed, depending on the mechanism used, either one-direction winding with a click wheel or two-direction winding with the help of an additional click wheel.

This movement was later rebuilt as a fast beat Clinergic 21 anchor escapement.

While AS is not the first manufacturer of a rotor automatic with winding in one direction, the firm did set off a new discussion as to the most effective winding principle, which after a scarcely questionable series of tests, surprisingly favored one-direction winding. It was defined by AS as "winding with dynamic free running", which was meant to express that the rotor turning in its free-running direction could turn more easily and thus farther against pressure and thus be able to wind more effectively in the other direction. The advantage of simpler and space-saving drive construction, though, was opposed by the sprung locking click which consumed energy and was liable to wear.

The Caliber 1712, introduced about 1966, was made by the modular principle and also wound in only one direction via a sprung gear lever. It was made until the 1970s and was thus the last development of a simple automatic caliber by the AS firm.

AS set an absolute record in 1967 with the smallest automatic ever built, the Caliber 1775, which was also used by Mido in the "Minimido" model. It too wound in only one direction. In the alarm automatic (Caliber 5008) fully developed in 1974, the sound mechanism was also wound automatically, unlike the well-known "Memovox".

In retrospect, the firm of AS not only holds the distinction of being the first raw-movement producer to develop automatics, but along with the introduction of the switching rocker it also produced the smallest automatic caliber in series production. Not all the designs introduced by AS proved to be successful, but no other manufacturer built so many different calibers.

Caliber Notes

Caliber 1171
D = 24.8 mm (11.25 lignes), 15 jewels
AUTOMATIC MODULE screwed to basic movement, the hammer limited by buffer springs, winding in one direction via winding click on hammer, centrally mounted winding wheel, two reduction wheels, the first with sprung click, on the locking wheel.
BASIC MOVEMENT: two-shank Glucydur screw balance, self-compensating flat hairspring, 18,000 BPH, Incabloc shock resistance, without crown wheel, seconds bit.

Caliber 1191
D = 28.0 mm (12.25 lignes), 17 jewels
AUTOMATIC MODULE screwed to basic movement, hammer limited by buffer springs, winding in one direction via winding click on hammer, central winding wheel, two reduction wheels, the first with a spring click under a separate plate, on the locking wheel.
BASIC MOVEMENT: two-shank Glucydur screw balance, self-compensating flat hairspring, 18,000 BPH, Incabloc shock resistance, seconds bit.

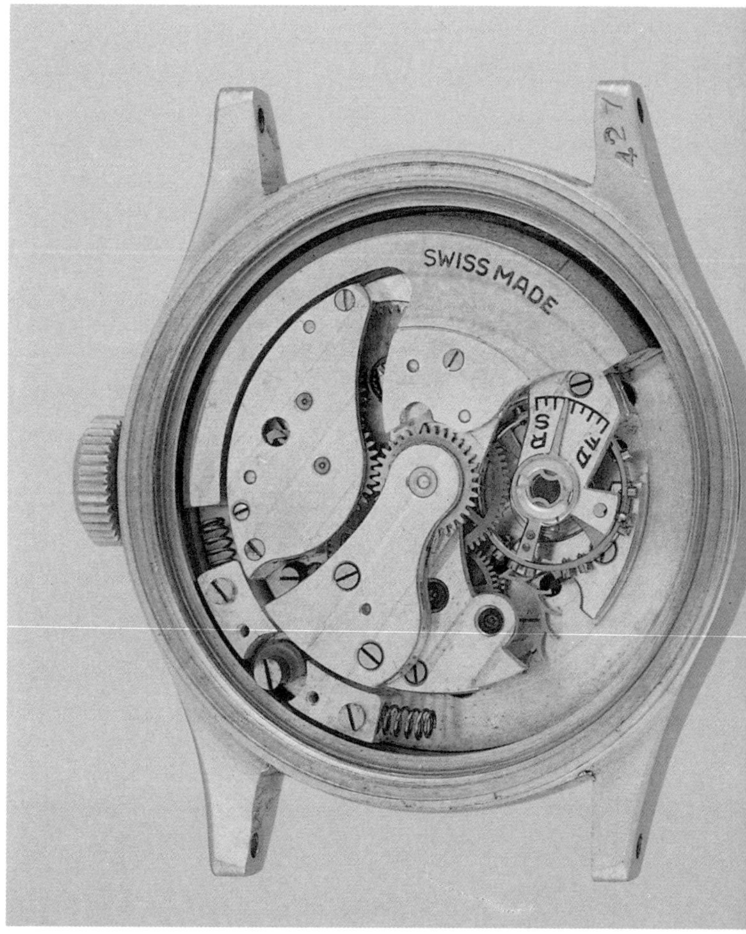

AS Caliber 1171, "Rodana"

Men's automatic of the "Rodana" brand of 1941, with AS Caliber 1171 raw movement.
Chrome plated Staybrite case with stainless steel screw bottom, crown only for hand-setting. Hammer winding in one direction over a Harwood type drive. Glucydur screw balance, Nivarox hairspring, Incabloc shock resistance.

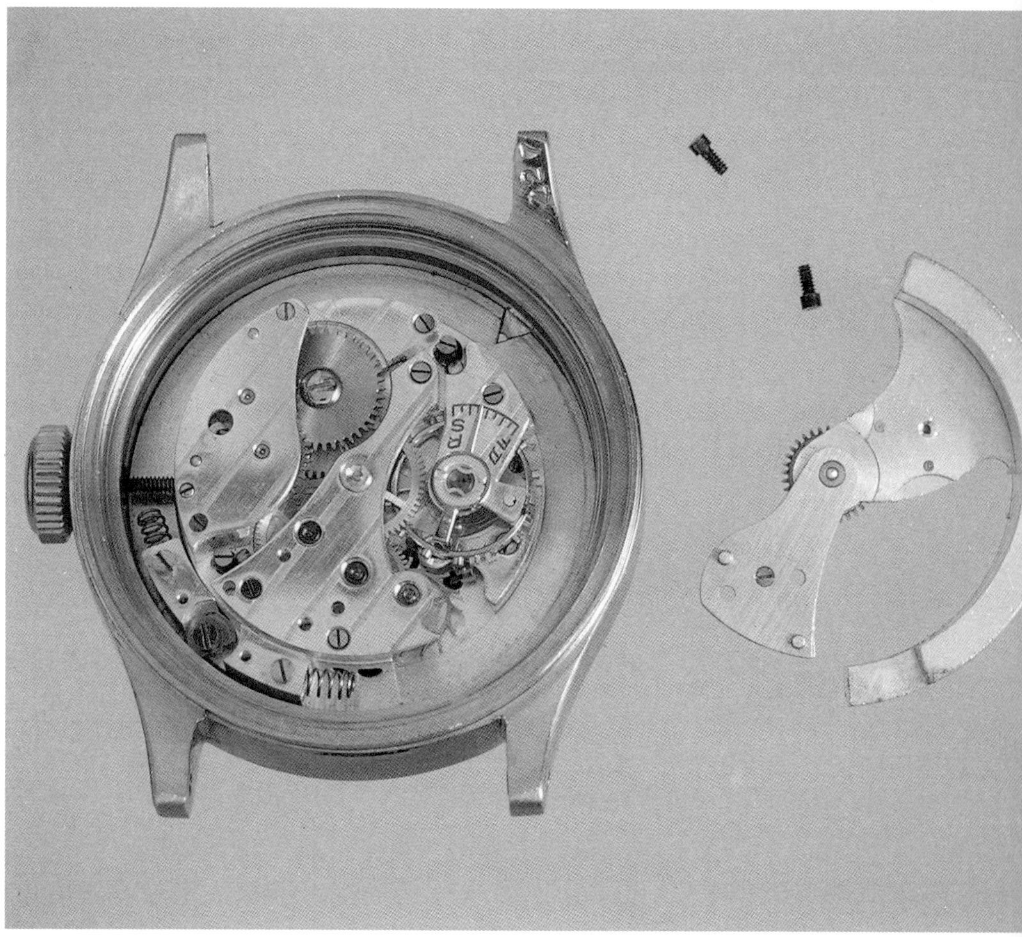

Opposite Page: AS Caliber 1191, "BWC"

Men's automatic of the "BWC" brand (by Richard), circa 1942, with an AS Caliber 1191 raw movement.
Gilded case with stainless steel screw bottom and additional dust cover.
Hammer winding in one direction over a Harwood type drive with additional uncoupling of the hand-winding mechanism, separate locking-click plate (by the 9).
Glucydur screw balance, Nivarox hairspring, Incabloc shock resistance.

B.W.C.
SUISSE
Automatic
SWISS MADE

RICHARD
AUTOMATIC
17 JEWELS
SWISS

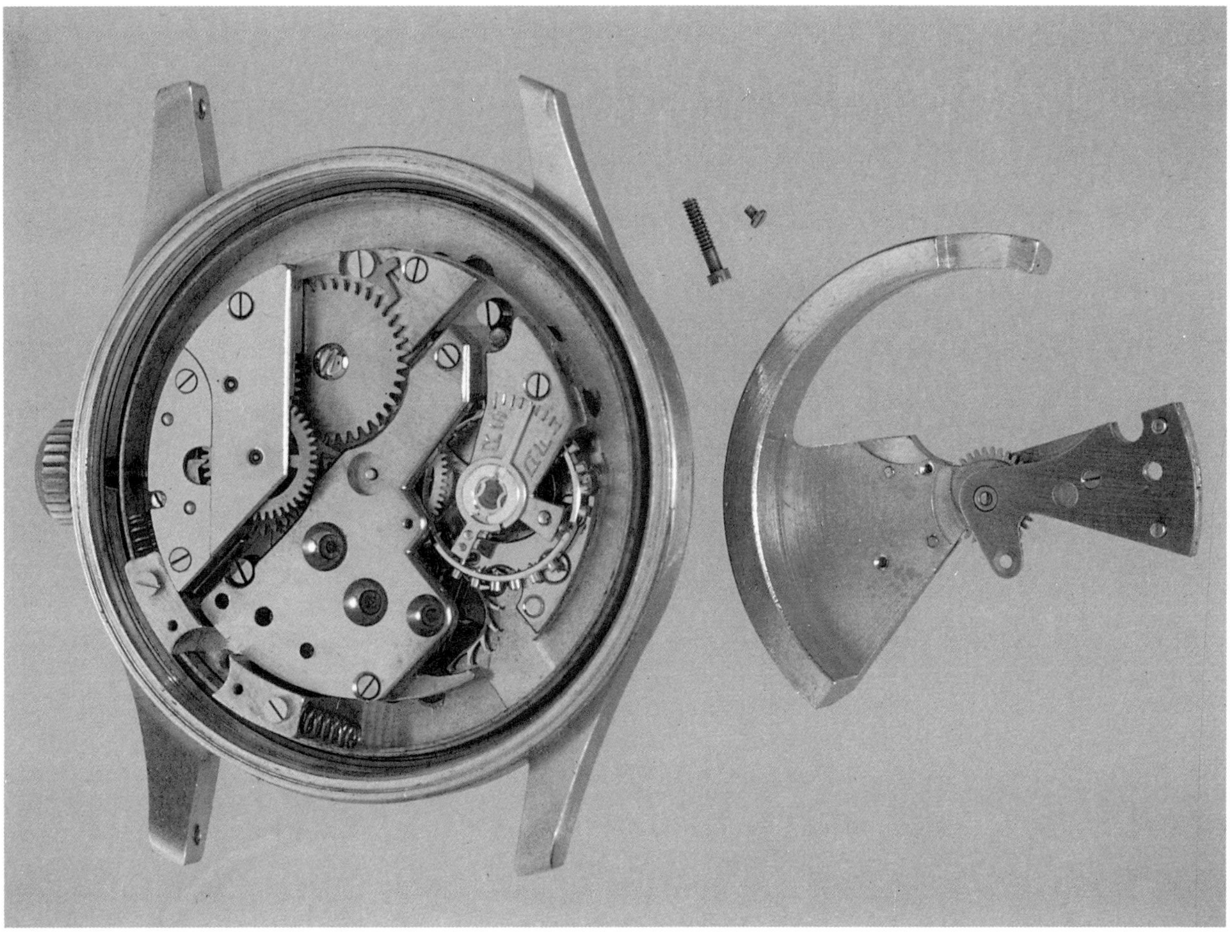

AS Caliber 1250, "Semos"

Men's automatic of the "Semos" brand, made in 1944, with an AS Caliber 1250 raw movement.

Chrome plated Staybrite case with stainless steel screw bottom.

Hammer winding in one direction over a Harwood type drive with additional uncoupling of the hand-winding mechanism, separate locking click plate (at 7:00).

Glucydur screw balance, Nivarox hairspring, Incabloc shock resistance.

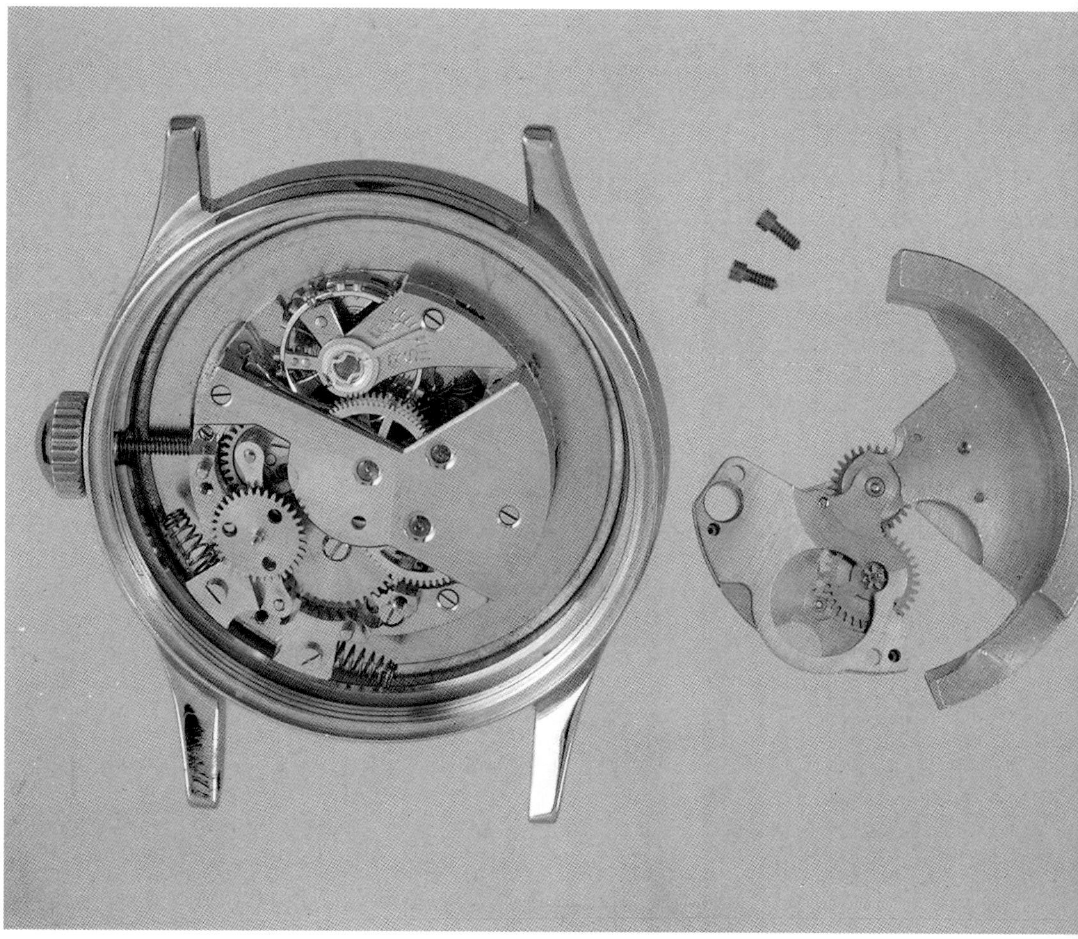

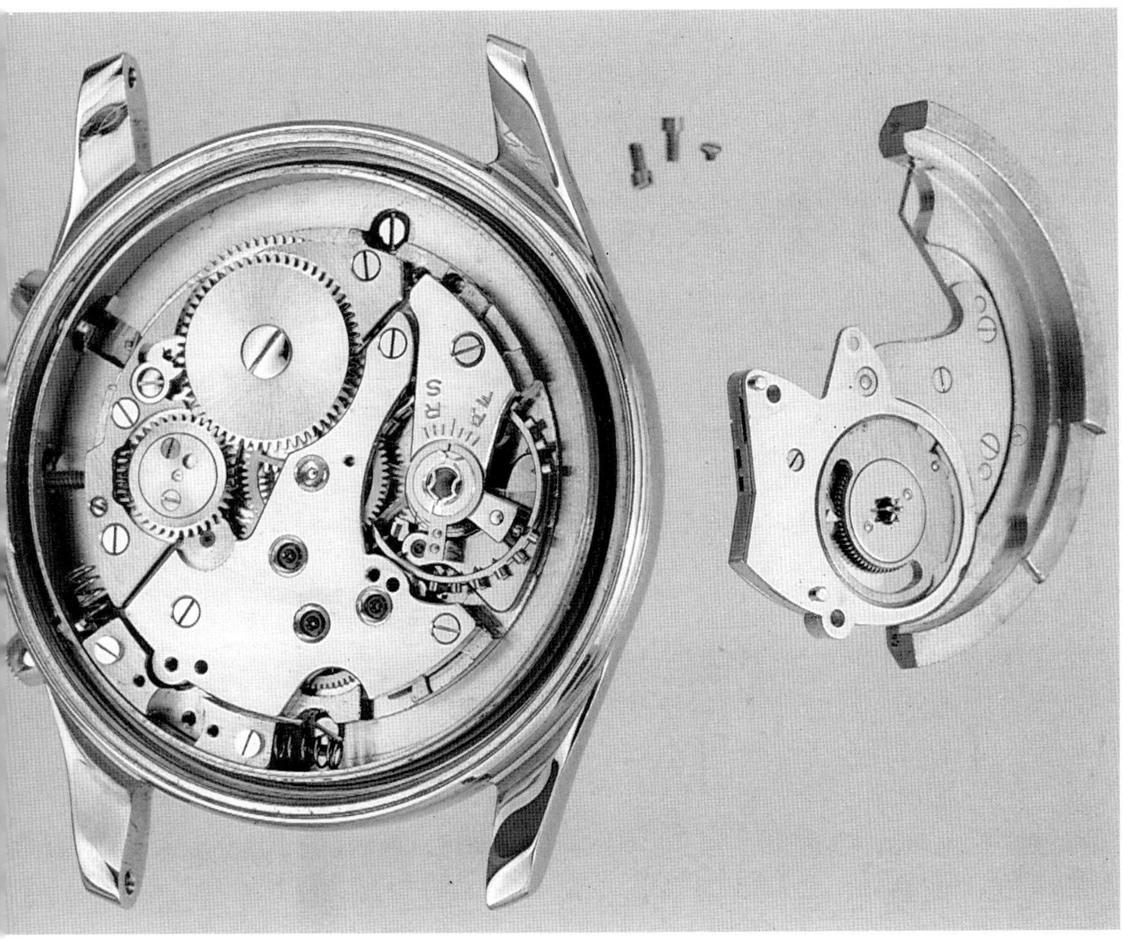

AS Caliber 1315, "Regatta"

Men's automatic of the "Regatta" brand, made in 1950, with an AS Caliber 1315 raw movement.
Stainless steel case with screw bottom.
Simple calendar with date indicator, side buttons for fast correction of the month by the 2 and the date by the 4, the week-day can be changed only by turning the hands.
Complete unscrewable automatic unit, hammer winding in one direction via a partly toothed click-carrier disc, separate locking-click plate (by the 6).
Glucydur screw balance, Nivarox hair-spring, Incabloc shock resistance.

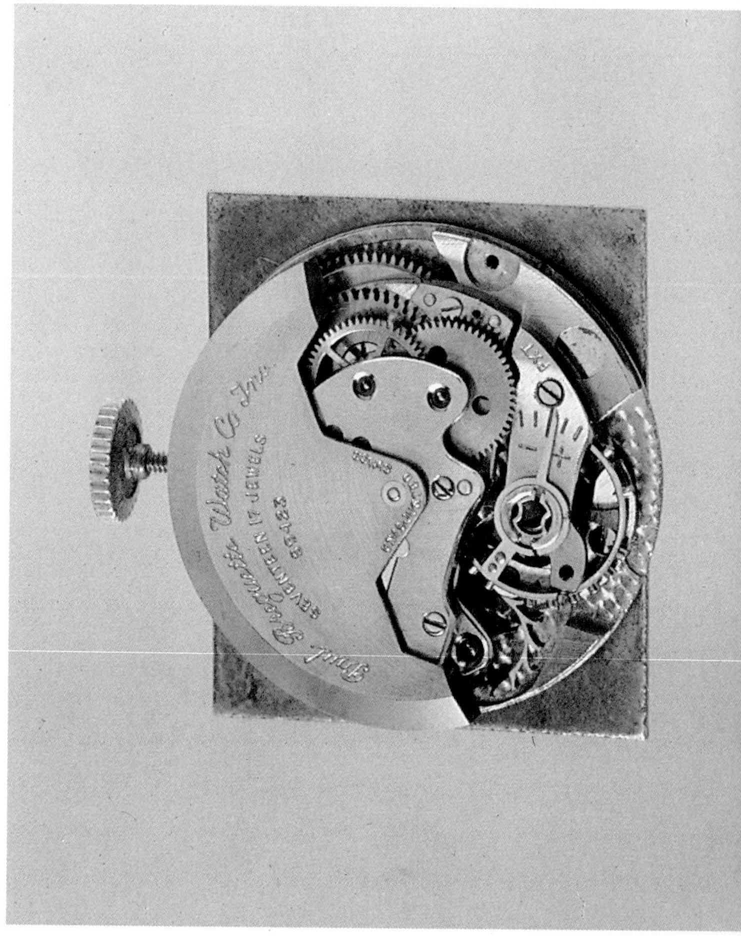

AS Caliber 1323, "Paul Breguette"
Ladies' automatic of the "Paul Breguette" brand, made circa 1952, with the first small-rotor Caliber 1323 by AS (first made in 1950). Rectangular 10-karat gold-filled case.
Rotor with heavy metal ring, winding in both directions via a switching rocker, uncoupling during hand-winding by a Breguet coupling, Glucydur screw balance, Nivarox hairspring, Incabloc shock resistance (Breguet coupling spring at 1:30 position is missing).

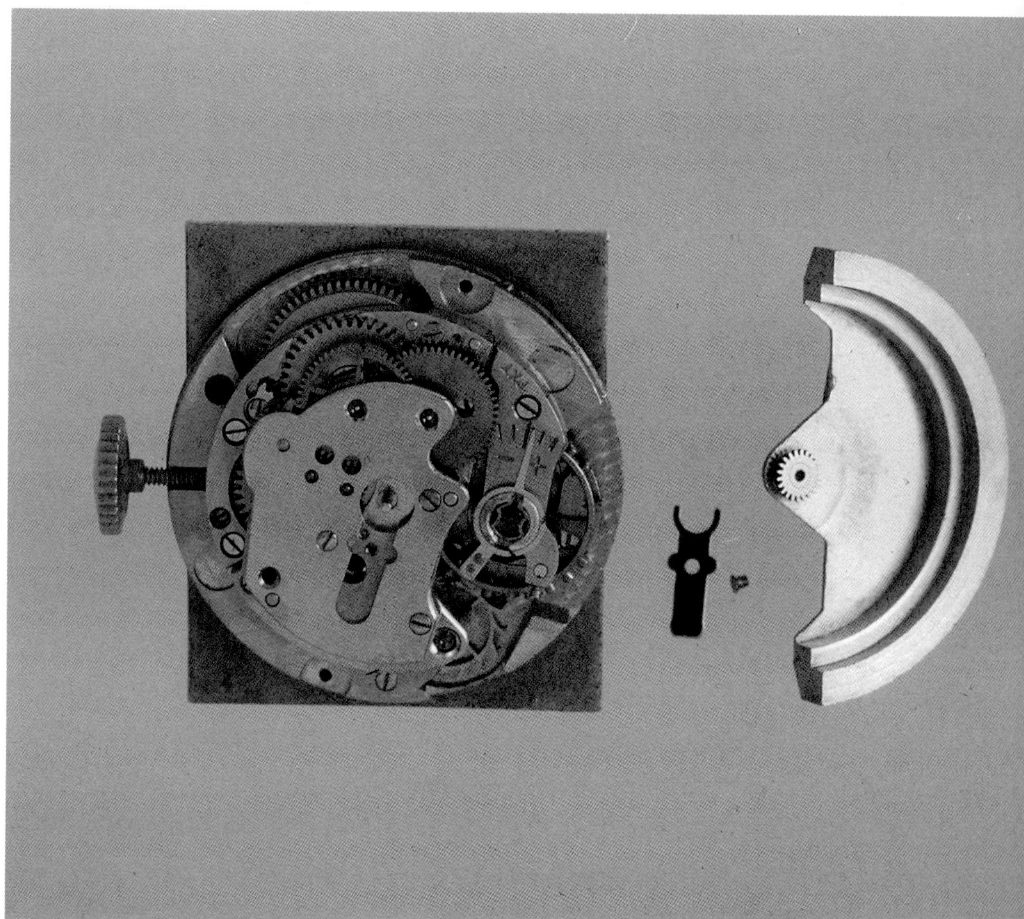

AS Caliber 1382/N "Nivada"

Men's automatic of the "Nivada" brand of 1952, with an AS Caliber 1382/N raw movement.

Gilded case with stainless steel screw bottom.

Auxiliary dial under the 12 to indicate running reserve. Rotor with heavy metal ring, winding in both directions via a switching rocker (partly concealed at left beside the balance), separate locking-click plate (at 6:00), uncoupling of the hand-winding drive by a Breguet coupling (Breguet coupling spring at 7:00).

Nickel screw balance, Nivarox hairspring, Incabloc shock resistance (case shows strong signs of use).

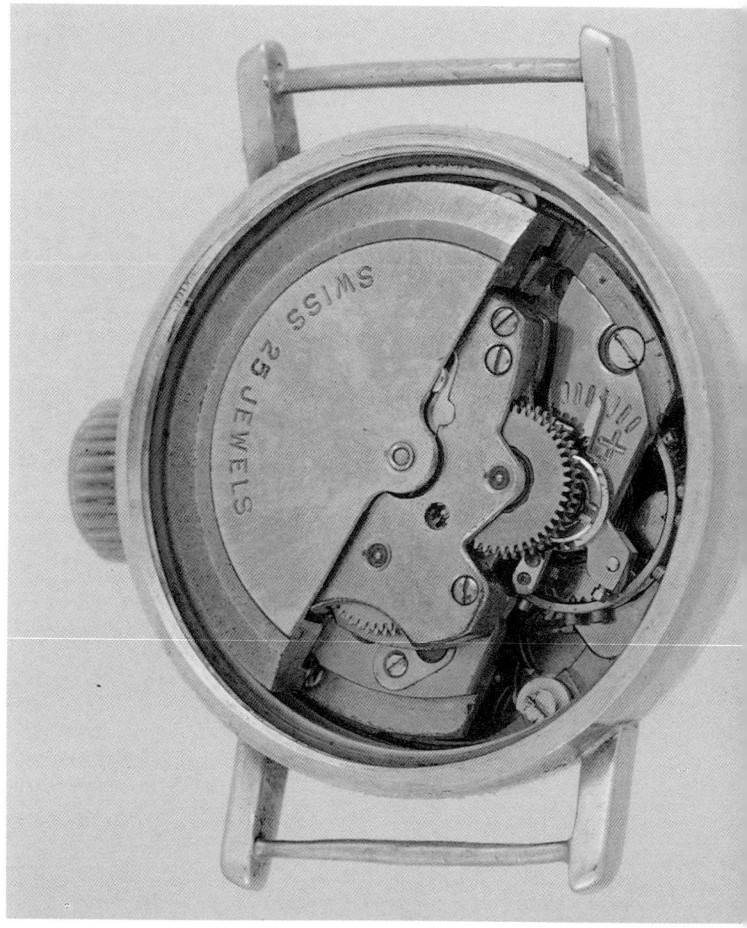

AS Caliber 1477, "Stowa"

Ladies' automatic of the German "Stowa" brand, circa 1955, with AS Caliber 1477 raw movement, the smallest automatic of its time. 14-karat gold case with screw bottom.

Rotor with heavy metal ring, winding in both directions via a click-wheel switcher, completely removable automatic module.

Glucydur screw balance, Nivarox hairspring, Incabloc shock resistance (winding crown replaced, dial and upper surface of movement corroded).

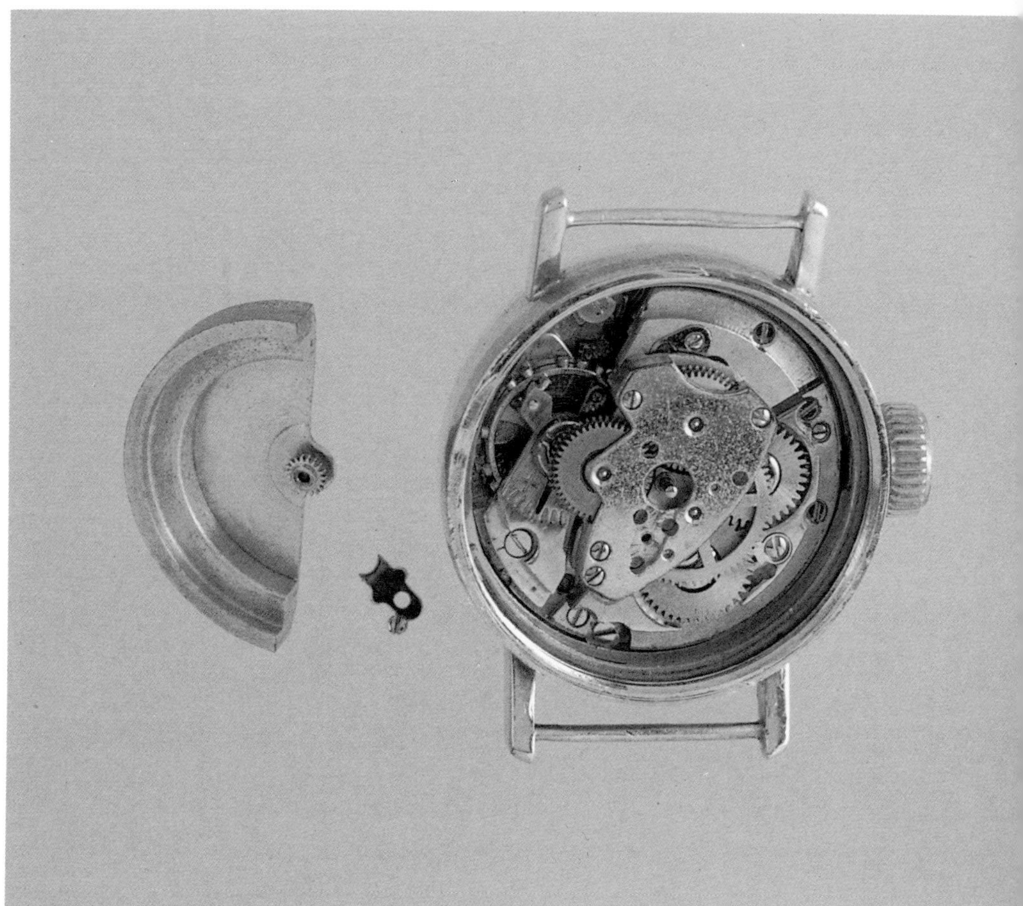

AS Caliber 1478 N, "Wittnauer"

Men's automatic of the US "Wittnauer" brand, circa 1954, with AS Caliber 1478 N raw movement.

Running reserve indication in a dial aperture under the 12.

Stainless steel case with screwed bottom lid.

Caliber identical to Caliber 1382 N (page 33), but with modified rotor and upper plate, plus Kif-Flector shock resistance (dial corroded, band attachments show signs of use).

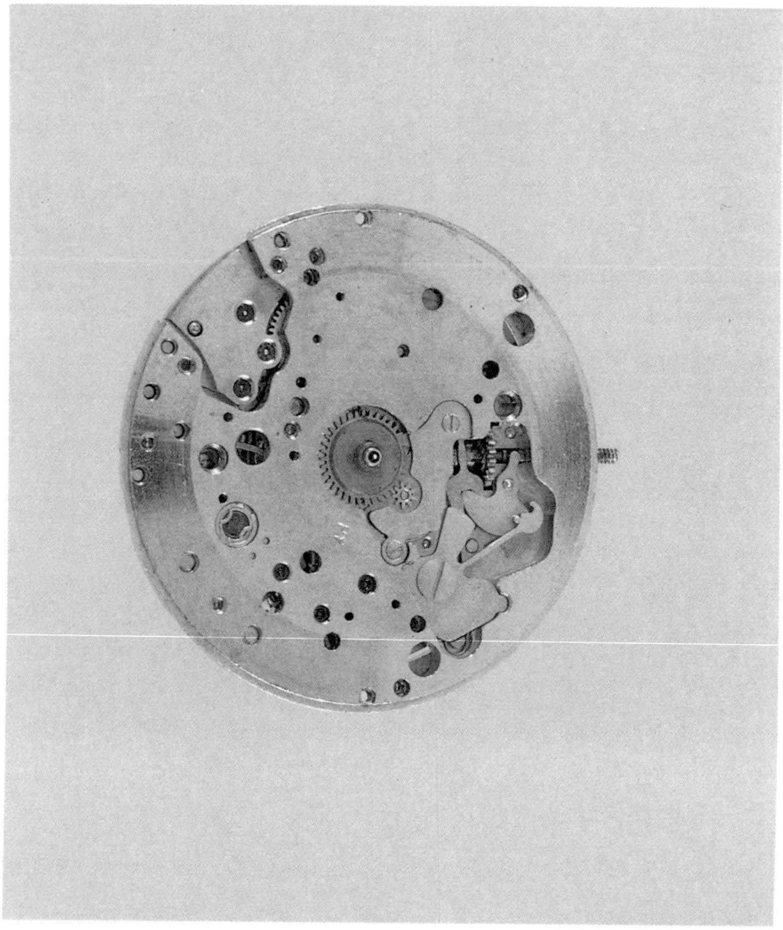

AS Caliber 1656 Movement

AS rotor Caliber 1656, circa 1958. Large rotor, mounted somewhat out of center, with heavy metal ring, winding in both directions via click-wheel switcher. Automatic drive integrated in the basic movement under a separate bridge (at 1:00).
Glucydur screw balance, Nivarox hairspring (missing), Incabloc shock resistance.

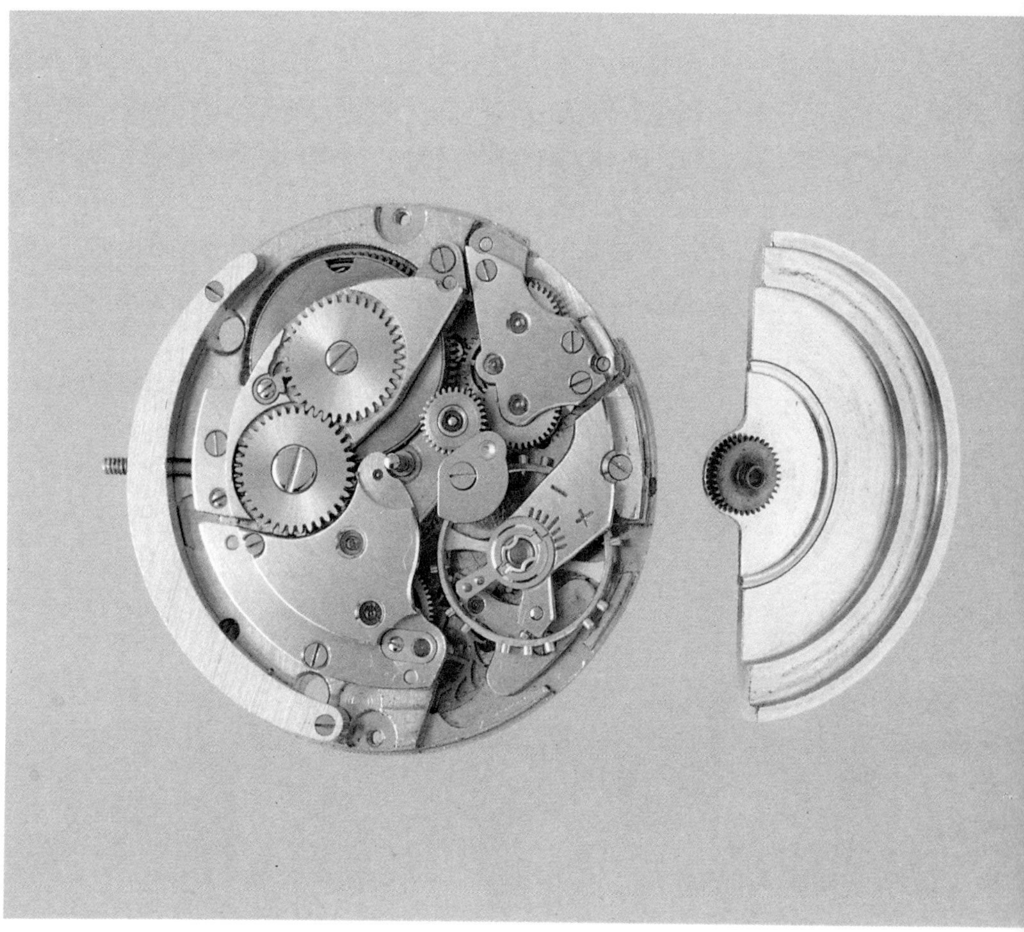

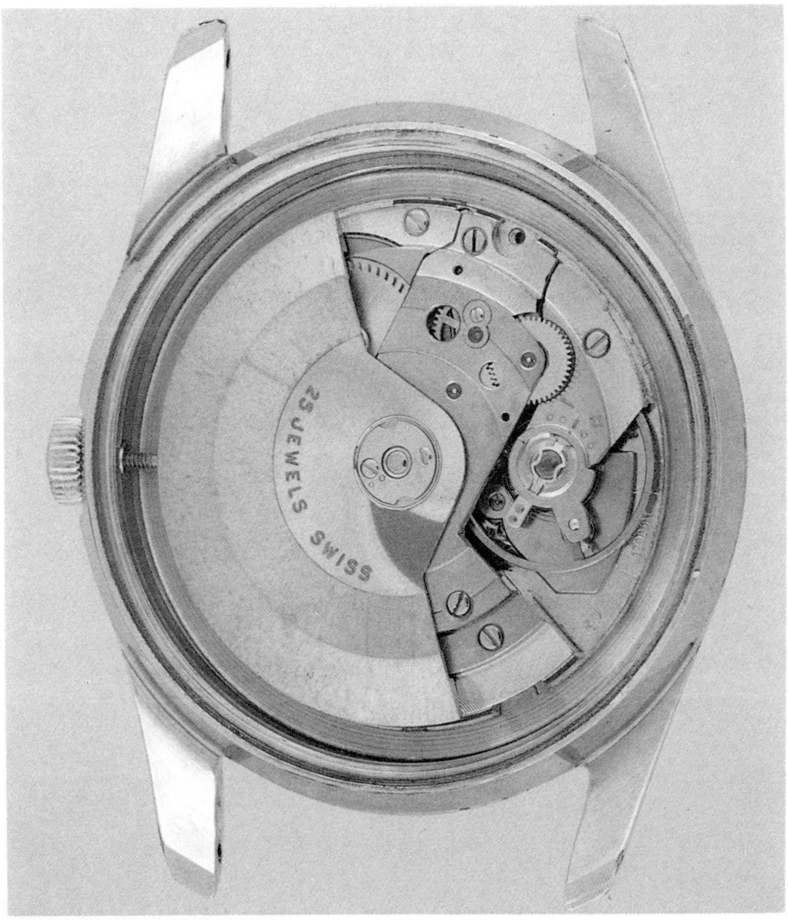

AS Caliber 1700/01, "Lovary"
Men's automatic of the "Lovary" brand, circa 1958, with AS Caliber 1700/01 ("Maximatic") raw movement.
Gilded case with stainless steel screw bottom, winding crown with Lovary emblem. Rotor with heavy metal ring, winding in both directions via a click-wheel switcher, uncoupling device for hand-winding. Partly integrated, completely removable automatic module.
Screwless nickel balance, Nivarox hairspring, Incabloc shock resistance (traces of wear on the band attachments).

AS Caliber 1895, "Felca"

Men's automatic of the Felca brand, "Skymaster", circa 1965, with AS Caliber 1895 raw movement.

Stainless steel case with screw bottom, winding crown with Felca emblem. Date indication with quick correction, and weekday indication.

Rotor with ball bearings and heavy metal ring, winding in both directions via click-wheel switcher. Oval automatic drive recessed in the basic movement (at 6:00), can also be exchanged for a one-direction drive.

Screwless Glucydur balance, Nivarox hairspring, Incabloc shock resistance.

AS Caliber 2063, Islamic Prayer Watch

Islamic prayer watch with automatic winding, made in 1971, with AS Caliber 2063 raw movement.

Gilded case with screw bottom, stainless steel, signed "Monte Carlo", with engraved directions in Arabic, mineral glass with compass.

The bezel with world cities is turnable by the crown by the 2 to the center (Mecca). The inner bezel with verses from the Koran can be set by the crown by the 4. Rotor with heavy metal ring winding in one direction via lever wheel (at left near center of movement).

Screwless nickel balance, Mivarox hairspring, Trishock shock resistance.

AS Caliber 5008, "Arctos"

Alarm automatic of the "Arctos" brand of 1974, with AS Caliber 5008. Stainless steel case with press-in bottom. Hand-winding of the movement by the crown by the 2, quick set of date and weekday as well as hand-setting (three positions). A rotor winds the movement in one direction, the alarm mechanism in the other. Alarm hammer (under the balance) strikes against the bottom plate. Screwless nickel balance, Nivarox hairspring, Incabloc shock resistance, fine regulator adjustment by eccentric lever.

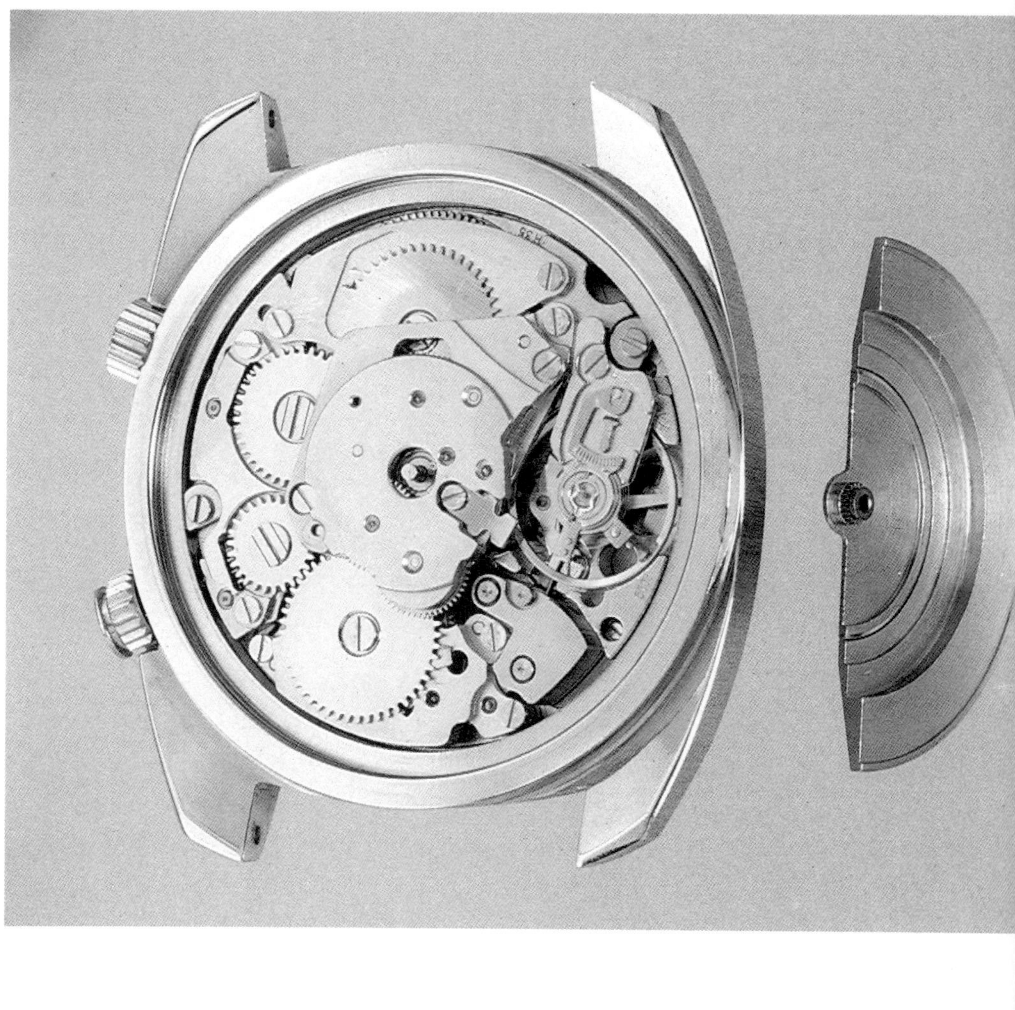

Caliber List: AS

Caliber	Diameter Lignes	mm	Height mm	Other	BPH	Jewels	Running Reserve	Notes	Year	Photo
(475)	5 1/2 x 8 1/2				18.000	7		Universe (see early AUT)	ca. 32	(xx)
796	6 1/2 x 11				18.000	15		Autorist (s. early AUT)	1931	(xx)
913	11 1/4			s	18.000				1935	
1034	12 1/4			s	18.000					
1049 A	11 1/4		5,2	s	18.000				1938	
1081	10 1/2			s	18.000			See under Mido 817		(xx)
1171	11 1/4	24,8		s	18.000	15		See Caliber 1049 A	1941	xx
1173	"	"	5,75	SCI	"	17		"	1941	
1191	12 1/4	28,0		s	18.000			See Caliber 1034	ca. 42	xx
1192	"	"		SCI	"					
1250	11	24,8	5,5	SCD	18.000	17			1944	xx
1298	12 1/4	28,0	5,15	SCD	18.000	17	36	See Zodiac 1225	ca. 46	(xx)
1302	12 1/4	28,0	4,85	s	18.000	17			ca. 48	
1315	"	"	6,25	s, QA, JG, MG	"	"			1950	xx
1320	9 1/4			SCI	18.000				1950	
1323	"			s	"				1950	xx
1336 /N	11 1/4	24,8	5,8	SCD	18.000	17		1st as Rotor		
1361 /N	11 1/2	25,6	"	SCD	"	"			1950	
1362 /N	"	"		SCD, RMA	"	"				
1364	10 1/2			SCI	18.000					
1365	"				"					
1382 /N	11 1/2	26,0		SCD, RMA	18.000	21	36	Basic Caliber 1336	1952	xx
1393	8 3/4			SCD	18.000					
1394	9 1/4				"					
1395 /N	11 1/2	26,0		SCD, QG, JG	18.000			Basic Caliber 1336		
1396 /N	"	"		SCD, QG	"					
1398	12 1/4	28,0		SCD, QG	18.000			Basic Caliber 1298		
1402 /N	11 1/2	26,0		SCD, QG, JG, MG	18.000			Basic Caliber 1336		
1410	9 1/4	20,4	5,35	SCI	18.000	17				
1412	"	"	"	SCI, QG	"	"				
1413	10 1/2	23,3	"	SCI, QG	"	"				
1422	9 1/4	20,4	5,35	SCI	18.000			Basic Caliber 1330		
1423	11 1/2	26,0	5,8	SCD	18.000	17		Basic Caliber 1336		
1424 /N	"	"	6,0	SCD, RMA	"	"		" s. Zodiac Autographic	1952	(xx)
1427	11 1/4	24,8	5,8	SCD	"	"		"		
1447	11 1/2	26,0		SCD, QG	"	"		"		
1448	"	"		SCD, RMA	"	"		"		
1457	"	"		SCD	"	"		"		
1463	"	"		SCD, RMG	"	"		"		
1468	"	"		SCD, QG, JG	"	"		"		
1476 /N	"	"		SCD, QG	"	"		"		

Caliber List: AS

Caliber	Diameter Lignes	mm	Height mm	Other	BPH	Jewels	Running Reserve	Notes	Year	Photo
1477	7	16	5,3		18.000	25			ca. 54	xx
1478 /N	11 1/2	26,0		SCD, RMG	18.000	17	36	Basic Caliber 1336		xx
1479	"	"		SCD, RMA	"	"	"	"		
1488	"	"		SCD, RMG	"	"	"	"		
1490	"	"		SCD	"	"		"		
1493	"	"		SCD, QG, JG, MG	"	"		"		
1512	"	"		SCD, RMA	"	"		"		
1513	7				21.600					
1515	11 1/2	26,0		SCD, QG	18.000			Basic Caliber 1336		
1519	"	"		SCD	"			"		
1520	"	"		SCD, RMA	"			"		
1523	"	"		SCD, RMA	"			"		
1533	6 3/4 x 8				18.000					
1535	8 3/4			SCD	18.000			Basic Caliber 1393		
1555 /N	11 1/2	26,0		SCD, QG	18.000			Basic Caliber 1336		
1563	7	16	5,3		18.000			Basic Caliber 1477		
1571	11 1/2	26,0		SCD, QG	18.000					
1572	"	"		SCD, QG	"					
1574	"	"		SCD, RMG	"					
1578	"	"		SCD, QG	"					
1580	11 1/2		5,2	SCD	18.000			Maximatic	1958	
1581	"			SCD, QG	"					
1583	"			SCD	"					
1584	"			SCD	"					
1585	"			SCD	"					
1590	13				18.000					
1591	"			QG	"					
1607	9 1/4			SCI	18.000			Basic Caliber 1320		
1624	11 1/2			SCD	18.000					
1626	13			SCD	18.000			Basic Caliber 1590		
1628	11 1/2			SCD, QG	18.000					
1635	7			SCD	18.000					
1636	13			SCD, QG	18.000			Basic Caliber 1590		
1639	6 3/4 x 8				21.600					
1641 /N	11 1/2			SCD, QG	18.000					
1654	7			SCD	21.600					
1656	11 1/2		4,2	SCD	18.000	25				xx
1657	7			SCD	21.600			Basic Caliber 1654		
1659	11 1/2			SCD, QG	18.000					
1664	"			SCD	"					

Caliber List: AS

Caliber	Diameter Lignes	mm	Height mm	Other	BPH	Jewels	Running Reserve	Notes	Year	Photo
1665	11 1/2			SCD, QG	18.000					
1673	"		4,5	SCD	"				1960	
1674	"			SCD, QG	"					
1675	13			SCD	"					
1678	"			SCD, QG	"	17		s. Mido Cal 1117 PC		(xx)
1680	11 1/2			SCD	18.000			Basic Caliber 1580		
1681	"			SCD, QG	"					
1695	"			SCD	"					
1700	11 1/2		5,2	SCD	18.000	25	44	Basic Caliber 1580	1958	xx
1701	"		"	SCD, QG	"		"	"	1958	
1709	"			SCD, QG, JG	"		"	"		
1712	11 1/2	25,6	5,2	SCD	18.000				ca. 66	
1713	"	"	"	SCD, QG, RDR	"					
1716	11 1/2	25,6	4,0	SCD	18.000				ca. 64	
1717	"	"	"	SCD, QG, RDR	"					
1739	11 1/2			SCD	18.000					
1740	"			SCD, QG	"					
1748	11 1/2	25,6	4,5	SCD	18.000			Basic Caliber 1716		
1749	"	"	"	SCD, QG, RDR	"					
1767	11 1/2			SCD, QG, JG	18.000			Basic Caliber 1580		
1775	6		4,8		18.000			Smallest auto built	1967	
1782	8 3/4			SCD	18.000					
1783	"			SCD, QG	"					
1788	11 1/2			SCD	18.000					
1789	"			SCD, QG	"					
1798	11 1/2	25,6	4,5	SCD	18.000					
1799	"	"	"	SCD, QG, RDR	"			Basic Caliber 1748		
1806	11 1/2			SCD, QG, JG	18.000					
1816	11 1/2	25,6	5,95	SCD, QG, JG, RDR	18.000			Basic Caliber 1712		
1820	11 1/2	25,6	4,75	SCD, QG, JG, RDR	18.000			Basic Caliber 1716		
1824	"	"	5,25	SCD, QG, JG, RDR	"			" " 1748		
1825	13 3/4	31,0	"	SCD, QG, JG, RDR	"			" " "		
1826	11 1/2	25,6	"	SCD, QG, JG, RDR	"			" " "		
1827	13 3/4	31,0	"	SCD, QG, JG	"			" " "		
1831	11 1/2	25,6	4,75	SCD, QG, JG	"			" " 1716		
1839	"	"	4,0	SCD, QG	"			" " "		
1847	"	"		SCD	"					
1848	"	"		SCD, QG	"					
1849	"	"		SCD, QG, JG	"					
1857	"	"		SCD	"					
1858	"	"		SCD, QG	"					
1859	"	"		SCD, QG, JG	"					
1862	11 1/2	25,6	4,0	SCD	21.600			Basic Caliber 1716		
1863	"	"	"	SCD, QG, RDR	"			"		

Caliber List: AS

Caliber	Diameter Lignes	mm	Height mm	Other	BPH	Jewels	Running Reserve	Notes	Year	Photo
1866	11 1/2	25,6	4,75	SCD, QG, JG, RDR	21.600			Basic Caliber 1716		
1873	"	"	4,0	SCD, QG	"			"		
1876	"	"	4,75	SCD, QG, JG, RDR	"			"		
1882	"	"	4,5	SCD	"			"		
1883	"	"	"	SCD, QG, RDR	"			"		
1885	13 3/4	31,0	5,25	SCD, QG, JG, RDR	"			"		
1886	11 1/2	25,6	"	SCD, QG, JG, RDR	"			"		
1892	"	"	4,5	SCD	"			"		
1893	"	"	"	SCD, QG, RDR	"			"		
1895	13 3/4	31,0	5,25	SCD, QG, JG, RDR	"			"		xx
1896	11 1/2	25,6	5,25	SCD, QG, JG, RDR	"			"		
1902	11 1/2	25,6	5,2	SCD (digital)	21.600	17		Basic-Cal 1712 (s.Revue)	1966	(xx)
1903	"	"	"	SCD, QG, RDR	"			"		
1906	"	"	5,95	SCD, QG, JG, RDR	"			"		
1913	"	"		SCD, QG	"			"		
1916	"	"		SCD, QG, JG	"			"		
1920	12 1/2			SCD, QG, JG	36.000			Basic Caliber 1716		
1985	13 3/4			SCD, QG, JG	21.600			"		
1986	11 1/2			SCD, QG, JG	"			"		
1995	13 3/4			SCD, QG, JG	"			"		
1996	11 1/2			SCD, QG, JG	"			"		
2063	11 1/2		5,2	SCD, QG	21.600	17		Basic Caliber 1712	1971	xx
2066	"		"		"				1971	
2073	"			SCD, QG (digital)					1974	
5008	13 3/4	31,0	7,6	SCD, QG, ST, REV	28.800	25			1974	xx
5103	8 3/4			QG					1975	
5106	"		5,35						1975	
5203	12	6,5							1976	

Caliber 1250
D = 24.8 mm (11 lignes), H = 5.5 mm, 17 jewels
AUTOMATIC MODULE screwed onto the basic movement as a unit, hammer swing limited by buffer springs, winding in one direction via click on hammer, central winding wheel, two reduction wheels, the first with a spring click, the second with Breguet coupling to disconnect during hand-winding, on crown wheel.
BASIC MOVEMENT: two-shank Glucydur screw balance, self-compensating flat hairspring, 18,000 BPH, Incabloc shock resistance, direct center seconds.

Caliber 1315
D = 28.0 mm (12.25 lignes), H = 6.25 mm, 17 jewels.
AUTOMATIC MODULE as a complete unit screwed to the basic movement, hammer limited by buffer springs, winding in one direction via partly toothed click-carrier disc with combined winding and reduction wheel and sprung locking click under separate plate, on crown wheel.
BASIC MOVEMENT: two-shank Glucydur screw balance, Nivarox hairspring, 18,000 BPH, Incabloc shock resistance, seconds bit.

Caliber 1323
D = 20.4 mm (9.25 lignes), 17 jewels
AUTOMATIC MODULE screwed to the basic movement as a complete unit, central rotor with heavy metal ring, winding in both directions via switching rocker, two reduction wheels, the first with spring click and the second with Breguet coupling to disconnect during hand-winding, on the locking wheel.
BASIC MOVEMENT: two-shank Glucydur screw balance, Nivarox hairspring, 18,000 BPH, Incabloc shock resistance, seconds bit.

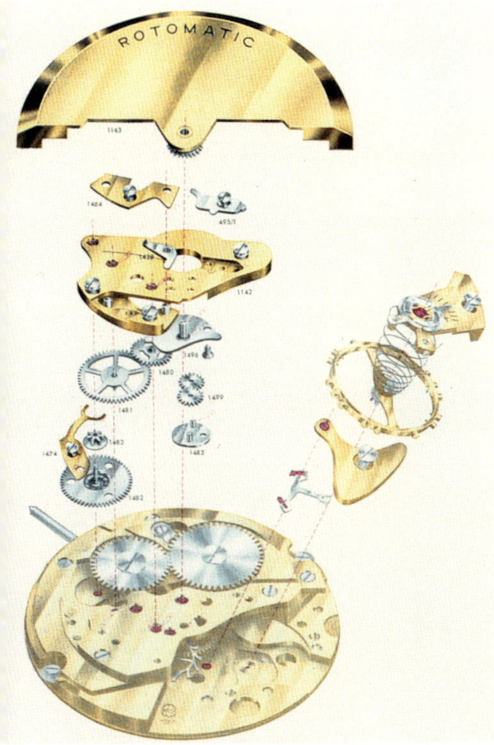

Exploded view of an automatic module (AS Caliber 1361) as oiling guide: 1485—switching rocker with two switching wheels 1499, 1480—tension wheel with locking click (1428) under separate plate (1464), 1481—first reduction wheel, 1482/83—combined reduction and Breguet coupling wheel with Breguet coupling spring (1474).

Caliber 1382 N
D = 26.0 mm (11.5 lignes), 21 jewels
AUTOMATIC MODULE screwed to basic movement, central rotor with heavy metal ring, winding in both directions via switching rocker, one tension wheel with spring click under a separate plate, two reduction wheels, the second with Breguet coupling to uncouple during hand-winding, on the crown wheel.
BASIC MOVEMENT: two-shank nickel screw balance, Nivarox hairspring, 18,000 BPH, Incabloc shock resistance, direct center seconds. Running reserve indication via ball-wheel differential drive.

Caliber 1424 N
D = 26.0 mm (11.5 lignes), H = 6.0 mm, 17 jewels
See Caliber 1382 N.

Caliber 1477
D = 16 mm (7 lignes), H = 5.3 mm, 25 jewels
AUTOMATIC MODULE screwed to the basic movement as a complete unit, central rotor with heavy metal ring, winding in both directions via unsprung twin click wheels, one reduction wheel on locking wheel.
BASIC MOVEMENT: two-shank Glucydur screw balance, Nivarox hairspring, 18,000 BPH, Incabloc shock resistance, no second.

Caliber 1478 N
D = 26.0 mm (11.5 lignes), 17 jewels
See Caliber 1382 N, but with Kif-Flector shock resistance.

Caliber 1656
D = 26.0 mm (11.5 lignes), H = 4.2 mm, 25 jewels
AUTOMATIC MODULE integrated into basic movement, large rotor with heavy metal ring mounted somewhat out of the center, winding in both directions via unsprung twin click wheels, three reduction wheels in all on a wheel lying under the dial and corresponding to the locking wheel.
BASIC MOVEMENT: two-shank Glucydur screw balance, Nivarox hairspring, 18,000 BPH, Incabloc shock resistance, direct center seconds.

Calibers 1700/01
D = 25.6 mm (11.5 lignes), H = 5.2 mm, 25 jewels
AUTOMATIC MODULE partly integrated into basic movement, central rotor with heavy metal ring, winding in both directions via unsprung twin click wheels, two reduction wheels on a wheel lying under the dial and corresponding to the locking wheel, with Breguet coupling to uncouple during hand-winding.
BASIC MOVEMENT: three-spoke screwless Glucydur balance, Nivarox hairspring, 18,000 BPH, Incabloc shock resistance, moveable hairspring block carrier, direct center seconds.

Calibers 1882/83, 1895
D = 31.0 mm (13.75 lignes), H = 5.25 mm, 25 jewels
AUTOMATIC MODULE integrated into basic movement, central rotor with ball bearings and heavy metal ring, winding in both directions via unsprung twin click wheels, two reduction wheels on crown wheel.
BASIC MOVEMENT: three-spoke screwless Glucydur balance, Nivarox hairspring, 18,000 BPH, moveable hairspring block carrier, direct center seconds.

Caliber 1902 (see page 270) D = 25.6, (11.5 lignes), H = 5.2 mm, 17 jewels
AUTOMATIC MODULE as a complete unit screwed to the basic movement, central rotor with heavy metal ring, winding in one direction via sprung lever wheel, two reduction wheels, the first with spring click, on crown wheel.
BASIC MOVEMENT: three-spoke screwless Glucydur balance, Nivarox hairspring, 18,000 BPH, Incabloc shock resistance, moveable hairspring block carrier, direct center seconds. Indication of hour, minute and second through dial apertures.

Caliber 2063
D = 25.6 mm (11.5 lignes), H = 5.2 mm, 17 jewels
See Caliber 1902, but with nickel balance, 21,600 BPH, Trishock shock resistance, without digital indication but with date indication and quick set.

Caliber 5008
D = 31.0 mm (13.75 lignes), H = 7.6 mm, 25 jewels
AUTOMATIC MODULE placed upon basic movement, central rotor with heavy metal ring, winding in one direction via one intermediate wheel, unsprung click wheel, one reduction wheel on locking wheel. Other rotor direction provides automatic winding of alarm via identical reduction drive but without intermediate wheel.
BASIC MOVEMENT: three-spoke screwless nickel balance, Nivarox hairspring, 28,800 BPH, Incabloc shock resistance, fine adjustment of regulator via eccentric lever, moveable hairspring block carrier, direct center seconds.
ALARM MECHANISM integrated into basic movement, alarm hammer strikes against the case.
Date and weekday indication with quick set.

Caliber Development

Jules Audemars and Edouard-Auguste Piguet founded a business in Le Brassus in 1875 that became renowned in particular for its fine watches with minute repeat and perpetual calendars. The firm made its first automatic around 1956. The raw movement for this Caliber 2498 came from Jaeger LeCoultre, which likewise produced the ebauches for the later automatics and reserved them especially for AP, but simultaneously also for Vacheron Constantin and in part for Patek Philippe. The rotor used in this design is weighted with a gold peripheral segment and is additionally supported by four ruby balls that run under cover in the central area. Winding is done in both directions via a robust switching rocker. In Caliber 2070, the four balls are replaced by ruby rollers but are now visible in the rotor cutouts. In 1960 the movement was fitted with a Gyromax balance and date indication (Caliber 2072).

With the new Caliber 2120 of 1967, AP was able to offer one of the slimmest of all automatics, its height measuring only 2.45 mm. With the low height of this movement, the secure mounting of the big central rotor weighted with an 18-karat gold segment is a special problem. Therefore an additional periphery brace with four ruby rollers mounted on the backplate became necessary.

The sporting watch model of 1972, named "Royal Oak" after the British battleship, was equipped with this movement. The slogan used when it was introduced: "steel outside—gold inside", was meant to refer to the rotor.

By adding a module 1.66 mm high between the movement and the dial, a slim automatic with a perpetual calendar (Caliber 2121/1) could be made in 1978.

In addition to this movement, the more robust Caliber 2122 ff was used, likewise utilized by Jaeger LeCoultre since 1967 as Caliber 888. However here a heavy metal segment was used instead of the 21-karat gold segment used by AP. In its last years, a smaller (nine-lignes) automatic movement by the "Frederic Piguet SA", likewise of Le Brassus, was used; its gold rotor wound in only one direction.

The automatic tourbillon completed in 1986 was developed completely by AP and thus is the firm's only independent design with automatic winding. In terms of design it is actually a throwback to the days of the pocket tourbillon and the early automatics. The production of this lightweight titanium turning frame, only 7.2 mm in diameter and 2.5 mm high, could be done only with the help of computer-aided electroerosion. The case back itself serves as a bottom plate, in the cutouts of which the swinging mass of the rotor is recessed. The rotor was made of a heavy platinum-iridium alloy and wound in only one direction. The bottom crown can be used only for setting, the winding drive had to be omitted for lack of space. The tourbillon carriage is visible in a dial cutout, and requires something less than one minute for a full rotation. There is no provision for the seconds to be read.

Caliber Notes

Caliber *2499*
Jaeger LeCoultre Ebauche
D = 29.5 mm (13 lignes), H = 5.4 mm
AUTOMATIC MODULE screwed to basic movement. Central rotor with 18-karat gold ring, additionally mounted on four ruby balls, winding in both directions by a switching rocker, two reduction wheels, the first with a locking click, on a doubled crown wheel (to disconnect during winding).
BASIC MOVEMENT: two-spoke Glucydur screw balance, self-compensating Breguet balance, 18,000 BPH, Jaeger LeCoultre shock resistance, swan's-neck fine regulation, center seconds.

Caliber 2071
D = 29.5 mm (13 lignes), H = 5.4 mm
See Caliber 2499, but the ruby balls are replaced by four visible ruby rollers, locking click under separate plate.

Caliber 2020/1
Jaeger LeCoultre Ebauche
D = 28.0 mm (12.5 lignes), H = 2.45 mm, 36 jewels
AUTOMATIC MODULE integrated into the basic movement under the same plate. The large central rotor has ruby bearings, a 21-karat gold ring and peripheral beryllium ring for mounting support, and glides on four ruby rollers recessed in the plate. It winds in both directions with an intermediate wheel with ball bearings, one small reduction wheel, switching rocker, two additional reduction wheels (the first with a locking click). On the locking wheel there is a coupling wheel that disconnects during winding.
BASIC MOVEMENT: three-spoke screwless Glucydur balance, Nivarox hairspring, 19,800 BPH, Kif-Elastor shock resistance, moveable hairspring block carrier, anchor-wheel mount as combined jewel casing, eccentric second wheel.

Audemars Piguet Caliber 2499

Men's automatic made by Audemars Piguet circa 1956 with Caliber 2499 (exclusive raw movement by Jaeger LeCoultre). 18-karat gold case with press-in bottom, winding crown, dial inlays and hands of gold.

Ball-supported rotor with 18-karat gold ring, winding in both directions via switching rocker, uncoupling during hand-winding. Glucydur screw balance, self-compensating Breguet hairspring, JLC shock resistance, fine regulator adjustment.

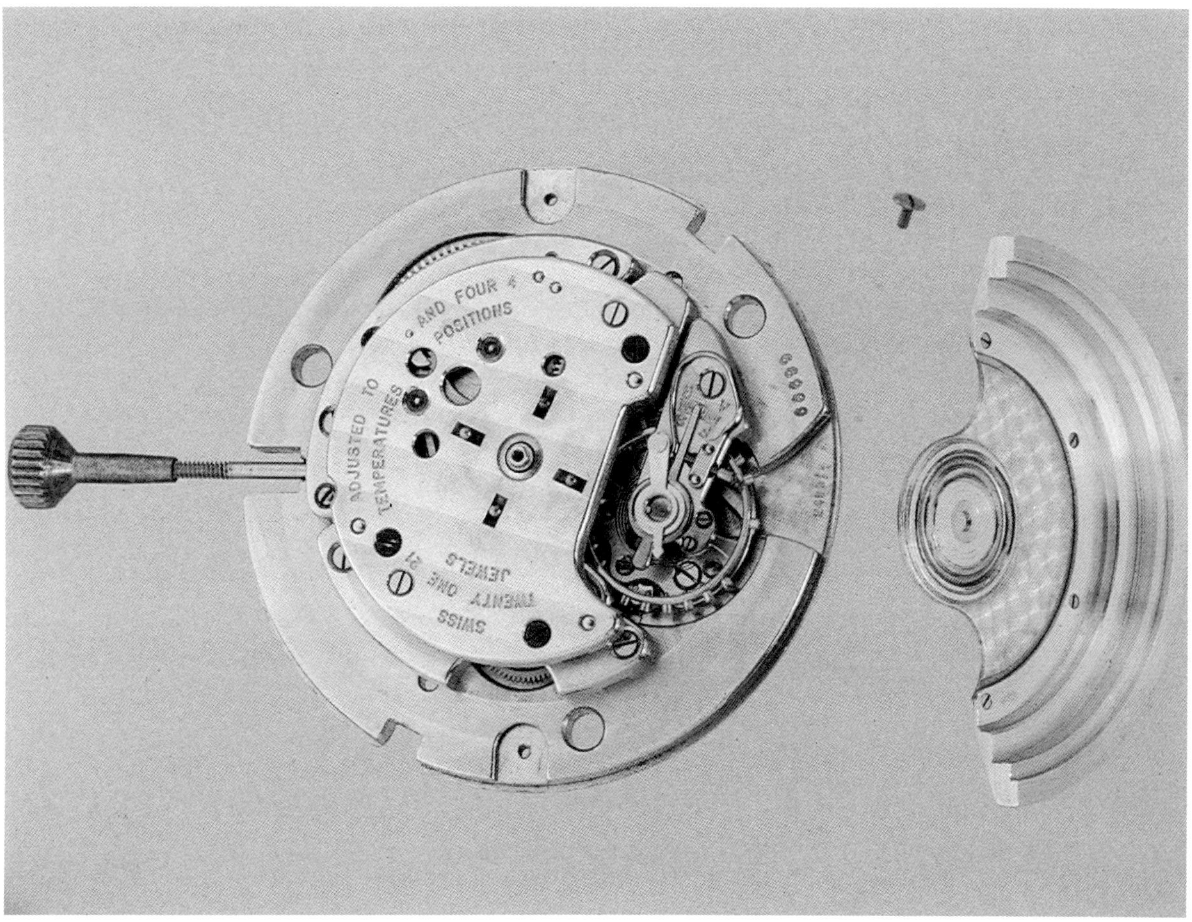

Audemars Piguet Caliber 2071

Men's automatic made by Audemars Piguet circa 1960 with Caliber 2071 (exclusive raw movement by Jaeger LeCoultre).

18-karat gold case with screw bottom and gold hands.

Centrally mounted rotor with 18-karat gold ring, supported by four ruby rollers, movement otherwise identical to Caliber 2499 (see page 47), but with Kif-Flector shock resistance and separate locking-click plate (at 12:00).

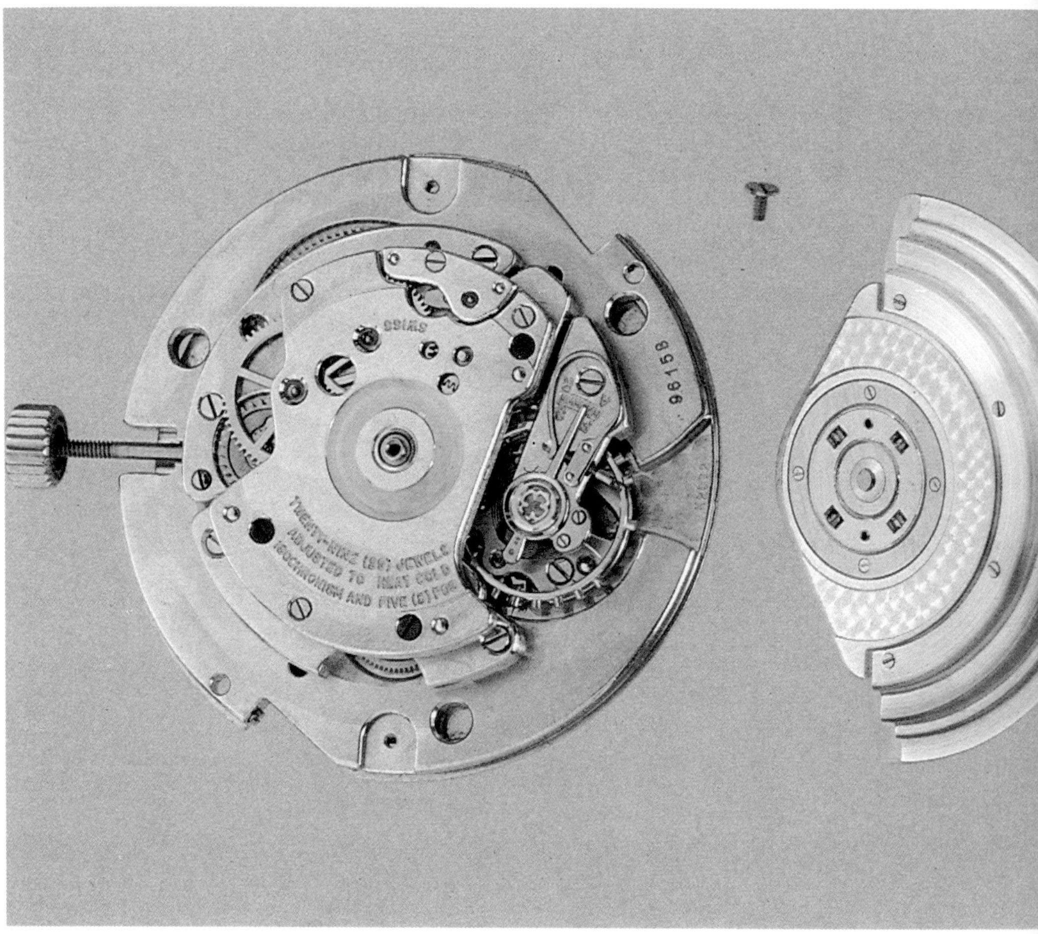

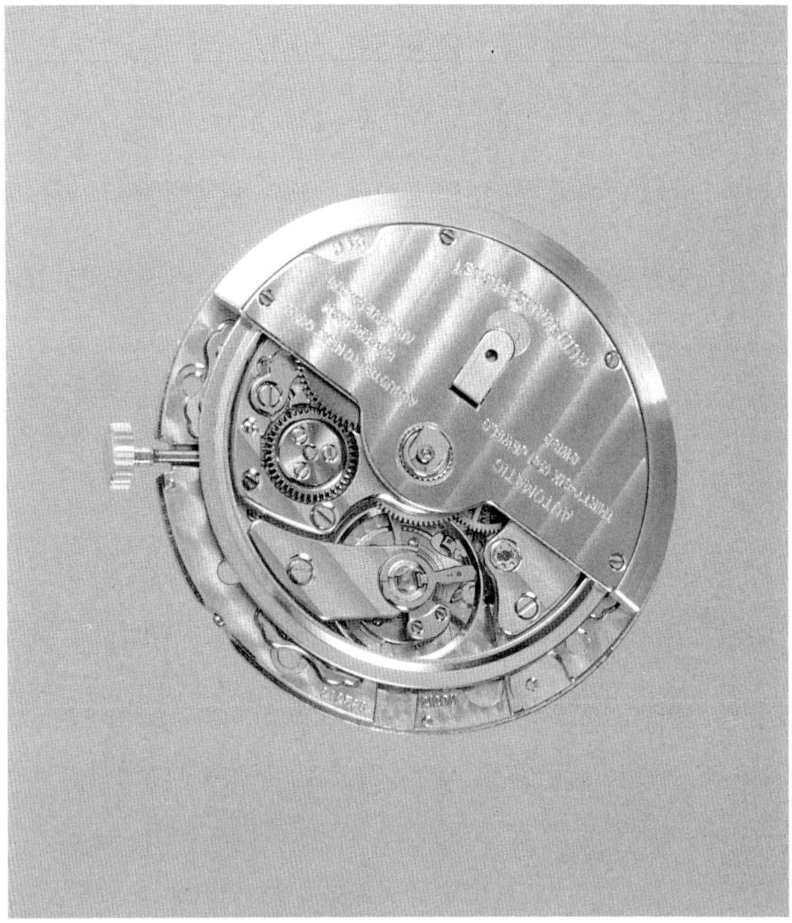

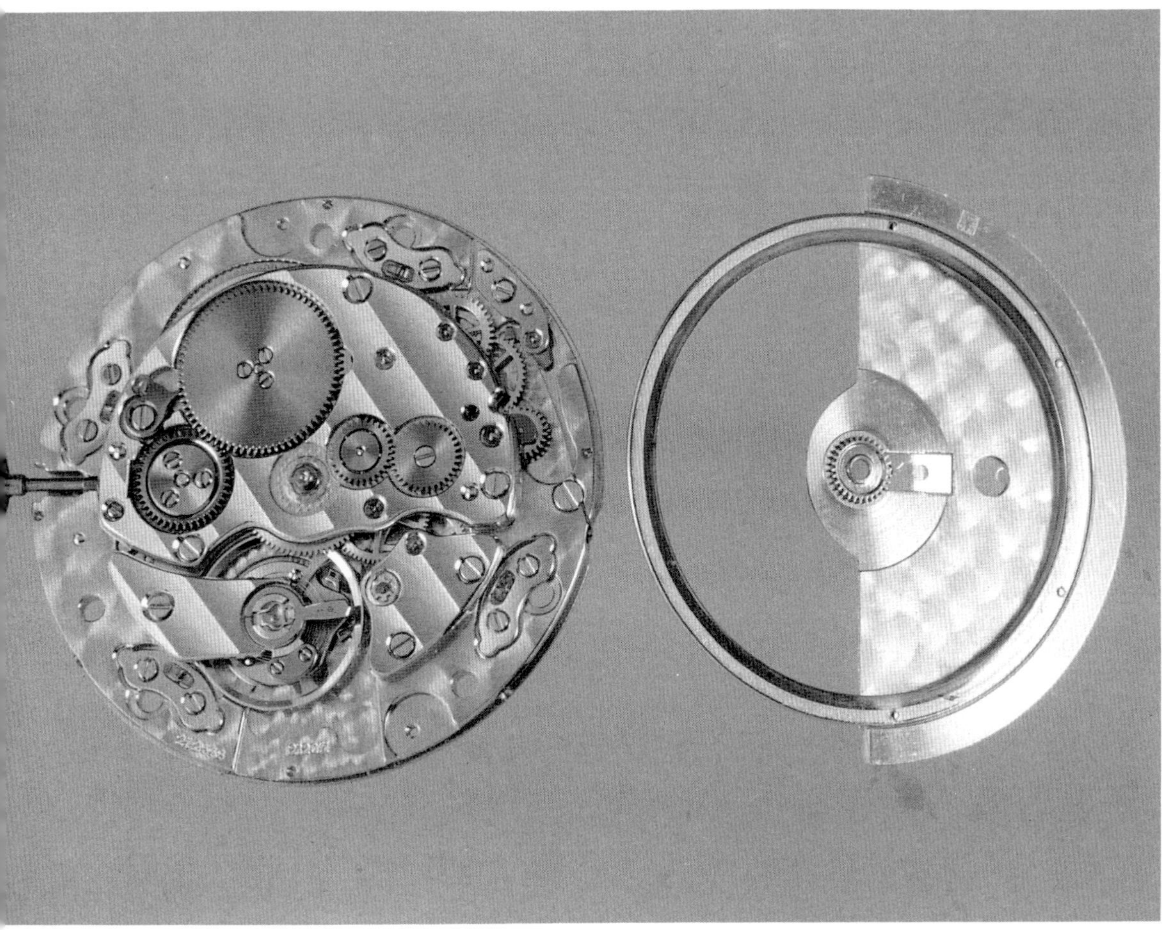

Audemars Piguet Caliber 2120/1

Very flat men's automatic by Audemars Piguet, circa 1970, with Caliber 2120/1 (exclusive raw movement by Jaeger LeCoultre).
18-karat gold case, dial inlays and hands of gold.
Centrally mounted rotor with 21-karat gold ring and outer ring, supported by four peripheral ruby rollers (at 1, 4, 7 and 10:00), winding in both directions via switching rocker (at 2:30), uncoupling of the hand-winding drive.
Screwless Glucydur balance, Nivarox hairspring, Kif-Ultraflex shock resistance.

Audemars Piguet Caliber 2062

Automatic Caliber 2062 for a lady's watch, made by Audemars Piguet circa 1970, with similar construction to Omega Caliber 685 (see page 235).

Rotor with gold ring, winding in both directions via double click-wheel switcher, uncoupling of the hand-winding drive.

Screwless Glucydur balance, Nivarox hairspring, Kif-Flector shock resistance.

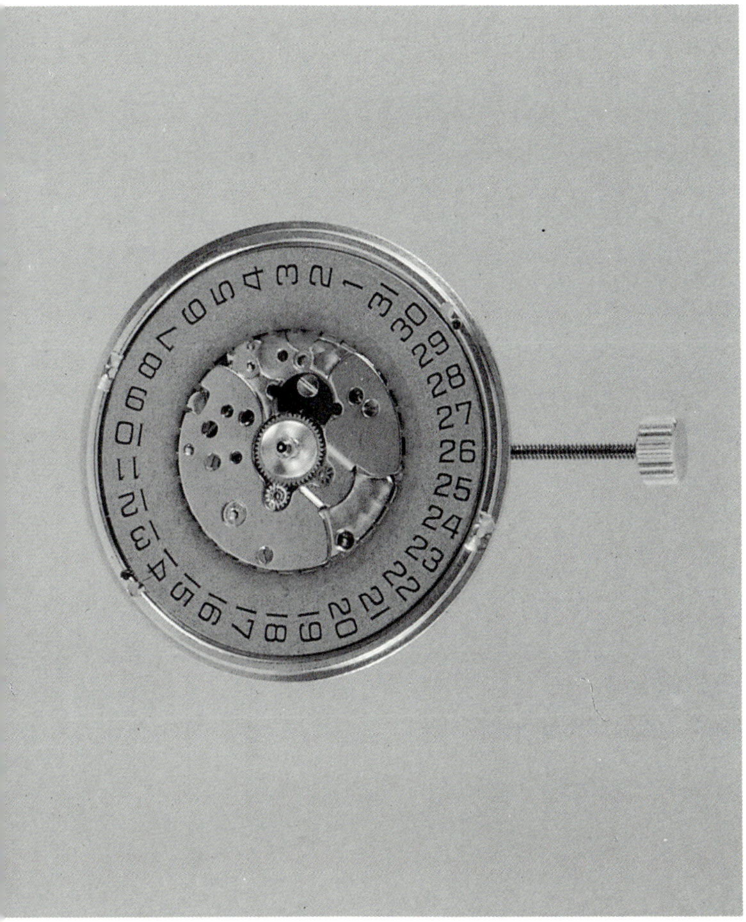

Audemars Piguet Caliber 2131
Automatic Caliber 2131 in medium size, made by Audemars Piguet circa 1988, with a raw movement by Fr. Piguet.
12-karat gold rotor with ball bearings, winding in one direction via click wheel (at 5:00 in photo B).
Screwless Glucydur balance, self-compensating flat hairspring, Kif-Ultraflex shock resistance.

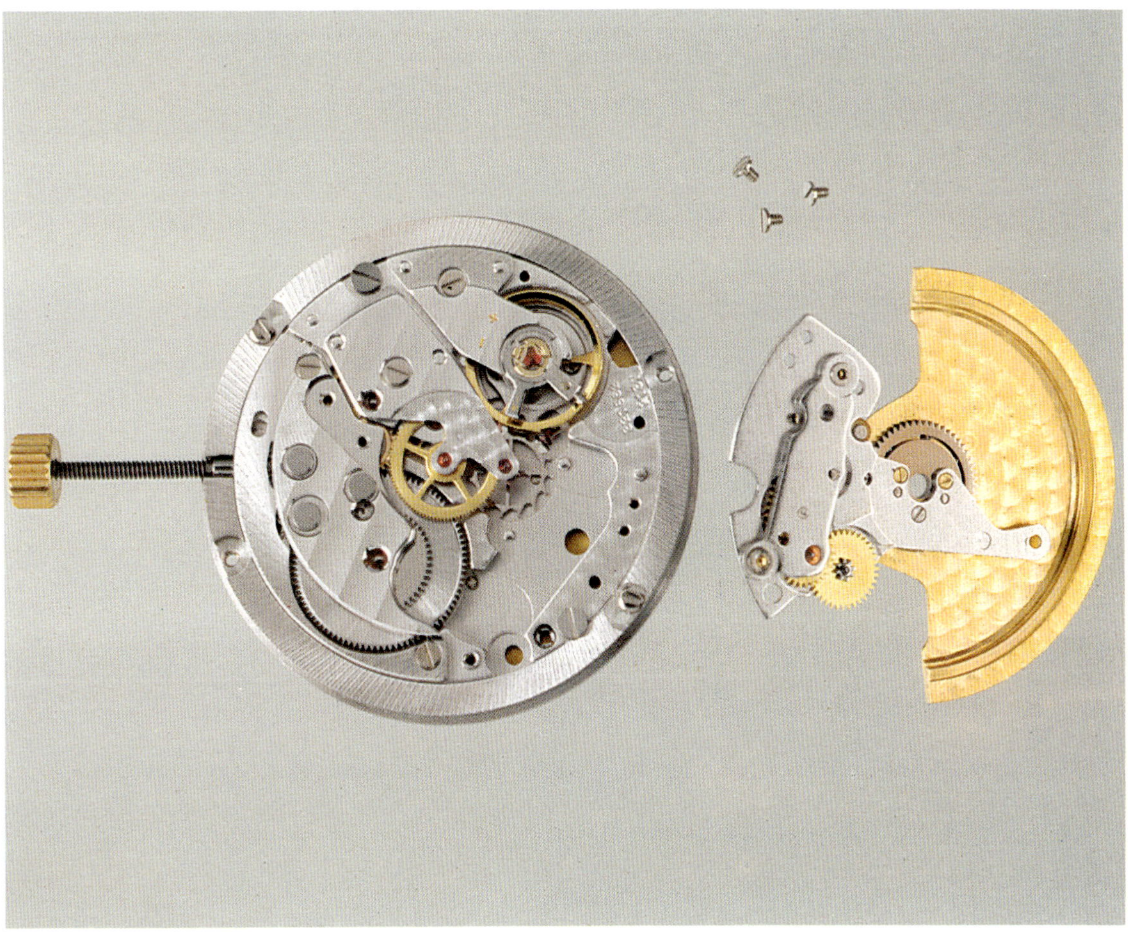

Caliber 2062
D = 15.3 mm (6.75 lignes), H = 4.4 mm, 24 jewels
AUTOMATIC MODULE partially integrated into the basic movement. Central rotor with gold ring and roller bearings, winding in both directions, each with one intermediate pinion on the unsprung double click wheel, one reduction wheel on the locking wheel. Hand-winding drive uncoupled by a sprung gliding pinion (and reversing wheel) between the crown and locking wheels.
BASIC MOVEMENT: four-spoke screwless Glucydur balance, Nivarox hairspring, 21,600 BPH, Kif-Flector shock resistance, moveable hairspring block carrier, eccentric second wheel without second indication.

Caliber 2131
Ebauche by Frederic Piguet of Le Brassus
D = 23.3 mm (10.5 lignes), H = 3.2 mm, 19 jewels
AUTOMATIC MODULE integrated into the basic movement, 21-karat gold rotor with ball bearings, winding in one direction via an unsprung combined winding and locking wheel as gear lock, two reduction wheels on the locking wheel of the barrel.
BASIC MOVEMENT: three-spoke screwless Glucydur balance, self-compensating flat hairspring, 21,600 BPH, Kif-Ultraflex shock resistance, moveable hairspring block carrier, direct center seconds, date indication.

Caliber List: Audemars Piguet

Caliber	Diameter Lignes	mm	Height mm	Other	BPH	Jewels	Running Reserve	Notes	Year	Photo
2062	6 3/4	15,3	4,4		21.600	24			ca. 70	xx
2070	13	29,0	5,1	s	18.000	29		Eubache by J. Coultre	ca. 58	
2071	"	"	5,4	SCI	"	"		"		xx
2072	"	"		SCI, QG	"	"		"	1960	
2120	12 1/2	28,0	2,45		19.800	36		Eubache by J. Coultre	1967	xx
2120/1	"	"	"		21.600	"		"		
2121	"	"	3,05	QG	19.800	"		"	1977	
2121/1	"	"	4,0	Perpetual Calendar	"	38		"	1978	
2122	11 1/2	26,0	3,25	QG, RDR		33		Eubache by J. Coultre		(xx)
2123	"	"	"	SCI, QG, RDR	28.800	"		" (Caliber 888)		
2123/1	"	"	"	SCI, QG, RDR	"	"		"		
2124	"	"	"	QG, RDR	21.600	"	45	"		
2125	"	"	"	SCI, QG, RDR	"	"	"	"		
2130	9	20,4	3,2	SCD, QG	21.600	19	38	Ebauche Fr. Piguet	ca. 86	
2131	10 1/2	23,3	"	SCD, QG	"	"	45	"		xx
2498	13	29,0	5,1	s	18.000			Eubache by J. Coultre	ca. 56	
2499	"	"	5,4	SCI	"			"		xx
Tourbillon		28,5 x 32,5	4,8			32	50	Pendulum Winding	1986	

Caliber Development

The firm of "Baumgartner & Frères S.A." (BFG), founded at Grenchen in 1899, limited itself to the production of raw movements with pin lever escapements (Roskopf escapements). Opening additional production facilities in Switzerland (Moutier, Reconvillier, Rueti) and northern Italy (Verbania), the firm became one of Switzerland's largest producers of raw movements.

The firm's first automatic was Caliber 92, the "Bovimatic", introduced around 1950 and featuring a hammer as a swinging mass, winding in both directions. The hammer had no central pinion and made contact with the specially designed click-carrier rocker at both ends of its swing. In between, it moved completely freely and could use this distance to gain speed. The actual automatic drive could thus be kept very simple. When it was in action, the winding mechanism was uncoupled. Although the movement was composed only of stamped sheet-metal parts, all the important bearings were made of ruby jewels.

The next automatic movement was ready to go into production in 1960. This Caliber 1582 was equipped with a central rotor winding in both directions and built into the hand-wound Caliber 158, the drive of which was designed according to the most modern concepts. It was able to hold a large barrel with a correspondingly dependable up-down indicator.

Probably the last design for an automatic caliber was made in 1972 and produced in two sizes, for men's watches as Caliber 630 and for ladies' models as Caliber 921, the latter with fast beat escapement for its balance.

Caliber Notes

Caliber 92
D = 29.3 mm (13 lignes), H = 7.35 mm, 17 jewels
AUTOMATIC MODULE mounted on the basic movement. Monometallic hammer without central pinion, but having ruby bearings, striking against arched sheet-metal strips,and winding in both directions via a click-carrier rocker that is moved in one direction by the stroke of the outer hammer ring, in the other by a pin notch next to the hammer mount. There is a combined winding and reduction wheel on the locking wheel and a winding drive which is disconnected by a gear rocker between the crown and locking wheels.
BASIC MOVEMENT: three-spoke brass ring balance, Nivarox hairspring, 18,000 BPH, Kif-Protechoc shock resistance, pin lever escapement, anchor wheel between cap jewels, center seconds.

Caliber 1582
D = 30.7 mm (13.5 lignes), H = 6.6 mm, 21 jewels
AUTOMATIC MODULE screwed to the basic movement. Monometallic central rotor with sliding bearings, winding in both directions via gliding pinion with "swimming" mount, tension wheel, two reduction wheels (the second serving as a coupling wheel to disconnect during winding) on the locking wheel. The winding mechanism is uncoupled by a moveable crown wheel.
BASIC MOVEMENT (Caliber 158): two-spoke brass ring balance, Nivarox hairspring, 18,000 BPH, Novodiac shock resistance, pin lever escapement, direct center seconds.

Caliber List: BFG

Caliber	Diameter Lignes	mm	Height mm	Other	BPH	Jewels	Running Reserve	Notes	Year	Photo
92	13	29,3	7,35	SCI	18.000	17		Bovimatic	ca. 50	xx
630	12 1/2	28,6	5,15		18.000				1972	
921	9 1/2	21,9	5,55		21.600	1/17/21/25	50		1972	
921	"	"	"		"	"	"			
921	"	"	6,25	SCD, QG, RDR	"	"	"			
921	"	"	6,95	SCD, QG, JG, RDR	"	"	"			
1582	13 1/2	30,7	6,6	SCD	18.000	17/21		Basic Movement 158	1960	
1582	"	"		SCD,QG	"	"		" "		xx

BFG Caliber 92, "Agora"
Men's automatic of the "Agora" brand, circa 1950, with BFG Caliber 92 movement.
Gilded case with stainless steel screw bottom.
Hammer winding in both directions via click-carrier rocker, uncoupling of the hand-winding drive.
Brass ring balance, flat hairspring, Kif-Protechoc shock resistance, Roskopf hook-anchor escapement.

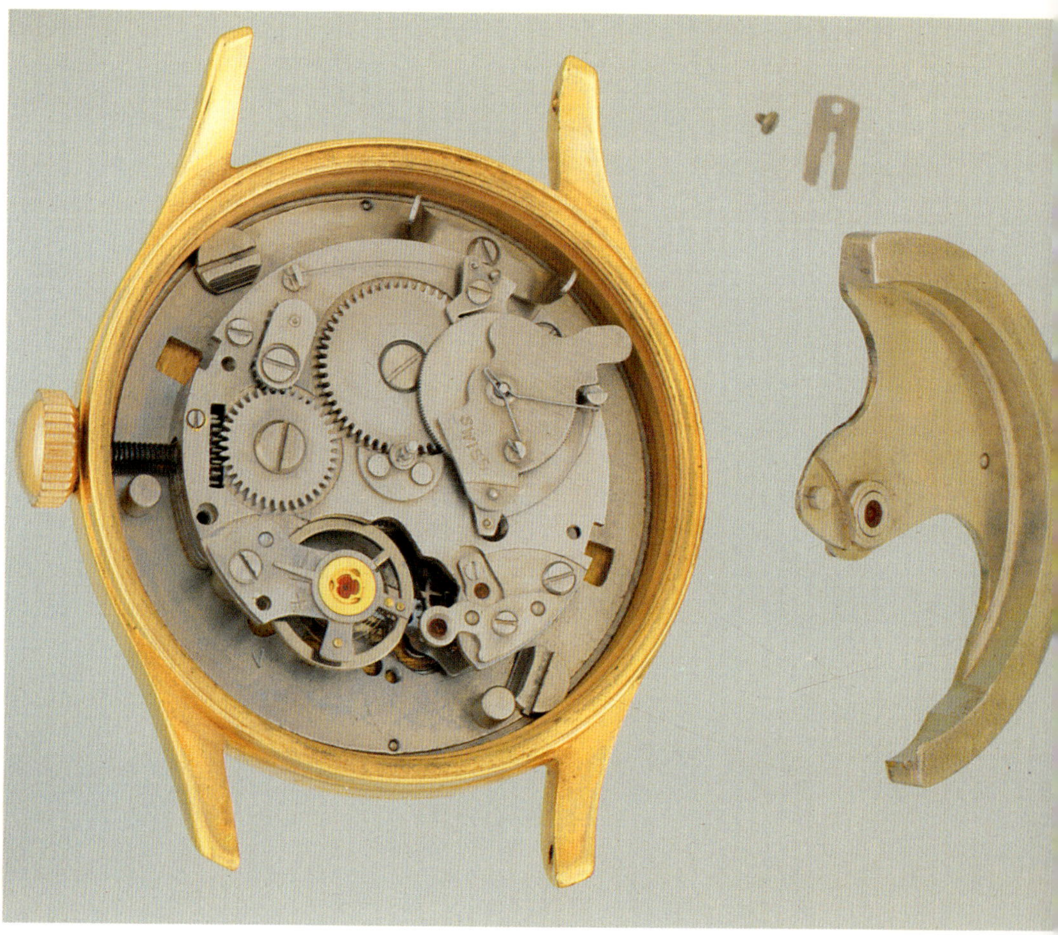

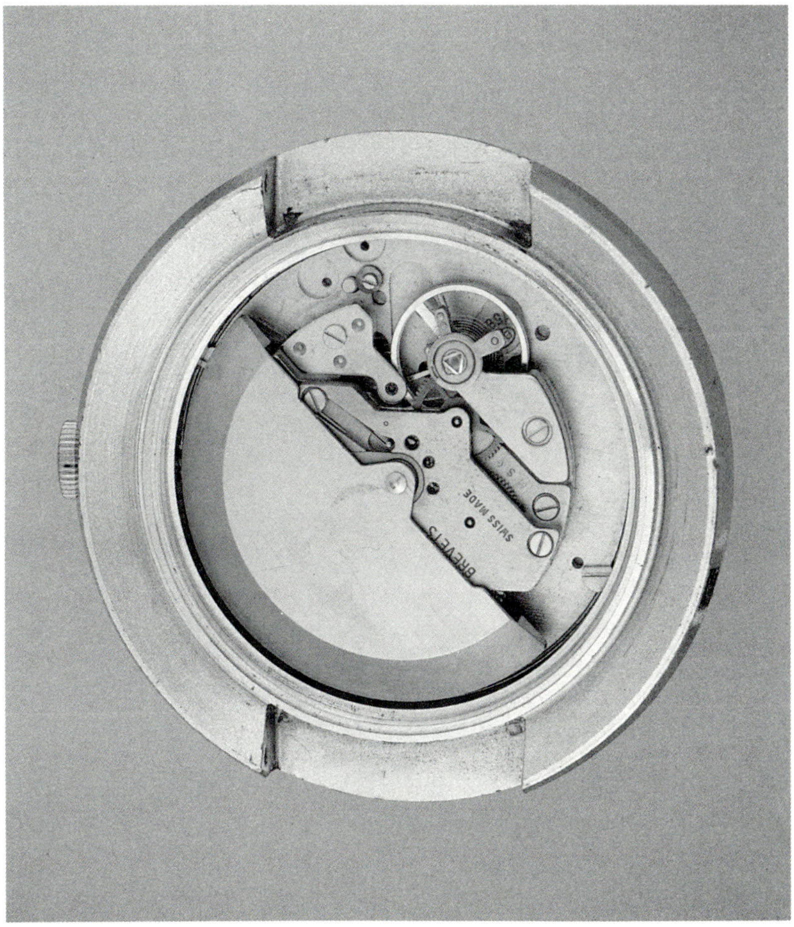

BFG Caliber 1582, "Geneva"

Diver's automatic of the "Geneva" brand, circa 1962, with BFG Caliber 1582.

Chromed brass case with stainless steel screw bottom, turning bezel with minute scale.

Monometallic rotor winding in both directions via gliding-pinion switcher, complete reciprocal uncoupling between the automatic and hand-winding drives.

Brass ring balance, flat hairspring, Novodiac shock resistance, Roskopf hook-anchor escapement.

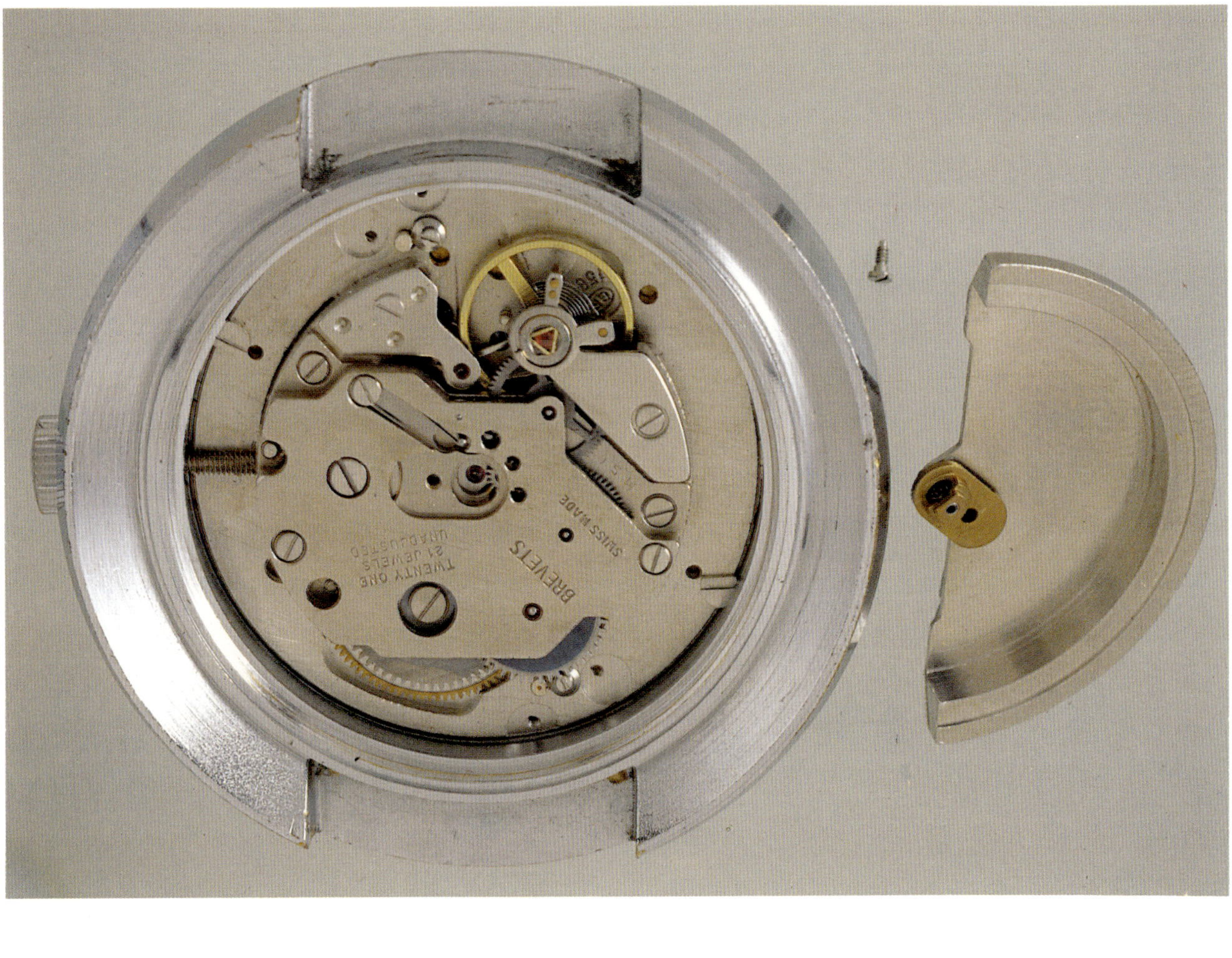

Caliber Development

The "Bulova Watch Co." firm was founded in Biel in 1885 and later established not only distribution but its own production facility in the USA. Their movements from the firm's Swiss headquarters were installed in American-made cases.

At the beginning of the 1930s, the firm showed a keen interest in the development of the automatic wristwatch and applied for Swiss patents in 1933 for both a model with pendulum winding (Patent 166843) and one with roller winding (Patent 169089).

The mass of the winding rotor was a flat plate over the movement. The entire caliber (13 AP) had been developed especially for this purpose and integrated the complete automatic drive. Winding was done in one direction via an eccentric click—which also uncoupled the winding mechanism and thus avoided additional friction loss during automatic operation (see p. 21).

This movement may have been produced by the firm of "La Champagne Ls. Müller & Cie", also located in Biel, which sold the "Aster" brand of watches and also produced the "Wig-Wag" roller automatic.

The roller winding design patented by Bulova, on the other hand, probably never went into series production, though one of similar design appeared under the brand name of Bucherer of Lucerne at the beginning of the 1930s (see p. 19).

Until the 1950s, the Bulova's activities in the automatic realm went no further, using movements from other sources. In the 1950s the firm began to design its own automatic again at almost the same time as the development of the electric wristwatch. This rotor movement, introduced as Caliber no. 11 ACAC, was equipped with a typical Bulova switching control. It used twin Breguet couplings combined with a V-shaped Breguet spring. This drive element was also found in the subsequent and slim automatic caliber made in the 1960s.

Caliber List: Bulova

Caliber	Diameter Lignes	mm	Height mm	Other	BPH	Jewels	Running Reserve	Notes	Year	Photo
6 BPA	6 3/4 x 8									
6 CHAC	6 3/4 x 7 3/4			SCI						
6 CMA	6 1/2 x 7				18.000	17		AS-Ebauche (Modified)		
6 CMAC	"			SCI	"	"		"		
6 COAC	"			SCI	"	"		"		
7 BEAC	7 3/4			SCD						
7 BEACD	"			SCD, QG						
7 EDA	7 3/4	17,2	3,9		21.600	30	45	Buren-Ebauche 04	1968	
7 EDAD	"	"	4,3		"	"	"	Buren-Ebauche 06	"	
8 BBA	8 3/4 x 9 1/4									
9 AR	9 1/4									
9 ARC	"			SCI						
9 AS	"									
9 ASC	"			SCI						
10 BOAC	10 1/2			SCI						
10 BP	"									
10 BPAC	"			SCI						
10 BPAD	"			QG						
10 BRC	9 1/4			SCI						
10 CBAC	10 1/2			SC						
10 CMC	"			SC						
10 CSC	9 1/4			SCD	18.000	17		AS-Ebauche 1320	ca. 50	
11 ACAC	11 1/2			SCD	18.000	23			ca. 55	xx

Caliber List: Bulova

Caliber	Diameter Lignes	mm	Height mm	Other	BPH	Jewels	Running Reserve	Notes	Year	Photo
11 AFA	11 1/2				18.000			Watch for the blind		
11 AFAC	"			SCI	"					
11 AFACD	"			SCI, QG	"					
11 AFCD	"			SCI, QG	"	17				
11 ALA	11 1/2				18.000					
11 ALAC	"			SCI	"					
11 ALACD	"			SCI, QG	"					
11 ALLA	"				"					
11 ANAC	11 1/2			SCD	19.800	23			ca. 65	
11 ANACB	"			SCD, QG, JG	"	"				
11 ANACD	"			SCD, QG	"	"				
11 AOACD	11 1/2			SCD, QG	21.600				1968	xx
11 BKACB	11 1/2			SCI, QG, JG	21.600	23				
11 BLA	11 1/2				18.000					
11 BLAC	"			SCI	"					
11 BLACD	"			SCI, QG	"	23			ca. 63	xx
11 BLLA	"				"					
12 EBA	12 1/2	28,2	2,85		19.800	30	54	Buren-Ebauche 1280 (Ambassador)	ca. 68	xx
12 EBAC	"	"	3,15	SCI,	"	"	"	" 1320		
12 EBACD	"	"	3,6	SCI, QG	"	"	"	" 1331		
12 EBAD	"	"	3,3	QG	"	"	"	" 1281		
12 ECACD	12 1/2			SCI, QG	19.800					
12 ECAD	"			QG	"					
13 AP	8 3/4 x 12			s	18.000	17		Pat. 166843 See Aster Cal 28-20	ca. 35	(xx)

The electric "Accutron" tuning-fork watch that had been developed to be ready for production by this time was manufactured in Bulova's American factories and thus went into competition with the firm's own automatic watch made in its Swiss factories. The firm has shown little interest in modernizing it, and makes wide use of movements from outside. For example, the very elegant-looking "Ambassador", introduced around 1968, utilizes a Buren Intramatic movement.

Bulova Caliber 11 ACAC

Men's automatic by Bulova, circa 1955, with Caliber 11 ACAC.

Chrome plated case with stainless steel screw bottom, winding crown with Bulova monogram, brass rotor winding in both directions via patented Breguet coupling switcher (the V-shaped Breguet coupling spring can be seen in the triangular plate cutout at 9:00).

Glucydur screw balance, Nivarox hairspring, Kif-Flector shock resistance (case with strong marks of use, lower left band attachment bent).

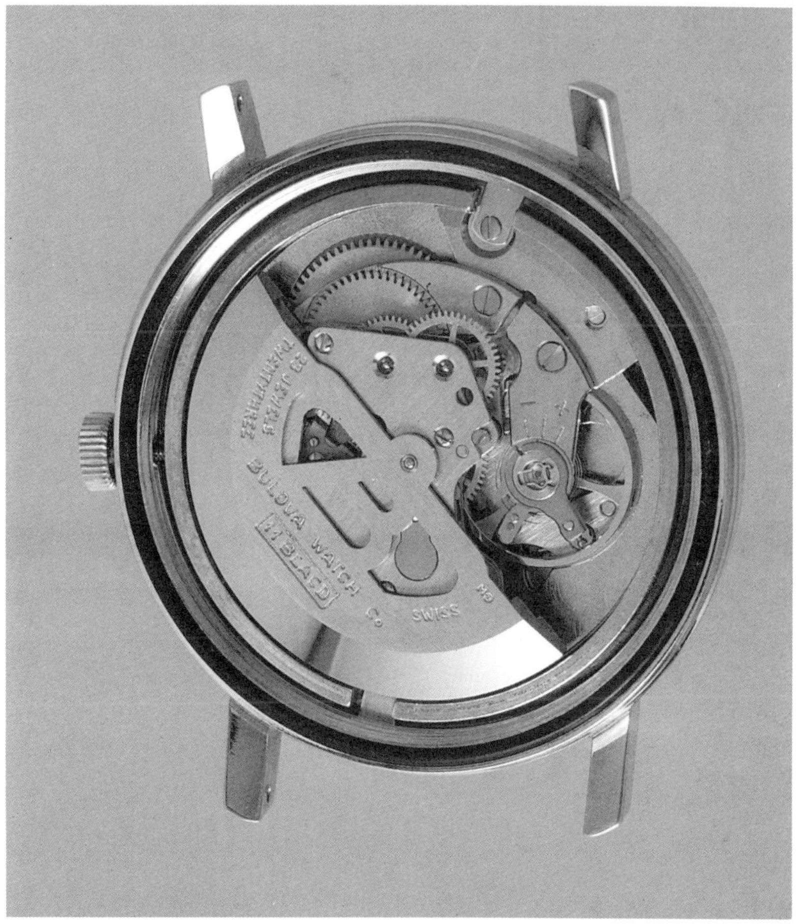

Bulova Caliber 11 BLACD

Man's automatic by Bulova, circa 1963, with Caliber 11 BLACD.

Gilded case with stainless steel screw bottom, winding crown with Bulova monogram.

Brass rotor winding in both directions via patented Breguet coupling switcher (the V-shaped Breguet coupling spring can be seen in the triangular plate cutout at 9:00).

Screwless Glucydur balance, Nivarox hairspring, Kif-Ultraflex shock resistance.

Bulova Caliber 11 AOACD

Men's sport automatic made by Bulova, circa 1968, with Caliber 11 AOACD.
Stainless steel case with screw bottom and screw crown, plus turning bezel with minute scale.
Rotor with heavy metal ring winding in both directions via Breguet coupling switcher.
Screwless Glucydur balance, Nivarox hairspring, Kif-Ultraflex shock resistance.

Bulova Caliber 12 EBA, "Ambassador"

Men's flat automatic, "Ambassador" by Bulova, circa 1968, with Intramatic movement made by Buren.
Gilded bowl case (40 u), winding crown with Bulova monogram.
Heavy metal microrotor winding in both directions via gliding-pinion switcher ("swimming" gliding-pinion mount under large ruby jewel). Complete reciprocal uncoupling between the automatic and hand-winding drives.
Screwless Glucydur balance, Nivarox hairspring, Incabloc shock resistance, "Isochron" fine adjustment.

Drawing of the automatic module of a Bulova watch:
Turning-moment transmission from the rotor, turning clockwise, via the direct switching wheel, Breguet coupling and two reduction wheels to the locking wheel of the barrel.
The Breguet coupling of the indirect switching wheel is running free here, as the upper pinion is driven by the direct switching wheel, the lower one by the first reduction wheel, in opposite directions.

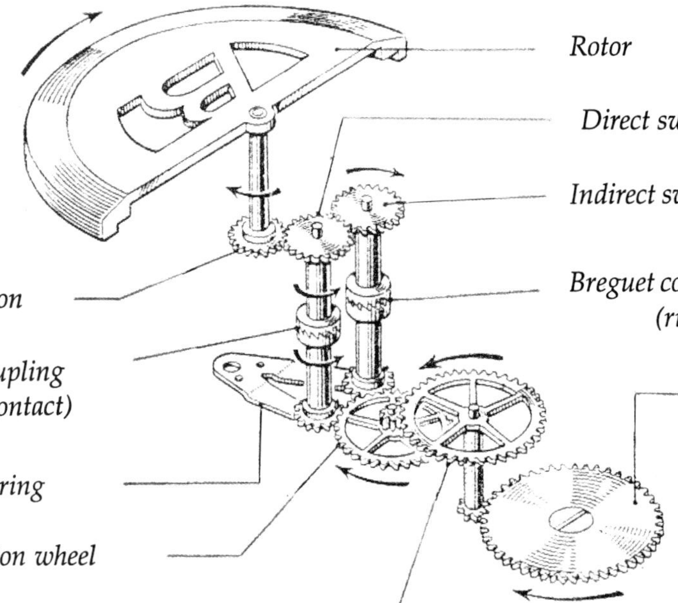

Caliber Notes

Caliber 13 AP
(same as Aster Caliber 28-20, see page 21)
8.75 x 12 lignes, 18 jewels
AUTOMATIC MODULE integrated into basic movement. Lead-weighted flat pendulum, winding in one direction via eccentric click, combined winding and reduction wheel on locking wheel, crown-wheel rocker to uncouple winding mechanism.
BASIC MOVEMENT: two-spoke screw balance, flat hairspring, 18,000 BPH, no shock resistance, jewel anchor with lateral anchor fork.

Caliber 11 ACAC
D = 25.6 mm (11.5 lignes), 23 jewels
AUTOMATIC MODULE screwed to the basic movement as a complete unit, central brass rotor winding in both directions via twin Breguet coupling wheels, two reduction wheels on the locking wheel.
BASIC MOVEMENT: two-spoke Glucydur screw balance, Nivarox hairspring, 18,000 BPH, Kif-Flector shock resistance, center seconds.

Caliber 11 BLACD
D = 25.6 mm (11.5 lignes), 23 jewels
AUTOMATIC MODULE: see Caliber 11 ACAC
BASIC MOVEMENT: three-spoke screwless Glucydur balance, Nivarox hairspring, 18,000 BPH, Kif-Ultraflex shock resistance, moveable hairspring block carrier, center seconds, date indication.

Caliber 11 AOACD
D = 25.6 mm (11.5 lignes), 23 jewels
AUTOMATIC MODULE screwed to basic movement, central rotor with heavy metal ring, winding in both directions via twin Breguet coupling wheels, two reduction wheels on crown wheel.
BASIC MOVEMENT: three-shank screwless Glucydur balance, Nivarox hairspring, 21,600 BPH, Kif-Ultraflex shock resistance, moveable hairspring block carrier, direct center seconds, date indication.

Caliber 12 EBA
Buren Ebauche (see page 68)
D = 28.2 mm (12.5 lignes), H = 2.82 mm, 30 jewels
AUTOMATIC MODULE integrated into the movement with a small heavy-metal planetary rotor recessed in the movement and mounted outside the center. It winds in both directions via a guiding pinion mounted swimming between ruby discs (formed as roller drive). There is one tension wheel with a spring click, three reduction wheels on the locking wheel, reciprocal uncoupling of hand- and automatic winding by a sliding coupling in the second reduction wheel or gliding pinion between the crown and locking wheels.
BASIC MOVEMENT: three-shank screwless Glucydur balance, Nivarox hairspring, 19,800 BPH, Incabloc shock resistance, double fine adjustment by regulator and hairspring block carrier (Buren isochronic regulator), center seconds, play-free hand-setting drive.

Caliber Development

This firm was founded at Buren on the Aare in 1873 by F. Suter and until 1972, operated under the names of H. Williamson (1898-1932), the Buren Watch (1932-1966) and Hamilton-Buren (1966-1972). The first Buren automatic watch, the Caliber 525 of 1945, had pendulum winding with its swinging mass located not over the movement, as in other manufacturers' designs, but recessed inside, thus not increasing the height. For the first time in this type of winding, the pendulum wound in both directions. Before developing this caliber, Buren used raw movements produced by the Felsa firm, calling them "Extra-Automatic".

Although Buren developed pendulum winding to its highest state of perfection in the Caliber 525, its winding performance could not compete with that of the rotor and this line of development thus had to be abandoned. In 1952 the manufacturer's first rotor automatic, the "Rotowind" (Caliber 541), appeared. Its effectiveness could be increased by the following measures:

Over a small-caliber basic movement (9.25 lignes) there swung a large and powerful central rotor, winding in both directions. Between the automatic and hand-winding drives there was a mutually effective and thus complete uncoupling mechanism installed, so that the friction losses in the hand-winding mechanism would have no influence on the automatic winding. The small basic movement actually offered no favorable conditions for outstanding running performance, but the developers were able to compensate for this weakness through the use of what was then the most modern watch technology, using a Glucydur balance, Nivarox hairspring, etc., as well as a high degree of manufacturing precision. On the basis of this caliber, the smallest automatic watch with up-down indicator, Caliber 538, was introduced in 1953. Thus the public, still skeptical about the automatic watch, was supposed to be offered personally confirmable evidence of reliability.

Buren attracted the attention of the watchmaking world in 1957 with the "Super Slender", an automatic design based on completely new concepts and patented in 1954. For the first time, all the functioning parts, including the rotor, automatic drive and the basic movement, were arranged not one above the other as before, but set beside each other on the plane of the movement. Thus one of the flattest automatic calibers of the time was created. The smaller arc potential of the small swinging mass of the planetary rotor would lead one to expect it requires being made of heavy metal as well as its winding in both directions over needed a particularly efficient drive. Here Buren used a so-called roller-bearing drive as a direction changer. The mutual uncoupling of the winding and automatic winding drives decreased further power losses.

In this type of construction, the functioning of all moving parts can be checked without disassembling the movement, which offers a distinct advantage for repairs.

With the "Intramatic" of 1962 (Caliber 1280), a further decrease in height could be achieved. Here too, as in the "Super Slender", a planetary rotor winds in both directions. Because of the low movement height, its arc of swing had to be increased, and it now spread over the middle of the movement. For that reason, the minute wheel had to be located outside the center and drive the minute hand indirectly. Both the "Intramatic" and the "Super Slender" calibers were used by other firms in their flat watch models, such as the Bulova "Ambassador", the Dugena "Super" and the Hamilton "Thin-o-matic".

By increasing the beat to 19,800 BPH in Caliber 1320, a further increase in running reliability was achieved in 1965.

This firm's only ladies' automatic was ready for series production in 1968, and was marketed with caliber numbers 04 and 06.

In cooperation with the firms of Breitling-Leonidas, Dubois-Depraz, Hamilton and Heuer, one of the first automatic chronograph watches—the "Chronomatic"—was created in 1969. The chronograph module 8510 was developed by the Dubois-Depraz firm, while Buren created the adapted "Intramatic" movement that underlies it.

In the firm's last automatic design, the Caliber 82 of 1970, two barrels work together. Their locations, flanking the center minute-wheel pinion, was chosen to guarantee a particularly uncritical and even transmission of the torque, and this fast beat escapement's high degree of chronometric quality could be increased even more. The increased need for energy again required the use of a large central rotor and winding in both directions. A choice could be made between a swinging rate of 28,800 or 36,000 BPH, and experiments with 43,200 BPH were even made.

The essential disadvantage of this very interesting movement is its thickness of 6.2 mm, which made it appear uncompetitive at the time. Thus only 200 to 300 watches could be delivered complete, although individual parts were produced for about 100,000 raw movements; they were taken over by the Comor firm in 1971.

With these interesting automatic designs, the Buren firm provided new impetus for the general development of automatic winding in the wristwatch. The swinging mass completely recessed in the movement is a typical element of this firm's production, whether it is a pendulum or a planetary rotor. Hans Kocher, who developed the planetary rotor as Buren's chief designer, is also the author of the only scientific work on the theory of automatic winding.

Buren Caliber 525

Men's automatic made by Buren in 1945, with Caliber 525.

Gilded case with stainless steel screw bottom.

Pendulum swinging mass winding in both directions via click wheels (at 5 and 6:00), automatic module integrated into the basic movement as a complete unit.

Glucydur screw balance, Nivarox hairspring, Incabloc shock resistance.

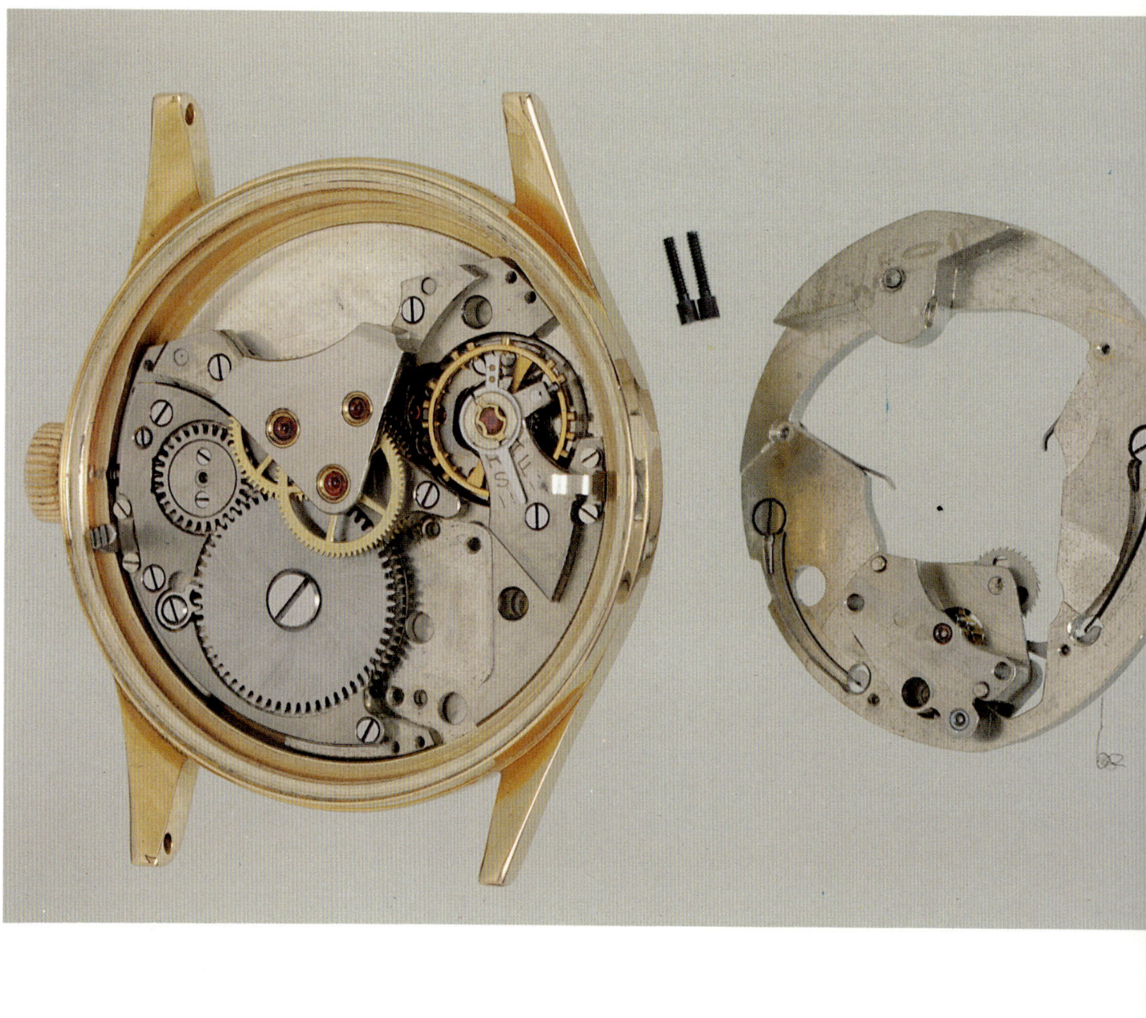

Buren Caliber 539

Men's automatic made by Buren in 1954, with Rotowind Caliber 539, smallest movement made with up and down indicator.

Gold and stainless steel case with screw bottom, winding crown with Buren emblem. Up and down indicator on a rotating auxiliary dial under the 12.

Rotor with heavy metal ring, winding in both directions via switching rocker, complete reciprocal uncoupling of automatic and hand-winding drives.

Glucydur screw balance, Nivarox hairspring, Incabloc shock resistance.

Buren Caliber 543

First rotor automatic made by Buren in 1952, with the "Rotowind" Caliber 543. Gilded case with stainless steel screw bottom, winding crown with Buren emblem.

Monometallic rotor winding in both directions via switching rocker (at 5:00 under rotor axle), complete reciprocal uncoupling between automatic and hand-winding drives.

Glucydur screw balance, Nivarox hairspring, Incabloc shock resistance.

Buren Caliber 1001, "Super-Slender"

First Buren automatic with a microrotor, made in 1958, with the "Super-Slender" Caliber 1001.
18-karat gold case with screw bottom, gold markings and hands, winding crown with Buren emblem.
"Planetary rotor" of heavy metal, winding in both directions via patented gliding pinion, complete reciprocal uncoupling between automatic and hand-winding drives.
Glucydur screw balance, Nivarox hairspring, Incabloc shock resistance. All components are located side by side, easy to service: Microrotor—regulatory organ—clockwork—barrel bridge—automatic drive (in clockwise order). Winding crown worn.

Buren Caliber 1322, "Intramatic"

Flat men's automatic made by Buren in 1965, with the "Intramatic" Caliber 1322.

Square stainless steel case, "Westminster" model, winding crown with Buren emblem.

"Planetary rotor" of heavy metal winding in both directions via gliding pinion ("swimming" gliding pinion mount under large ruby jewel), complete reciprocal uncoupling between automatic and hand-winding drives.

Screwless Glucydur balance, Nivarox hairspring, Incabloc shock resistance, "Isochron" fine adjustment.

Lower right:

Patented gliding-pinion switcher as roller-bearing drive (here from Buren Caliber 06): under each of the four wheels belonging to the switcher a disc-shaped roller is attached, its diameter such that the optimal meshing of the "swimming" gliding pinion (in the center) with the rotor pinion (left) and the tension wheel (upper right) or, with the rotor direction changed, with the first reduction wheel (lower right), is guaranteed.

Buren Caliber 06, "Hamilton"
Flattest ladies' automatic of its time, made in 1968, with Buren caliber 06, called "Intramatic" model by Hamilton but "Slendermatic" by Buren.
Gilded case with stainless steel screw bottom. Heavy metal rotor winding in both directions via gliding pinion switcher ("swimming" gliding pinion mount under the large ruby jewel at 1:00 above the center).
Screwless Glucydur balance, Nivarox hairspring, Ultraflex shock resistance.

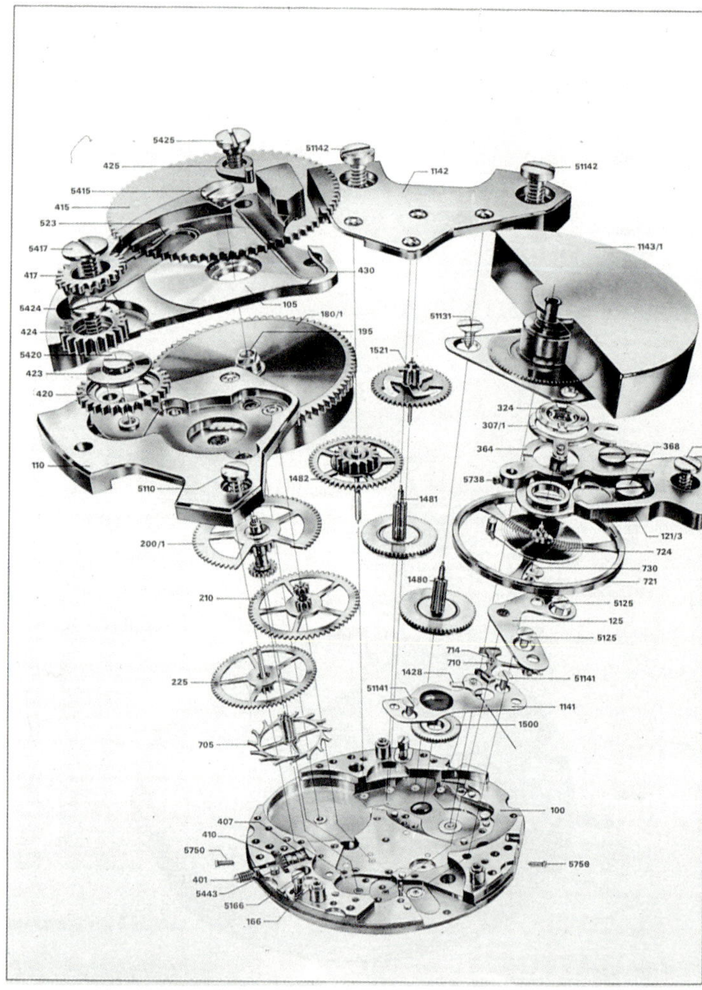

Buren Caliber 12

Automatic chronograph model, "Transocean", made by Breitling circa 1970, with "Chronomatic" Caliber 12 as joint development with Buren.

Case with stainless steel screw bottom and matching band, winding crown with Breitling symbol by the 9.

Modular construction of the basic movement—as adapted Buren "Intramatic" caliber (see page 68)—and chronograph unit 8510 with coulisse-lever switching, made by Dubois-Depraz.

Exploded view of the basic movement:

1143/1 — Microrotor

1480 — Tension wheel

1481 — 1st reduction wheel

1500 — Gliding pinion with "swimming" mount between two ruby cups, in alternate contact with 1480 and 1481.

Buren Caliber 82

Movement of a men's automatic made by Buren in 1970 as fast-swinging Caliber 82 with double barrel, of which only a small portion was covered.

Heavy metal rotor with ball bearings, winding in both directions via click-wheel switcher (at 4:00 and 5:30) parallel to both barrels (at 7 and 11:00).

Screwless Glucydur balance, Nivarox hairspring, Kif-Ultraflex shock resistance, Spirofin fine adjustment.

Caliber Notes

Caliber 525
D = 28.5 mm (12.5 lignes), H = 5.25 mm, 17 jewels
AUTOMATIC MODULE: pointed-oval shaped pendulum swinging mass recessed in the movement, with central cut-out and added weight, pendulum swing limited by leaf springs. Winding in both directions via two spring clicks, each on a click wheel, one reduction wheel on the locking wheel.
BASIC MOVEMENT: two-spoke Glucydur screw balance, Nivarox hairspring, 18,000 BPH, Incabloc shock resistance, direct center seconds.

Caliber 543
D = 25.6 mm (11.5 lignes), H = 5.4 mm, 21 jewels.
AUTOMATIC MODULE: Monometallic central rotor with broad ring, automatic drive mounted between two potences and upper movement plate, winding in two directions via switching rocker and three reduction wheels (the first with a spring click) on the locking wheel. Complete uncoupling of hand- or automatic winding, each by a gliding pinion.
BASIC MOVEMENT: Two-shank Glucydur screw balance, Nivarox balance, 18,000 BPH, Incabloc shock resistance, direct center seconds.

Caliber 539
D = 21 mm (9.25 lignes), H = 5.35 mm, 17 jewels
AUTOMATIC MODULE: Central rotor with heavy metal ring, drive as in Caliber 543.
BASIC MOVEMENT: Identical to Caliber 543 with additional running reserve indication by the relative movement of a hand (winding) and turning dial (running down).

Caliber 1001
D =28.1 mm (12.5 lignes), H = 4.8 mm, 21 jewels
AUTOMATIC MODULE: Small heavy-metal (planetary) rotor mounted outside the center, recessed in the movement, automatic drive integrated into the movement, winding in two directions via swimming guiding pinion (formed as roller drive), one tension wheel, two reduction wheels on the lower locking wheel, complete uncoupling of the hand- and automatic winding via a gliding pinion between the crown wheel and the upper locking wheel or by locking-wheel coupling.
BASIC MOVEMENT: Two-shank Glucydur screw balance, Nivarox hairspring, 18,000 BPH, Incabloc shock resistance, direct center seconds above double intermediate wheel on center seconds pinion. Date indication and correction possible.

Caliber 1322
D = 28.2 mm (12.5 lignes), H = 3.6 mm, 30 jewels
AUTOMATIC MODULE: Small heavy-metal (planetary) rotor recessed outside the center of the movement, automatic drive integrated into the movement, winding in both directions via a gliding pinion (formed as roller drive) mounted swimming between two ruby discs, one tension wheel with spring click, three reduction wheels on the locking wheel.
Mutual uncoupling of the hand and automatic winding by gliding pinion between the crown and locking wheels or sliding coupling in second reduction wheel, consisting of eloxied aluminum with plural-armed, axially working spring click.
BASIC MOVEMENT: three-spoke screwless Glucydur balance, Nivarox hairspring, 19,800 BPH, Incabloc shock resistance, patented Buren isochronic regulator with moveable hairspring block carrier, center seconds, play-free hand-setting pinion.

Caliber 06
D = 17.2 mm (7.75 lignes), H = 4.3 mm, 30 jewels
AUTOMATIC MODULE: Central rotor of heavy metal, automatic drive integrated into the basic movement, winding in both directions via a gliding pinion (formed as roller-bearing drive) mounted swimming between two ruby rollers, one tension wheel with spring click, two reduction wheels on locking wheel (under the dial), uncoupling of hand-winding mechanism.
BASIC MOVEMENT: three-spoke screwless Glucydur balance, Nivarox hairspring, 21,600 BPH, Kif-Ultraflex shock resistance, moveable hairspring block carrier, Nivaglide mainspring, slowly changing date indication.

Caliber 12
D = 31 mm (13.75 lignes), H = 7.7 mm, 17 jewels
Chronograph module 8510 with coulisse lever guidance, screwed onto the automatic movement.
AUTOMATIC MODULE: Adapted from Caliber 1322
BASIC MOVEMENT: Same as Caliber 1322 but with left-hand winding crown.

Caliber 82
D = 25.6 mm (11.5 lignes), H = 6.2 mm, 17 jewels
AUTOMATIC MODULE: Central rotor of heavy metal, with ball bearings, drive partly integrated into the basic movement, winding in both directions via unsprung twin click wheels, three reduction wheels on the locking wheel of the left barrel.
BASIC MOVEMENT: three-spoke screwless Glucydur balance, Nivarox hairspring, 28,800 BPH, Incabloc shock resistance, Spirofin fine regulation with moveable hairspring block carrier, direct center seconds, two barrels side by side, acting together on the centrally mounted minute wheel. Instantly changing date indication with quick correction, second-stopping mechanism.

Caliber List: Buren

Caliber	Diameter Lignes	mm	Height mm	Other	BPH	Jewels	Running Reserve	Notes	Year	Photo
04	7 3/4	17,2	3,9		21.600	30	45	Slendermatic	1968	
06	"	"	4,3	QG	"	"	"	"		xx
11	13 3/4	31	7,7	QG, CHR	19.800	17	40	Chronomatic	1969	
12	"	"	"	QG, CHR	21.600	"	"	"	ca. 72	xx
82	11 1/2	25,6	6,2	SCD, QG, RDR, ST	28.800	17	50	(2 Barrels)	1970	xx
525	12 1/2	28,5	5,25	SCD	18.000	17		(Pendulum Winding)	1945	xx
535	9 1/4	20,7	5,35	s	18.000	17	40	Rotowind	1952	
536	"	"	"	SCI	"	"	"	"		
537	"	"	"	SCD	"	"	"	"		
538	"	"	"	RMA	"	"	"	"	1953	
539	"	"	"	SCI, RMA	"	"	"	"		xx
540	"	"	"	SCD, RMA	"	"	"	"		
541	11 1/2	25,6	5,4		18.000		40	Rotowind	1952	
542	"	"	"	SCI	"		"	"		
543	"	"	"	SCD	"	21	"	"		xx
544	"	"	"	RM	"	"	"	"		
545	"	"	"	SCI, RMA	"	"	"	"		
546	"	"	"	SCD, RMA	"	"	"	"		
547	"	"	"	SCI	"	"	"	"		
549	"	"	"	SCI	"	"	"	"		
1000	12 1/2	28,1	4,2		18.000	21		Super Slender	1957	
1000 A	"	"	"	SCD	"	"		"		
1001	"	"	4,8	SCD, QG	"	"		"		xx
1001 A	"	"	"	SCD, QG	"	"		"		
1002	"	"	4,2	SCD	"	"		"		
1002 A	"	"	"	SCD	"	"		"		
1003	"	"	4,8	SCD, QG	"	"		"		
1003 A	"	"	"	SCD, QG	"	"		"		
1005	"	"	4,2	SCD	"	25		"	1958	
1006	"	"	"	QG	"			"		
1007	"	"	"		"			"		
1008	"	"	"	QG	"			"		
1280	12 1/2	28,2	2,85		18.000		54	Intramatic	1962	
1281	"	"	3,2	QG	"		"	"		
1282	"	"	"	QG	"		"	"		
1320	12 1/2	28,2	3,15	SCI	19.800	30	54	Intramatic	1965	
1321	"	"	3,6	SCI, QG	"	"	"	"		
1322	"	"	"	SCI. QG	"	"	"	"		xx

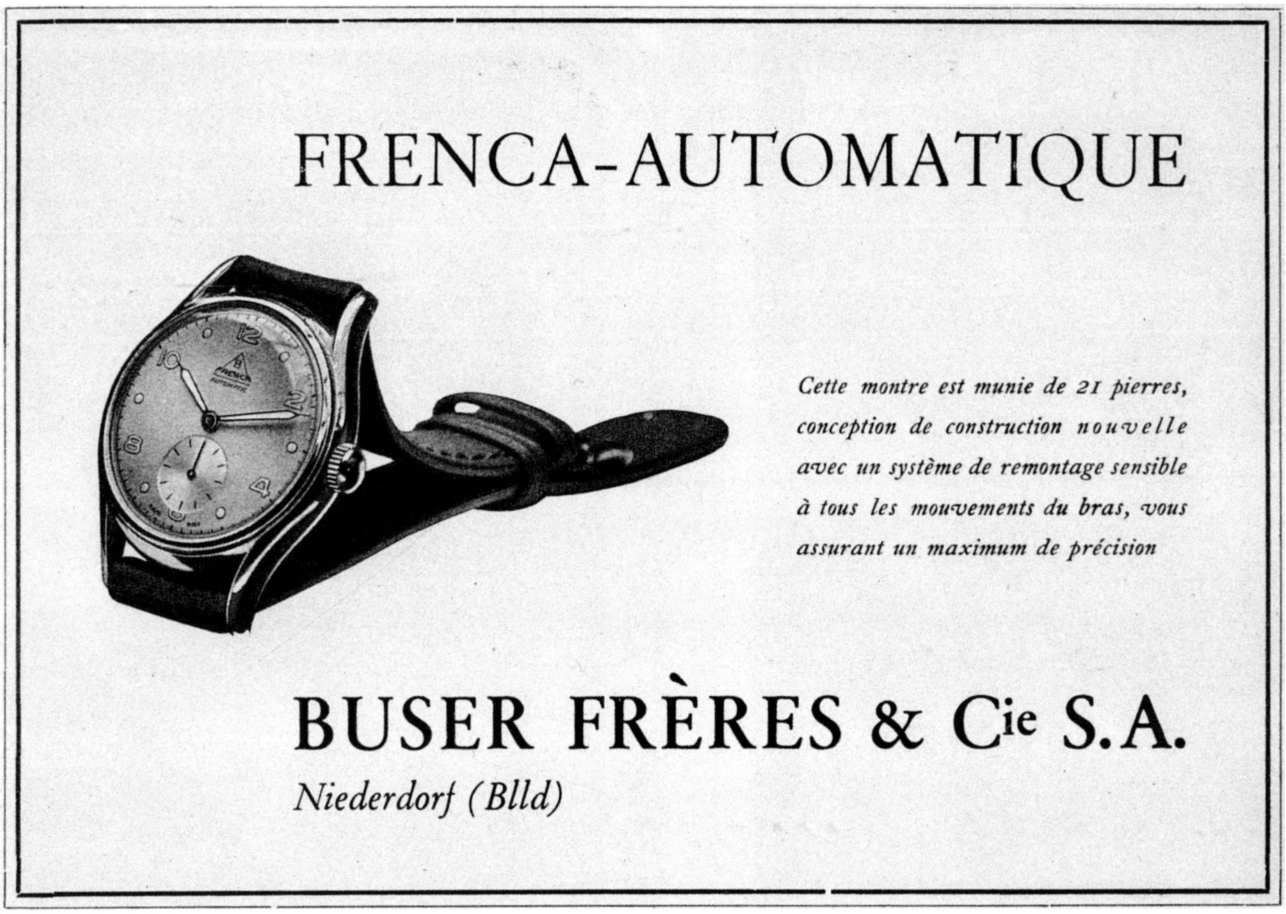

Magazine advertisement (from SJH, 1947)

Caliber Development

In 1893 Fritz and Albert Buser founded the "Buser Frères & Cie. S.A." watch factory in Niendedorf, near Biel. At first they produced only raw movements, but later manufactured complete watches. In 1962 the firm merged with the MSR (Manufactures d'Horlogerie Suisse réunies S.A.), which also included the firms of Revue, Vulcain, Studio and Phoenix.

The first Buser automatic was introduced as Caliber 20 in 1946. It was equipped with rotor winding that worked in only one direction via a sprung pusher pinion. The basic movement, measuring 8.75 lignes, was rather small and had a seconds bit. The successor type, Caliber 145, which was developed around 1950, used the same automatic component with slight modifications but it was mounted on a more modern basic movement with a direct center seconds and the entire surface under the rotor.

In Caliber 146, completely developed in 1953, the automatic component was replaced by a mechanism winding in both directions via sprung click wheels.

All Buser movements were equipped with an anchor with side fork.

Caliber List: Buser

Caliber	Diameter Lignes	mm	Height mm	Other	BPH	Jewels	Running Reserve	Notes	Year	Photo
20	12 1/4	28	5,55	s	18.000	21			1946	
145	12 1/4	28	6,25	SCD	18.000	21				
146	12 1/4	28	6,25	SCD	18.000	21			1953	

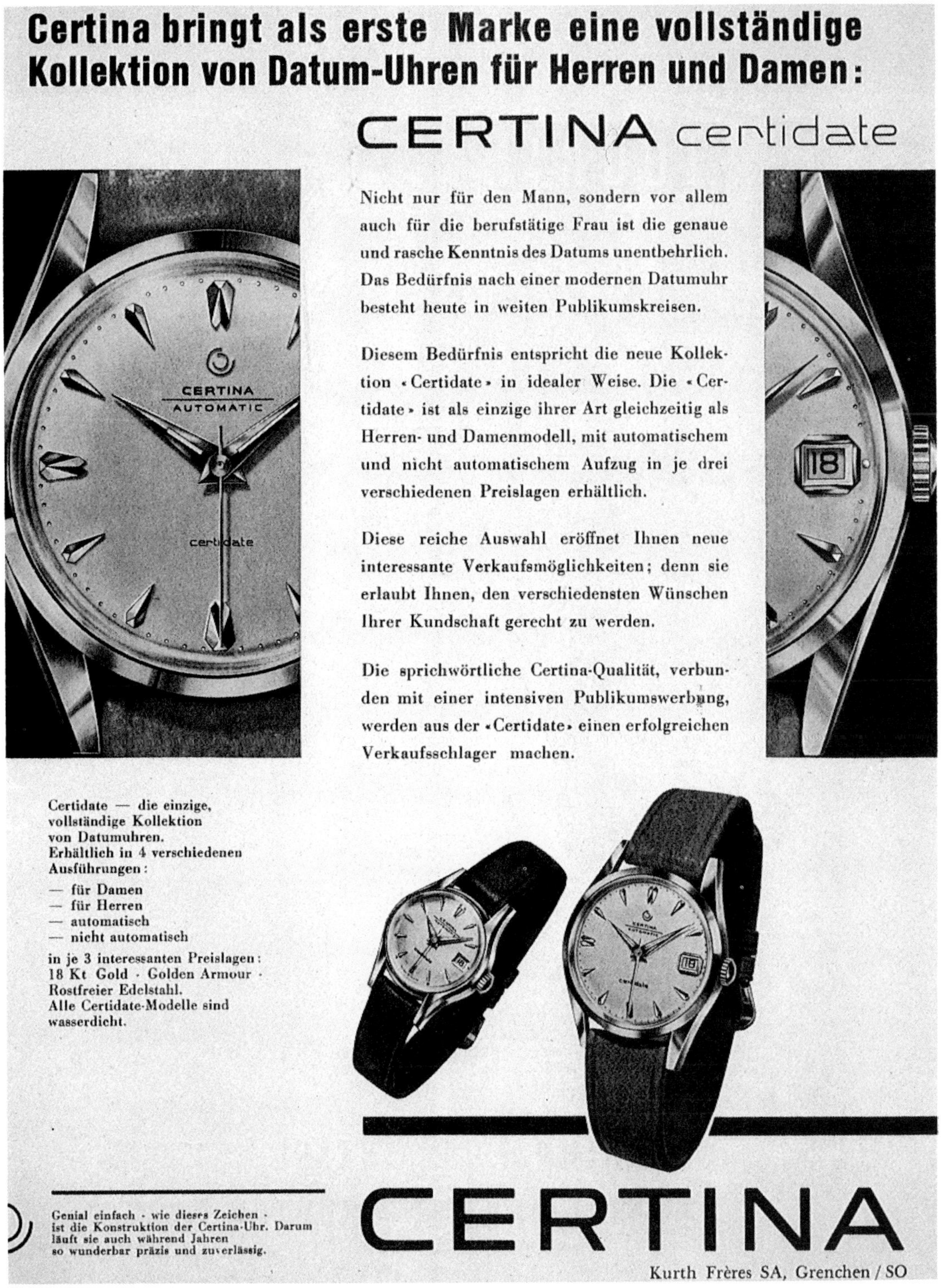

Magazine advertisement (from SUZ, 1959)

Caliber Development

This firm was founded by the Kurth brothers in Grenchen in 1888 and operated under the name of "Kurth Frères S.A." (KF). The first automatic it developed, Caliber 360, appeared in 1947, had hammer winding in one direction and differed little from its competitors.

When they designed a rotor automatic (Caliber 25-45) in 1951, the Certina engineers utilized the latest watch technology and created a very successful family of calibers, which was expanded in 1953 by a larger movement (Caliber 28-45) and in 1956 by a smaller one, Caliber 19-45, with its sides reversed.

A typical feature of this design was the double-lever switcher that worked in principle like a switching rocker and was used only by Certina. The coupling wheel, mounted ahead of the crown wheel to disconnect the automatic drive during hand-winding, is also built in an unusual manner.

A new rotor caliber, designated 25-65, was introduced in 1960; it featured a direction switcher of unsprung click wheels with anchor locks, and its movement was slimmer. The small movements for ladies' watches, Calibers 15-20 and 17-211, that were introduced in 1963 had the same winding principle.

The 288 model, a fast beat escapement modified from the 25-65 basic caliber, achieved chronometer quality. Its minute wheel with ball bearings deserves mention as a special feature.

In 1972 Certina developed a complex automatic movement, the Caliber 17-351, for the ladies' models. It was equipped with a heavy-metal rotor with ball bearings and a mutual uncoupling mechanism of the automatic and hand-winding mechanisms to guarantee safe winding. Like this model, the firm's last automatic design with rotor winding in two directions, built in 1976, was also made by the modu-

Certina Caliber 360

Men's automatic made by Certina in 1947, with Caliber 360.
Stainless steel case with press-in bottom.
Hammer swinging mass winding in one direction via click-carrier rocker (at 7:00).
Glucydur screw balance, Nivarox hairspring, Incabloc shock resistance.

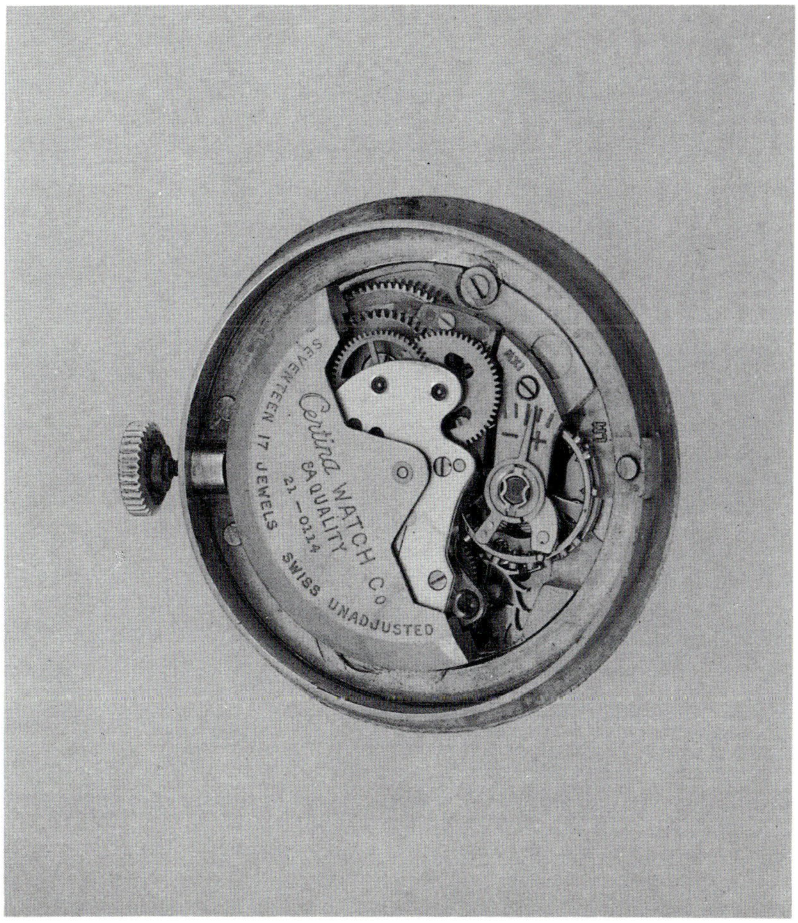

Certina Caliber 21-0114

Men's automatic made by Certina circa 1950, with medium-sized EA Caliber 21-0114.

14-karat gold case with press-in bottom. Movement substantially the same as AS Caliber 1323 (see page 32). This case is not original to the movement and would be called a "marriage".

Certina Caliber 28-451

Men's automatic made by Certina in 1953, with Caliber 28-451.

Gilded case with stainless steel screw bottom.

Ruby-mounted monometallic rotor with unbolting device, winding in both directions via double-lever switcher, separate locking-click plate, uncoupling of hand-winding drive.

Glucydor screw balance, Nivarox hairspring, Incabloc shock resistance, fine regulator adjustment by eccentric screw.

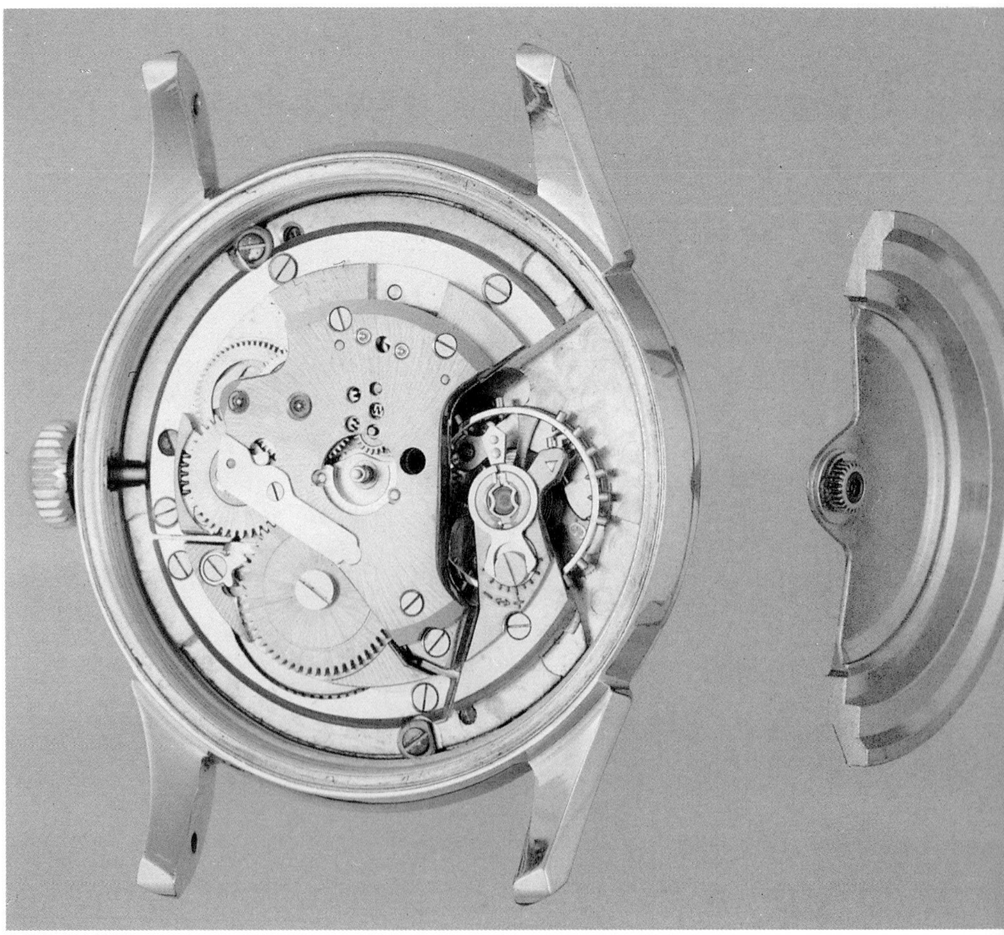

Certina Caliber 25-651 M, "Argonaut"

Men's automatic made by Certina in 1974, "Argonaut" model with Caliber 25-651 M.

Stainless steel case with screw bottom, winding crown with Certina emblem.

Ruby-mounted rotor with heavy metal ring and unbolting device, winding in both directions via click-wheel switcher.

Screwless nickel balance, Nivarox hairspring, Incabloc shock resistance, fine regulator adjustment by eccentric screw.

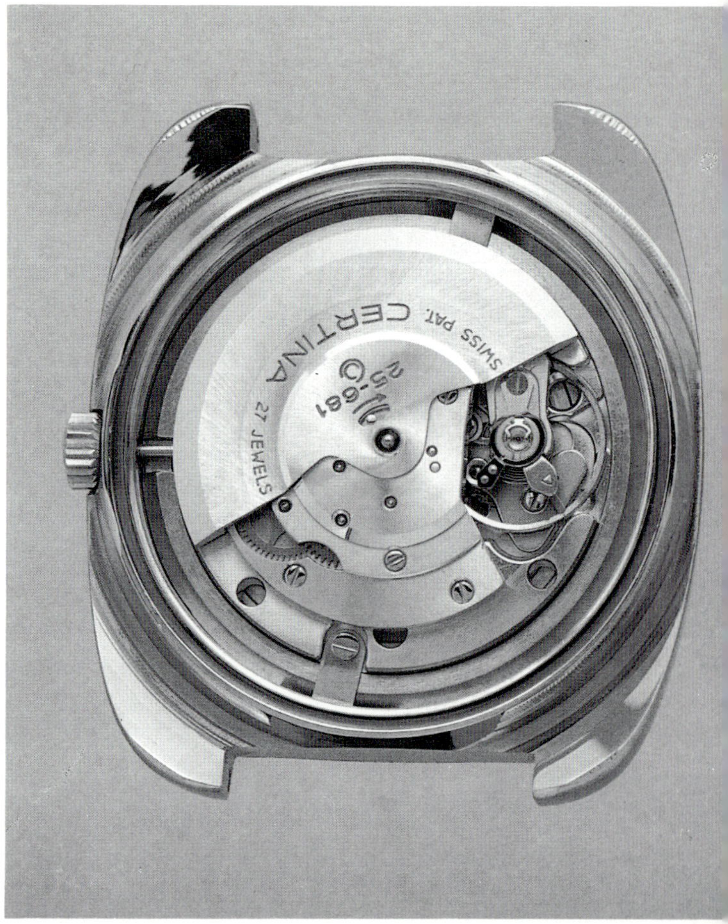

Certina Caliber 25-681 M, "288"

Men's automatic model "288" made by Certina circa 1970, with fast-swinging Caliber 25-681 M.

Massive stainless steel case with screw bottom, winding crown with Certina emblem.

Movement largely identical to Caliber 25-651 M (see page 79), but here with screwless Glucydur balance, Clinergic 21 anchor escapement, 28,800 BPH (hence "288"), minute wheel with ball bearings.

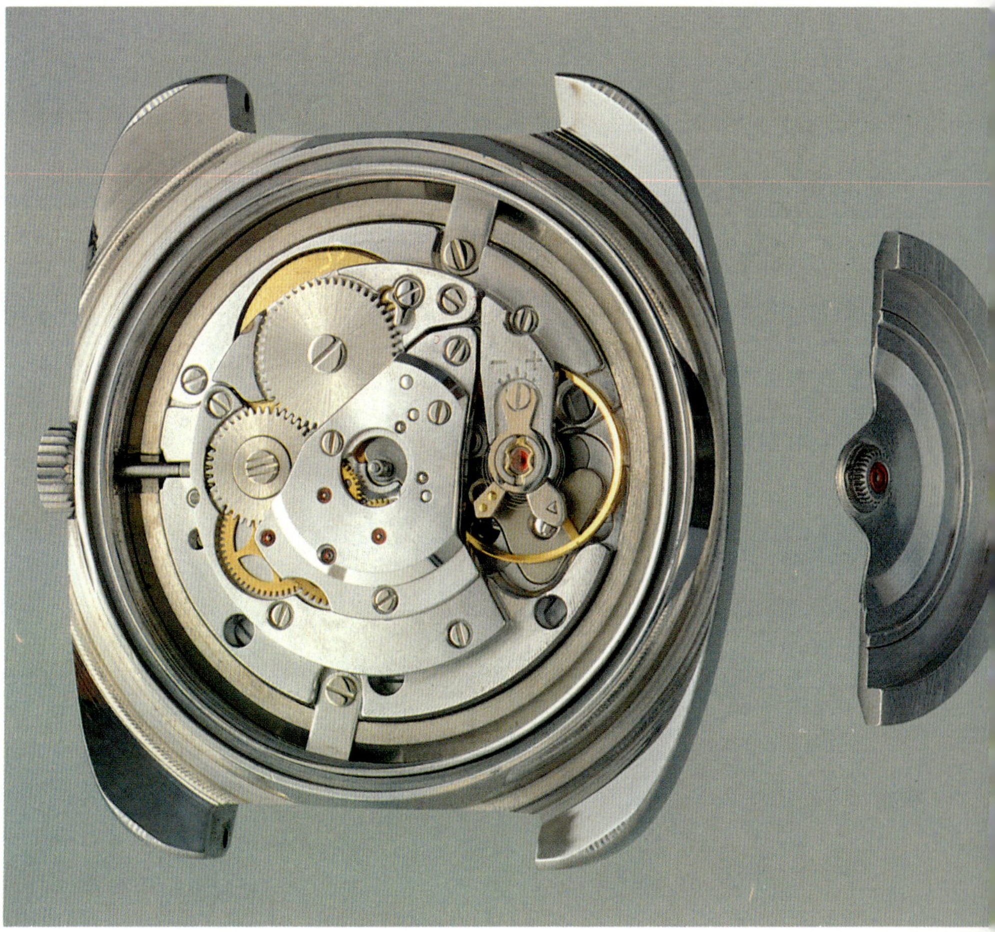

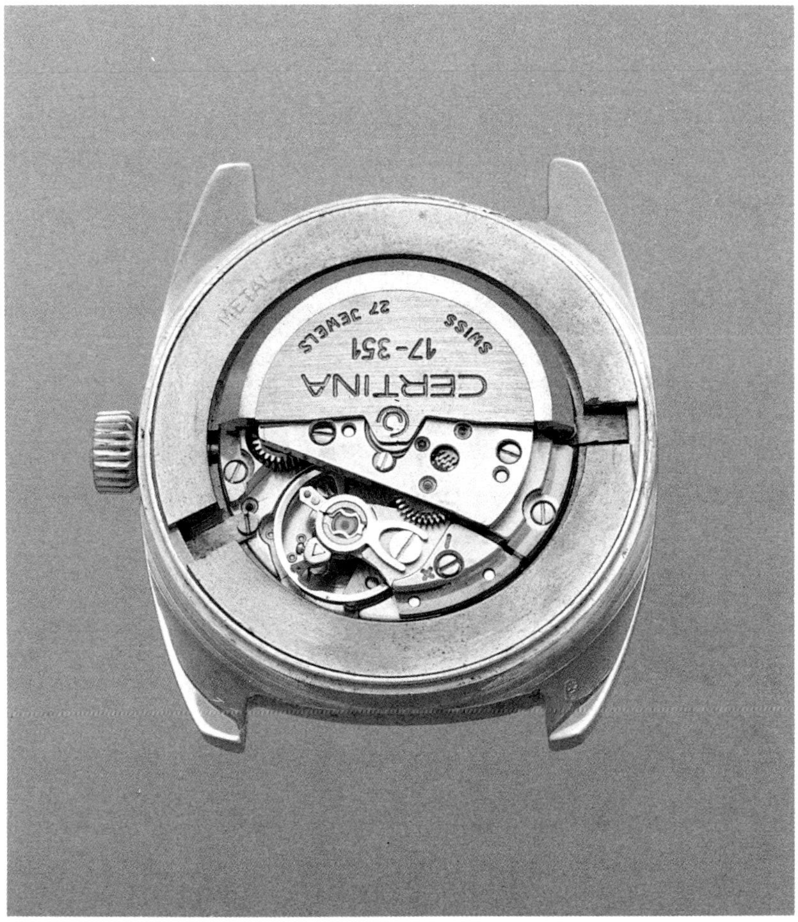

Certina Caliber 17-351, "Mayfair"
Ladies' automatic made by Certina in 1972, "Mayfair" model with Caliber 17-351.
18-karat gold case with press-in bottom, winding crown with Certina emblem.
Rotor with ball bearings and heavy metal ring, winding in both directions via click-wheel switcher, complete reciprocal uncoupling of automatic and hand-winding drives.
Screwless Glucydur balance, Nivarox hairspring, Incabloc shock resistance, fine regulator adjustment by eccentric screw.

lar technique, including a basic movement that could also be used in hand-wound models.

With the DS collection, introduced in 1960, Certina offered a very sturdy watch with the thickness of its case walls and glass doubled and its movement mounted in "suspended" fashion by an elastic, shock-absorbing ring. The crown seal and screwed bottom or single-piece case guaranteed watertightness to a diving depth of 200 meters.

Caliber Development

Caliber 360

D = 25.6 mm (11.5 lignes), H = 5.45 mm, 19 jewels
AUTOMATIC MODULE screwed to the basic movement, hammer winding in one direction via click-carrier rocker, one combined winding and reduction wheel on crown wheel.
BASIC MOVEMENT: two-spoke Glucydur screw balance, Nivarox hairspring, 18,000 BPH, Incabloc shock resistance, seconds bit.
Caliber 21-0114
AS Ebauche
D = approx. 21 mm (9.25 lignes), 17 jewels
AUTOMATIC MODULE screwed to the basic movement as a complete unit, central rotor with heavy metal ring, winding in both directions via switching rocker, two reduction wheels, the first with a spring click, the second with a Breguet coupling, on the locking wheel to disconnect during winding.
BASIC MOVEMENT: two-spoke Glucydur screw balance, Nivarox hairspring, 18,000 BPH, Incabloc shock resistance, center seconds.

Caliber 28-451
D = 28.2 mm (12.5 lignes), H = 6.5 mm, 21 jewels
AUTOMATIC MODULE screwed to basic movement, (monometallic) central rotor with ruby bearings, winding in both directions via double-lever switcher, two reduction wheels, the first under a separate plate and with a spring click, the second on the crown wheel and serving as a coupling to disconnect during winding.
BASIC MOVEMENT: two-spoke Glucydur screw balance, Nivarox hairspring, 18,000 BPH, Incabloc shock resistance, fine adjustment by regulator with eccentric screw, direct center seconds, slowly changing date indication.

Caliber List: Certina / KF

Caliber	Diameter Lignes	mm	Height mm	Other	BPH	Jewels	Running Reserve	Notes	Year	Photo
15-20	6 3/4	15,3	4,6		19.800	32			1963	
15-21	"	"	"		21.600	"			1964	
17-011	7 3/4	17,5	4,8	SCD, QG, RDR	21.600	17-25		Eta-Ebauche		
17-211	7 3/4	17,2	4,95	QG	21.600	32		Basic Caliber 15-21		
17-221	"	"	"	QG	"	"				
17-351	7 3/4	17,5	4,85	SCD, QG, ST	21.600	27			1972	xx
19-45	8 3/4	19,4	5,6	SCD	18.000	23			1956	
19-451	"	"	6,25	SCD, QG	"	"				
19-55	"	"	5,6	SCD	19.800	25				
19-551	"	"	6,25	SCD, QG	"	"			1956	
21-0114	9 1/4			SCI	18.000	17		AS-Ebauche 1320	1950	xx
25-011	11 1/2	26,0	5,2	SCD, QG, RDR	21.600	17-25		Eta-Ebauche		
25-012	"	"	5,9	SCD, QG, JG, RDR	"	21		"		
25-45	11 1/2	25,6	6,0	SCD	18.000	21				
25-451	"	"	6,5	SCD, QG	"	"		Certidate	1951	
25-65	11 1/2	25,6	5,0	SCD	19.800	27			1960	
25-651	"	"	5,35	SCD, QG	"	"		Certidate	1960	
25-651 M	"	"	"	SCD, QG	"	28	46		1974	xx
25-652	"	"	5,98	SCD, QG, JG	"	27		Certiday		
25-681	"	"	5,35	SCD, QG	28.800	"		288	ca. 70	xx
25-682	"	"	5,98	SCD, QG, JG	"	"				
28-45	12 1/2	28	6,0	SCD	18.000	21		Basic Cal 25-45		
28-451	"	"	6,5	SCD, QG	"	"		Basic Cal 25-451	1953	xx
28-55	"	"	6,0	SCD	"	"		Basic Cal 25-45		
360	11 1/2	25,6	5,45	S	18.000	19			1947	xx
384	"	"		SCI	"				1949	
681	13 1/4	30,0	7,6	SCD, QG, ST, REV	28.800	25		AS-Ebauche 5008	1975	
919-1	11 1/2	25,6	5,2	SCD, QG, JG, ST	28.800	25			1976	

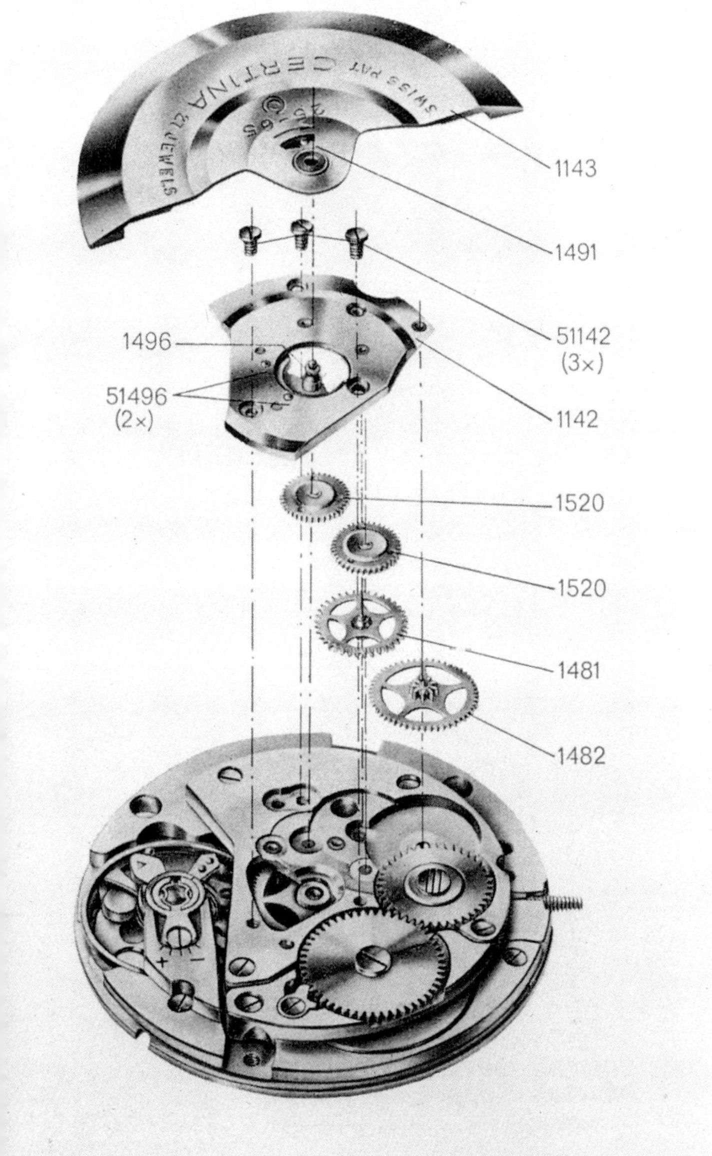

Click-wheel switcher (Certina patent): 1143—rotor drive, 1481—first reduction wheel.

Exploded view of Certina Caliber 25-65 (automatic module): 1520—twin click wheels, 1481—first reduction wheel, 1482—second reduction wheel.

Caliber 25-651 M
D = 25.6 mm (11.5 lignes), H = 5.35 mm, 28 jewels
AUTOMATIC MODULE screwed to basic movement, central rotor with heavy metal ring and ruby bearings, winding in both directions via twin click wheels with unsprung anchor clicks, two reduction wheels on crown wheel.
BASIC MOVEMENT: three-spoke screwless nickel balance, Nivarox hairspring, 19,800 BPH, Incabloc shock resistance, moveable hairspring block carrier, fine adjustment by regulator with eccentric screw, direct center seconds, Superflexo mainspring, date indication quick changes.

Caliber 25-681
D = 25.6 mm (11.5 lignes), H = 5.35 mm, 27 jewels
See Caliber 25-651, but with 28,000 BPH and central minute wheel with ball bearings.

Caliber 17-351
D = 17.5 mm (7.75 lignes), H = 4.85 mm, 27 jewels
AUTOMATIC MODULE screwed to basic movement, central rotor with heavy metal ring and ball bearings, winding in both directions via one intermediate wheel, unsprung twin click wheels, two reduction wheels on double locking wheel, mutual separation of automatic and winding by locking-wheel coupling or geared rocker between crown and locking wheels.
BASIC MOVEMENT: three-spoke screwless Glucydur balance, Nivarox hairspring, 21,600 BPH, Incabloc shock resistance, hairspring block carrier, fine adjustment by regulator with eccentric screw, direct center seconds, indirectly driven minute wheel, date indication with slow changing.

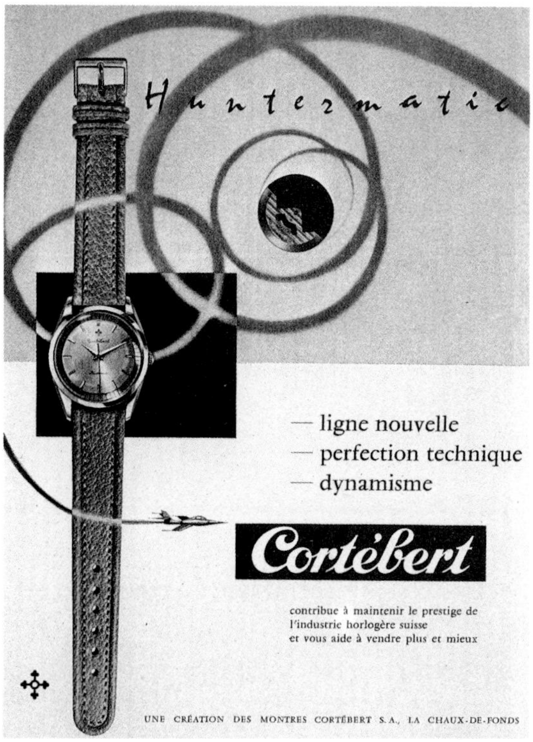

Magazine advertisement (from SUZ, 1960)

Caliber Development

The Cortebert Watch Co. S.A. was founded in Cortebert by A. L. Juillard in 1790, thus it ranks among the oldest Swiss watch manufacturers. It later opened a second production facility in La Chaux-de-Fonds.

The only automatic developed by the firm was the "Corterotor" (Caliber 700 ff), introduced in 1955.

This was a very solidly built rotor automatic winding in both directions, but without any unusual features. Even the disconnection of the reduction drive during hand-winding was done by the standard method of the time.

The other caliber numbers are those of movements from outside that were installed in Cortebert watch models.

Caliber Notes

Caliber 700
D = 25.6 mm (11.5 lignes), H = 6.05 mm, 21 jewels
AUTOMATIC MODULE screwed to the basic movement as a complete unit, central rotor with heavy metal ring, winding in both directions via switching rocker, three reduction wheels, the first with locking click and the third on the crown wheel as a coupling to protect the reduction drive during hand-winding.
BASIC MOVEMENT: two-spoke Glucydur screw balance, Nivarox hairspring, 18,000 BPH, Incabloc shock resistance, direct center seconds.

Caliber List: Cortebert

Caliber	Diameter Lignes	mm	Height mm	Other	BPH	Jewels	Running Reserve	Notes	Year	Photo
270	11 1/2			SCD						
"	"			SCD, QG						
272	7									
278	11 1/2			SCD						
279	"			SCD, QG						
284	7 3/4			SCD						
285	"			SCD, QG						
286	11 1/2			SCD, QG						
290	"			SCD, QG, JG						
297	11 1/2			SCD						
298	"			SCD, QG						
301	10 1/2			SCD						
305	7 3/4			SCD	21.600					
306	"			SCD, QG	"					
451	11 1/4	24,8	5,5	SCD	18.000	17		AS-Ebauche 1250	1944	
700	11 1/2	25,6	6,05	SCD	18.000	21	45	Corterotor	1955	xx
704	"	"	6,7	SCD, QG	"	"	"			
706	"	"								

Cortebert Caliber 700

Men's automatic made by Cortebert in 1955, with "Corterotor" Caliber 700. Stainless steel case with screw bottom, winding crown with Cortebert emblem. Rotor with heavy metal ring, winding in both directions via switching rocker (at 4:00 below the rotor axle), uncoupling of hand-winding drive.
Glucydur screw balance, Nivarox hairspring, Incabloc shock resistance.

Caliber Development

The Cyma Watch Co. S.A., founded in La Chaux-de-Fonds in 1862, merged with the Tavannes Watch Co. S.A. and was taken over by the Chronos Holding A.G. in 1966.

The construction of the first Cyma automatic movement, Caliber 420, in 1943 is reminiscent of the "Harwood" of 1929. Here too, the hammer strikes against spring buffers and moves around the outer rim of a relatively small movement. The automatic mechanism is almost identical, including its protection against overwinding. But hand-winding is possible by using the side crown, and the regulating organ consists of a modern self-compensating combination of balance and hairspring.

In the mid-1950s, two new calibers were introduced: the "Bijoumatic" (for ladies) and the "Autorotor" (for men). Both were rotor automatics with a very interesting eccentric lever drive and a centrally mounted winding wheel that is unique in this type. The central rotor and its axle are mounted movably and can be retained by either a ring around the movement or a specially turned case bottom when its outer rim is struck. During automatic operation, the hand-winding mechanism is uncoupled. The basic movement, extending out to the case rim, provides space for a large balance. The enlarged barrel is also moved far out toward the edge, but thus requires the non-central location of the minute wheel.

This movement is also used by Longines as Caliber 380, but it includes an almost necessary further development of the eccentric winding introduced by that manufacturer in 1945. Later, until the early 1950s, Cyma installed movements from outside in its watch models.

Caliber Notes

Caliber R 420
D = 23 mm (10.5 lignes), H = 5.9 mm, 17 jewels
AUTOMATIC MODULE: a complete unit attached to the basic movement by two screws, hammer with ruby bearings, limited by two buffer springs, winding in one direction via hammer click, centrally mounted click wheel, two reduction wheels, the first with locking click, on the locking wheel.
BASIC MOVEMENT: two-spoke Glucydur screw balance, self-compensating flat hairspring, 18,000 BPH Cymaflex shock resistance, center seconds, ruby bearings with chatons.

Caliber R 485
D = 28.0 mm (12.5 lignes), H = 5.7 mm, 27 jewels
AUTOMATIC MODULE: complete unit screwed onto the basic movement, central rotor with ruby bearings, heavy metal ring and axial shock resistance, winding in both directions via rotor eccentric, two winding levers with ruby-bearing rollers, alternately extended from the eccentric to move the centrally mounted winding wheel via spring clicks, one reduction wheel on the locking wheel. Winding uncoupled by toothed rocker between crown and reduction wheels.
BASIC MOVEMENT: two-spoke Glucydur screw balance, Nivarox hairspring, 18,000 BPH, Cymaflex shock resistance, direct center seconds via double intermediate wheel, turning on bored ruby axle, instantly changing date indication.

Caliber List: Cyma

Caliber	Diameter Lignes	mm	Height mm	Other	BPH	Jewels	Running Reserve	Notes	Year	Photo
420	11 1/2	23	5,8		18.000	17	36			
420 KT	"	"	5,9	SCI	"	"	"		1943	xx
425	8 3/4			SCI, QG	18.000					
470	8	18,0	5,4	SCD	18.000			Bijoumatic	ca. 54	
471	"	"			"					
480	12 1/2	28,0	5,2	SCD	18.000	27	40	Autorotor	ca. 55	
480.1	"	"			"	"	"	"		
481	"	"			"	"	"	"		
482	"	"		QG	"	"	"	"		
485	"	"	5,7	SCD, QG	"	"	"	"	ca. 57	xx
485.1	"	"			"	"	"	"		
504	7 3/4			SCD, QG	21.600					
804	11 1/2			SCD, QG	21.600			Eta- Ebauche		
814	"			SCD, QG	36.000			"		
824	"			SCD, QG, JG	21.600			"		
834	"			SCD, QG	36.000			"		

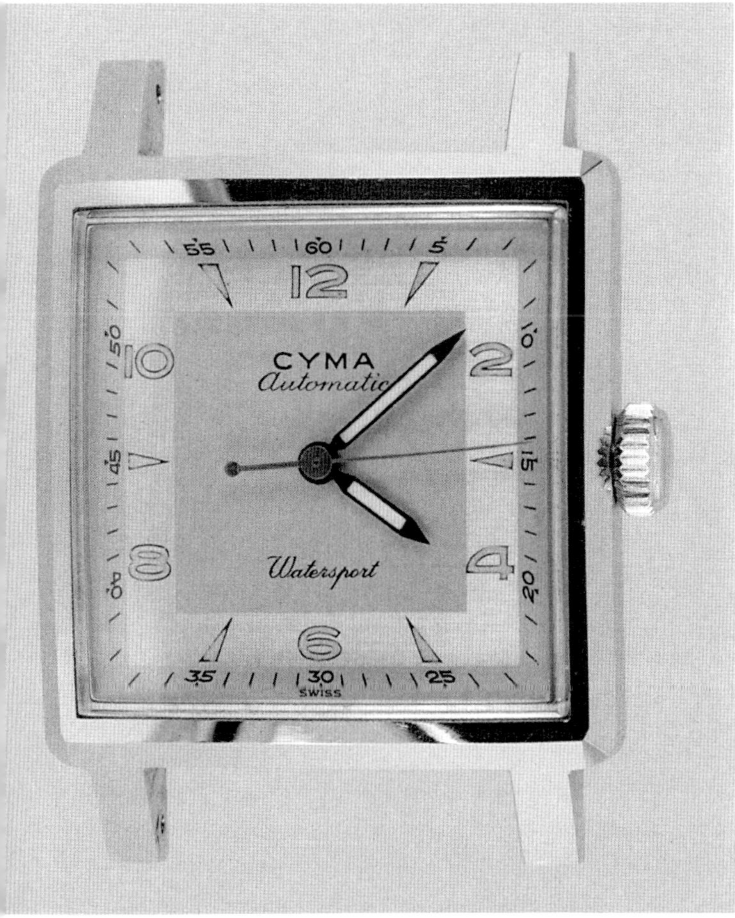

Cyma Caliber 420 KT

Men's automatic model, "Watersport", made by Cyma in 1943, with Caliber 420 KT.

Square stainless steel case with screwed bowl bottom and sectioned winding stem.

Ruby-mounted hammer winding in both directions via Harwood-type automatic drive.

Glucydur screw balance, Nivarox hairspring, Cymaflex shock resistance.

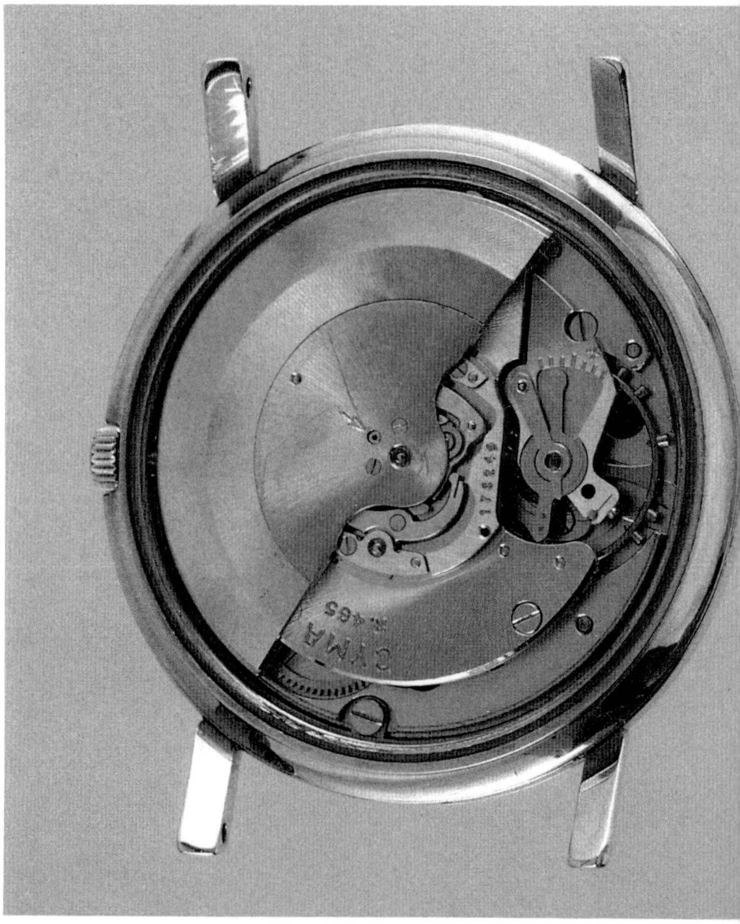

Cyma Caliber 485, "Autorotor"
Men's automatic made by Cyma circa 1957, "Navystar" model, with "Autorotor" Caliber 485.
Gilded case with stainless steel screw bottom, recessed winding crown.
Ruby-mounted rotor with heavy metal ring and unbolting device, plus shock resistance (by sprung axle and safety bottom), winding in both directions via eccentric switcher (in the center, directly under the rotor), uncoupling of hand-winding drive.
Glucydur screw balance, Nivarox hairspring, Cymaflex shock resistance (case with clear signs of wear).

Caliber Development

The Fabrique d' ébauches de Bettlach S.A. originated in 1932 to produce Roskopf raw movements, purchasing the Kummer S.A. in Bettlach.

The firm's first automatic pin lever movement, Caliber 1575, came on the market in 1958. As with all subsequent EB designs, the rotor winds in only one direction. The manufacturers of watch movements in the lower price ranges use the most modern technology when it leads to a simplification of the movement and guarantees a high degree of reliability. Thus one finds in the subsequent calibers of the mid-1960s an interesting version of the centrally located combined winding and locking wheel.

Caliber Notes

Caliber 8127
D = 25.6 mm (11.5 lignes), H = 5.4 mm, 21 jewels
AUTOMATIC MODULE screwed to the basic movement. Monometallic rotor with gliding bearings, winding in one direction via centrally mounted combined winding and locking wheel, two reduction wheels on the locking wheel of the barrel.
BASIC MOVEMENT: brass ring balance, Nivarox hairspring, 18,000 BPH, Kif-Trior shock resistance, pin lever escapement, direct center seconds, instantly changing date indication.

Caliber List: Ebauches Bettlach/EB

Caliber	Diameter		Height	Other	BPH	Jewels	Running Reserve	Notes	Year	Photo
	Lignes	mm	mm							
1575 /N	11 1/2	25,6	6,7	SCI	18.000				1958	
8100	8 3/4	19,4	4,8	s	18.000					
8101	"	"		SCI	"					
8103	"	"		SCD	"					
8104	"	"	5,4	s, QG	"					
8105	"	"		SCI, QG	"	21				
8107	"	"		SCD, QG	"					
8120	11 1/2	25,6	4,8	s	18.000					
8121	"	"	"	SCI	"					
8123	"	"	"	SCD	"					
8124	"	"	5,4	s, QG	"					
8125	"	"	"	SCI, QG	"					
8127	"	"	"	SCD, QG	"	21				xx
8131	11 1/2	25,6	4,8	SCD	18.000					
8135	"	"	5,4	SCD, QG, RDR	"					
8137	"	"	5,9	SCD, QG, JG, RDR	"					
8155	8 3/4	19,4	5,55		18.000	21			1967	
8175	11 1/2	25,6	5,55	SCD	18.000				1967	
8177	11 1/2	25,6		SCD, QG	18.000					

Ebauches Bettlach Caliber 8127, "Tirza"

Men's automatic of the "Tirza" brand, from the 1970s, with Ebauches Bettlach (EB) Caliber 8127.

Chrome plated case with stainless steel screw bottom, monometallic rotor, winding in one direction via Harwood-type drive.

Brass ring rotor, Nivarox hairspring, Kif-Trior shock resistance, pin lever escapement.

Caliber Development

Ariste Racine (AR) founded a factory in La Chaux-de-Fonds in 1913 and moved the firm's headquarters to Lengnau three years later.

The first Enicar automatics were equipped with AS and Eta movements. Around 1962, though, the firm prepared its own design for production. This Caliber 1124, known as "Rubirotor", operates with a central rotor mounted in ruby bearings and winding in both directions. Breguet free-running devices, similar to those made by Bulova (see Bulova), were used as switching controls. The turning moment of the swinging mass was so completely in tune with that of the gliding line in the barrel that the weight was balanced when the watch was fully wound, and premature damage to the sliding coupling could be avoided.

As of the mid-1960s, this caliber, in slightly modified form, was installed in over a hundred different "Sherpa" models.

Caliber Notes

Caliber 165

D = 26 mm (11.5 lignes), H = 4.9 mm, 24 jewels

AUTOMATIC MODULE integrated into the basic movement as a complete unit. Central rotor with ball bearings, heavy metal ring, winding in both directions via twin Breguet free-running units, one reduction wheel on lower locking wheel. Hand-winding mechanism uncoupled by sprung pusher pinion between crown wheel and upper locking wheel.

BASIC MOVEMENT: three-spoke screwless Glucydur balance, Nivarox hairspring, 21,600 BPH, Incabloc shock resistance, moveable hairspring block carrier, fine adjustment by regulator with eccentric screw, direct center seconds, date indication with quick correction.

Magazine advertisement (from ÖU, 1957)

Caliber List: Enicar / AR

Caliber	Diameter Lignes	mm	Height mm	Other	BPH	Jewels	Running Reserve	Notes	Year	Photo
122	11 1/2	26,0		s	21.600	29		Winding in one		
165	"	"	4,9	SCD, QG	"	24	44	direction	1967	xx
177	"	"		SCD, QG, JG	18.000			Sherpa- Star		
197	"	"		SCD, QG, JG	21.600					
257	13			SCD, QG	18.000					
297	13			SCD, QG, JG	21.600					
764	7 3/4	17,5	4,5	SCD	21.600			Ebauche Eta 2550	ca. 66	
765	"	"		SCD, QG	"			Ebauche Eta 2551		
1034	11 1/2	26,0	5,8	SCD	18.000			Ebauche AS 1423	ca. 52	
1035	"	"	6,0	SCD, QG	"			Ebauche AS 1447		
1124	11 1/2	26,0	4,9	SCD,	18.000	30	44	Rubirotor	ca. 62	
1125	"	"		SCD, QG	"	"	"	"		
1126	"	"		SCD, QG (24 STD.)	"	"	"	"		
1144	11 3/4	26,8	4,9	SCD	"	"	"	"		
1145 /B	"	"		SCD, QG	"	"	"	"		
1146	"	"		SCD, QG (24 STD.)	"	"	"	"		
1146 /B	"	"		SCD, QG	"	"	"	"		
1147 /B	"	"		SCD, QG, JG	"	"	"	"		
1177	"	"		SCD, QG, JG	"	"	"	"		
1650				SCD, QG	21.600	27				

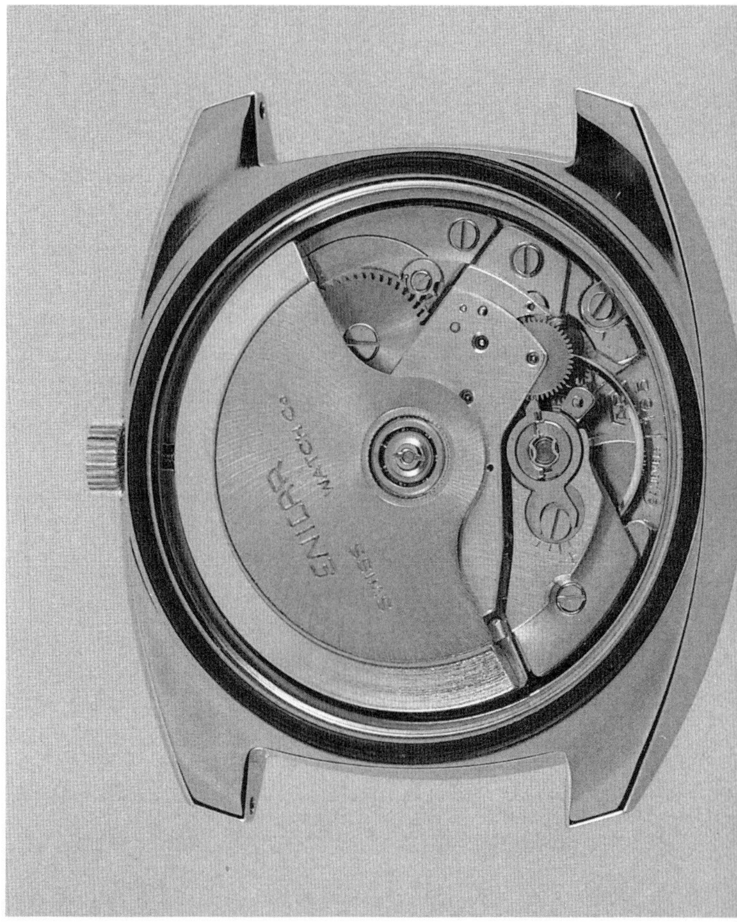

Enicar Caliber 165

Men's automatic model, "Sherpa Star", made by Enicar circa 1967, with Caliber 165.

Barrel-shaped, sharply angled stainless steel case with screw bottom and sapphire glass, winding crown with Enicar symbol, rotor with ball bearings and heavy metal ring, winding in both directions via Breguet-coupling switcher (one switching wheel visible at 2:00), uncoupling of hand-winding drive.

Screwless Glucydur balance, Nivarox hairspring, Incabloc shock resistance, fine regulator adjustment by eccentric screw.

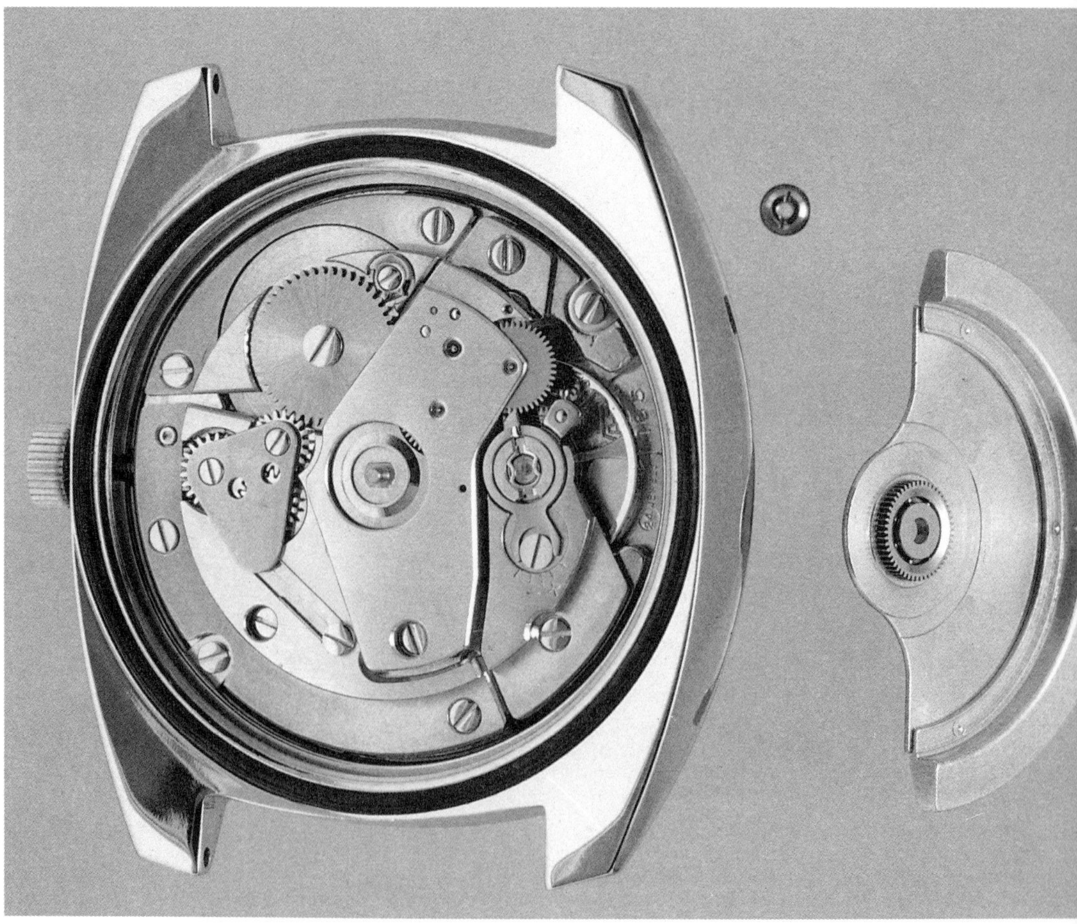

Caliber Development

In 1932 the Eterna firm separated from its producer of raw movements and had the latter, as an independent firm, the "ETA S.A.", join the Ebauches S.A. The most important Swiss producers of raw movements were included here (see AS). But there were still close associations in development and production between Eterna and ETA. This cooperation is also obvious in the first automatic movements produced by ETA in 1949-50, Caliber 1216 for ladies' and Caliber 1256 for men's watches. Both are equipped with a rotor that winds in both directions via the springless double click wheel patented by Eterna. Whereas the otherwise almost identical "Eterna-Matic" has a rotor running on miniature ball bearings, ETA used an adjustable conical gliding bearing.

The next caliber developments, made in 1952-53, included a modern drive mechanism with the indirectly driven minute hand beside the direct center seconds already used in the previous model, as well as improved click wheels now mounted side by side. The ladies' movement could be made smaller and attained production figures of three million pieces.

The subsequent ETA developments were based on these principles into the 1960s and can be identified easily by their characteristic twin click wheels. They differ only in the decreasing height of their movements (achieved through the use of flatter wheels of beryllium bronze, among other things), their rotor bearings, an anchor escapement with increasing rates of swing up to a fast speed of 36,000 BPH, and through added features such as instant date changing with fast correction, weekday indication and second stopping.

With the introduction of the "Flat line" Caliber 2890, the ETA engineers were able in 1976 to apply design and production technology to prepare what was then the flattest automatic movement with center seconds and date indication, the "Eterna-Matic 3000", for automated mass production. A reasonably priced, sturdy precision movement with dependable automatic winding could now be marketed. It can still be seen, with its noticeably large ball-bearing rotor mount, in most of today's wristwatches and is even used by renowned Swiss watch manufacturers.

The developmental history of the ETA automatic, extending more than twenty years, includes no essential caliber changes and thus gives a clear indication of the far-seeing original conception. ETA is, among other things, the most successful producer of raw movements for ladies' automatic wristwatches.

Caliber Notes

Caliber 1216
D = 20.5 mm (9.25 lignes), H = 5.35 mm, 17 jewels
AUTOMATIC MODULE screwed to the basic movement as a complete unit. Monometallic central rotor with conical gliding bearings, winding in both directions via patented springless double click wheel with switching wheel, two

Caliber 1256
D = 25.6 mm (11.5 lignes), H = 5.9 mm, 17 jewels, (running reserve indication)
AUTOMATIC MODULE screwed to the basic movement as a complete unit. Monometallic central rotor winding in both directions via springless (patented) double click wheel with switching wheel, two reduction wheels on the doubled crown wheel to uncouple the hand-winding mechanism.
BASIC MOVEMENT: two-spoke nickel screw balance, Nivarox hairspring, 18,000 BPH, Incabloc shock resistance, anchor wheel mounted between cap jewels, direct center seconds, differential drive.

Caliber 2472
D = 25.6 mm (11.5 lignes), H = 5.2 mm, 21 jewels
AUTOMATIC MODULE screwed to the basic movement as a complete unit, central rotor with heavy metal ring and conical gliding ruby bearings, winding in both directions via springless twin click wheels, two reduction wheels on the locking wheel, crown-wheel rocker to uncouple the hand-winding mechanism.
BASIC MOVEMENT: three-spoke screwless Glucydur balance, Nivarox hairspring, 18,000 BPH, Incabloc shock resistance, moveable hairspring block carrier, direct center seconds, date indication with immediate changing.

Caliber 2670
D = 17.5 mm (7.75 lignes), H = 4.8 mm, 21 jewels
AUTOMATIC MODULE screwed to the basic movement as a complete unit, central rotor with heavy metal ring, winding in both directions via springless twin click wheels, two reduction wheels on the locking wheel, movably mounted crown wheel to uncouple the hand-winding mechanism.
BASIC MOVEMENT: three-spoke screwless nickel balance, Nivarox hairspring, 28,800 BPH, Incabloc shock resistance, moveable hairspring block carrier, direct center seconds.

Caliber 2738
D = 25.6 mm (11.5 lignes)
AUTOMATIC MODULE screwed to the basic movement as a complete unit, central rotor with heavy metal ring and ball bearings, winding in both directions via springless twin click wheels, two reduction wheels on the locking wheel, movably mounted crown wheel to uncouple the hand-winding mechanism.
BASIC MOVEMENT: three-spoke screwless Glucydur balance, Nivarox hairspring, 36,000 BPH, Incabloc shock resistance, Triovis fine regulation with moveable hairspring block carrier, Clinergic-21 anchor escapement, direct center seconds, second stop, instantly changing date indication with fast correction, weekday indication.
reduction wheels on the locking wheel, crown-wheel rocker to uncouple the hand-winding mechanism.
BASIC MOVEMENT: two-spoke Glucydur screw balance, Nivarox hairspring, 18,000 BPH, Incabloc shock resistance, direct center seconds.

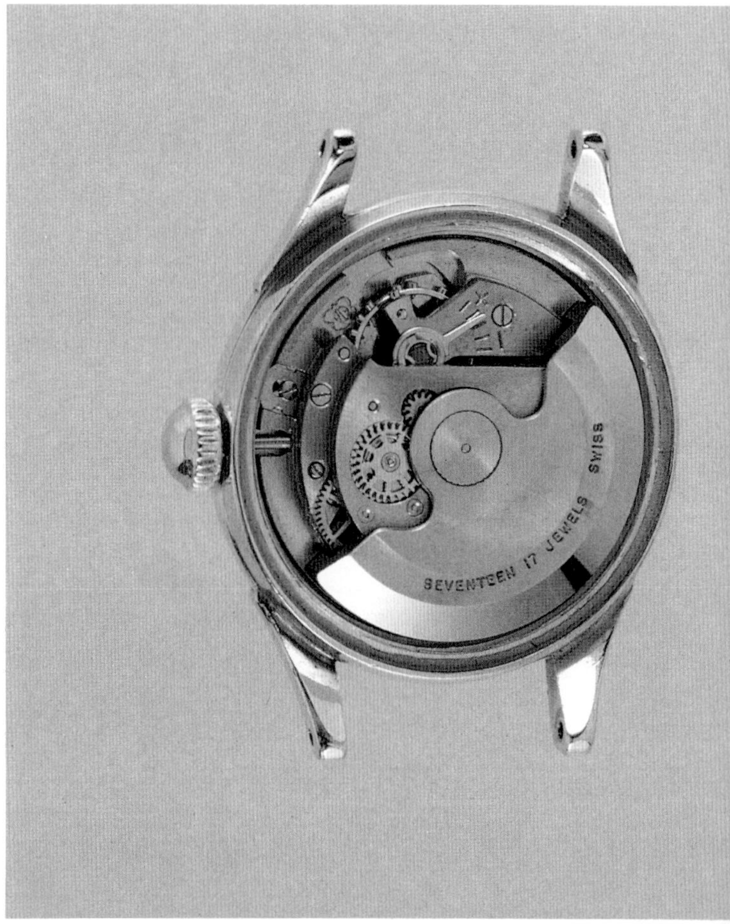

Eta Caliber 1216, "Paté"

Ladies' automatic of the "Paté" brand of 1950, with the first automatic Eta Caliber 1216.

Gilded case with stainless steel screw bottom.

Monometallic central rotor winding in both directions via a patented click-wheel switcher (at 9:00), uncoupling of the hand-winding drive. Automatic module removable as a complete unit.

Glucydur screw balance, Nivarox hairspring, Incabloc shock resistance (dial corroded, case shows signs of wear).

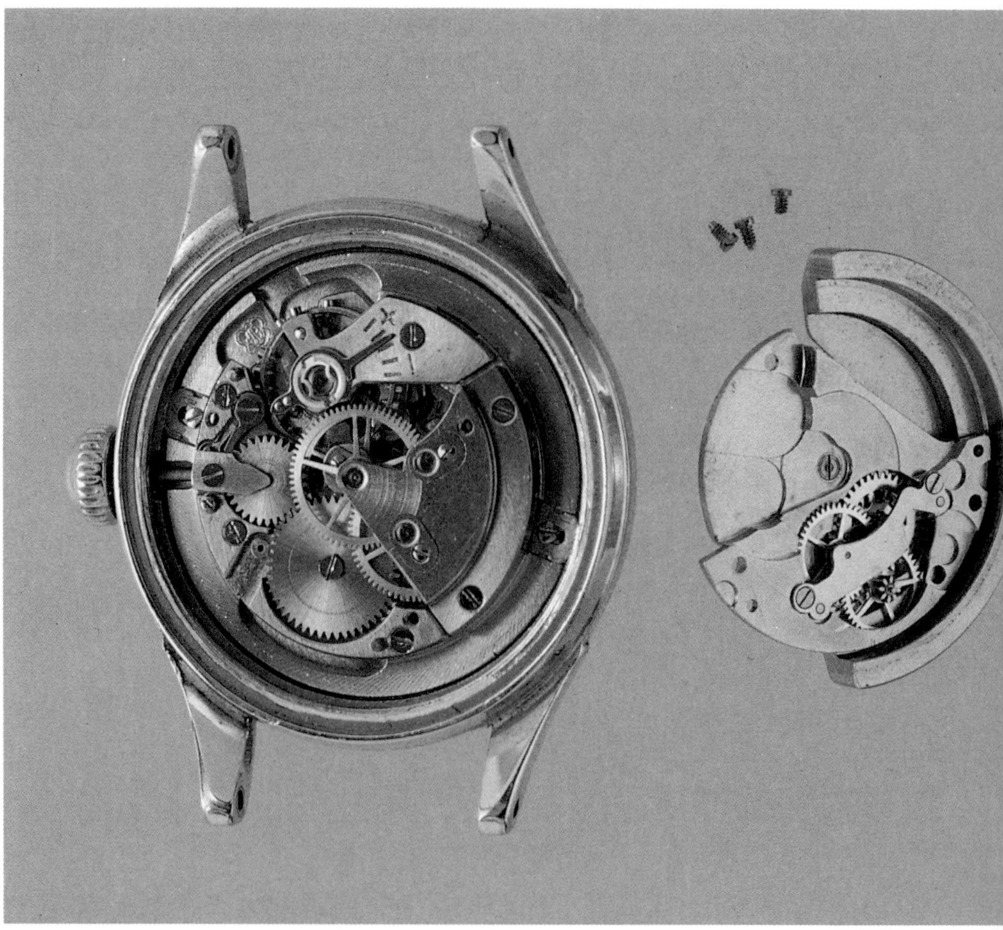

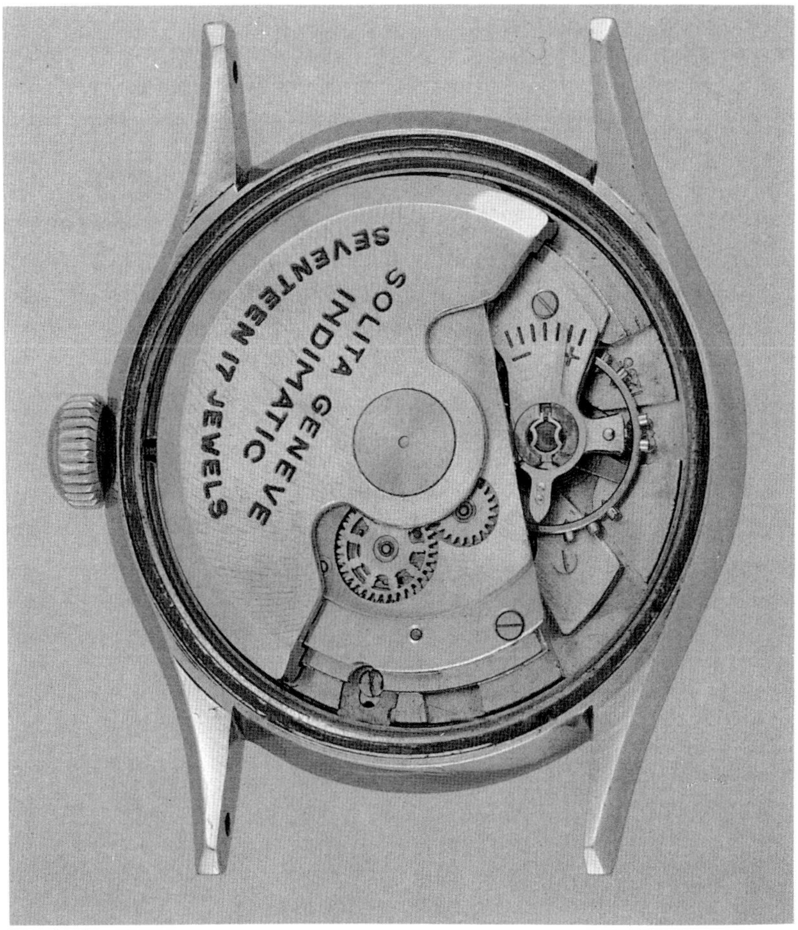

Eta Caliber 1256, "Indimatic"
Men's automatic of the "Solita" brand of 1951, with the first large automatic rotor Caliber 1256 by Eta, completed with running reserve indication in the "Indimatic" model. Chrome plate Staybrite case with stainless steel screw bottom. Monometallic rotor winding in both directions via click-wheel switcher (as patented double click wheel at 7:00), uncoupling of hand-winding drive. Nickel screw balance, Nivarox hairspring, Incabloc shock resistance.

Eta Caliber 2472, "Doxa"

Diver's automatic made by Doxa in 1967, "Sub 300 T searambler" model, with Eta Caliber 2472 (Doxa 118).

Massive stainless steel case with screw bottom and crown, turning bezel with one-sided lock, mineral glass.

Rotor with heavy metal ring, winding in one direction via click switcher (in typical Eta open fashion at 2:00), uncoupling of the hand-winding drive.

Screwless Glucydur balance, Nivarox hairspring, Incabloc shock resistance.

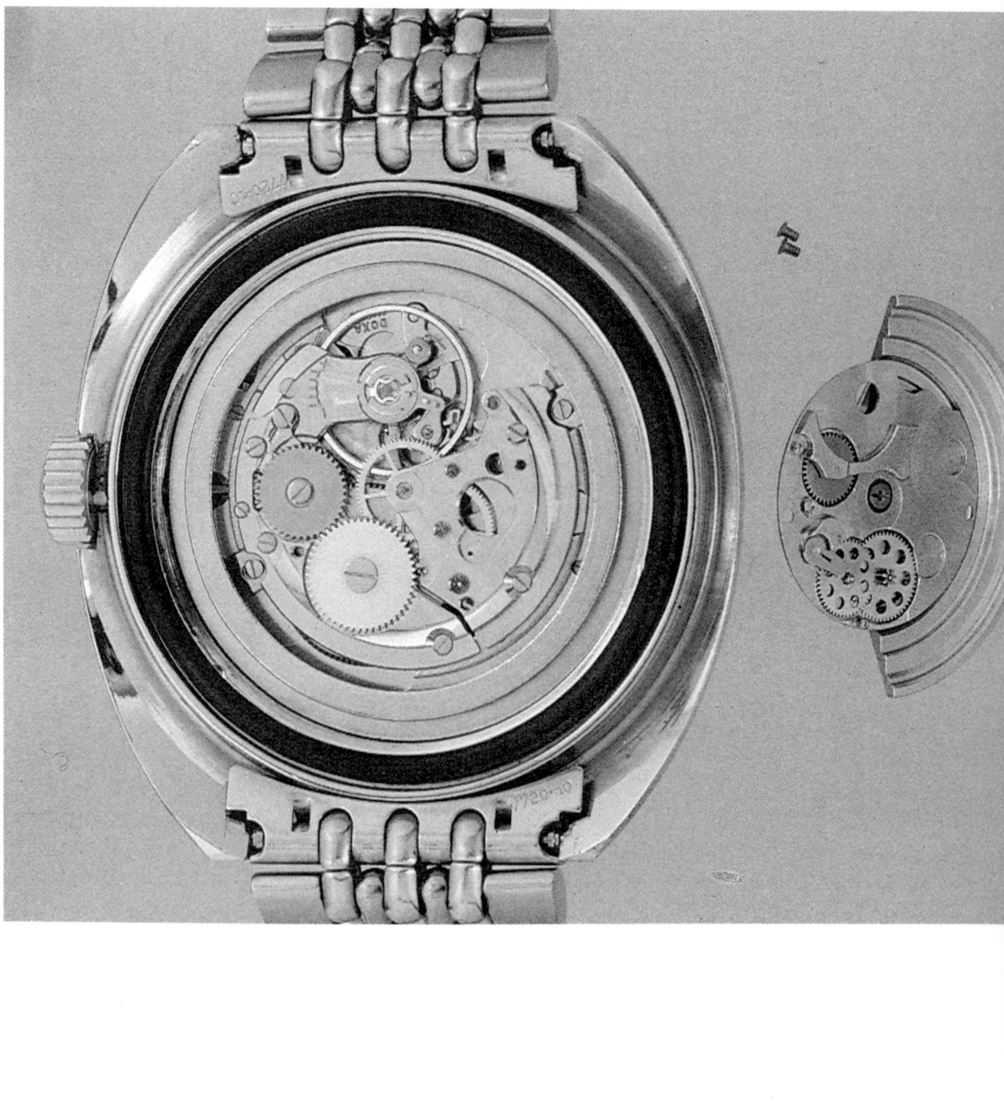

Eta Caliber 2670

Ladies' automatic of the German "Stowa" brand, circa 1970, with fast-swinging Caliber 2670 by Eta.
Gilded case with stainless steel screw bottom.
Rotor with heavy metal ring, winding in both directions via patented click-wheel switcher (at 2 and 3:00), uncoupling of hand-winding drive.
Screwless nickel balance, Nivarox hairspring, Incabloc shock resistance (winding crown replaced).

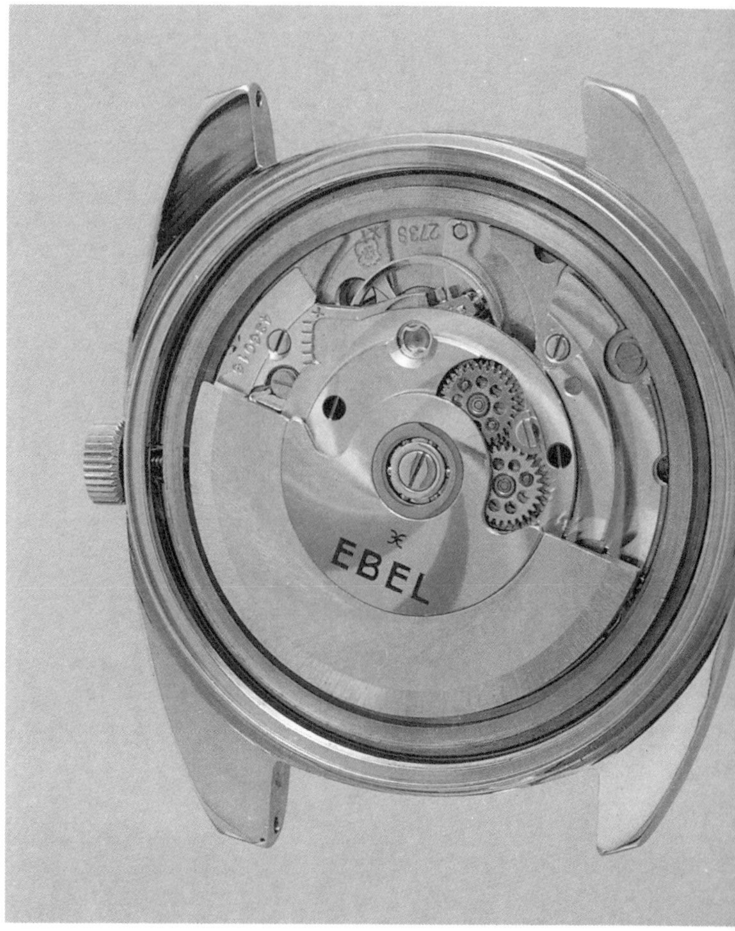

Eta Caliber 2738

Men's automatic as "Fast Beat" chronometer model, made by Ebel circa 1970, with fast-swinging Eta Caliber 2738.
Barrel-shaped stainless steel case with screw bottom.
Rotor with ball bearings and heavy metal ring, winding in both directions via patented click-wheel switcher (at 2 and 3:00), uncoupling of hand-winding drive.
Screwless Glucydur balance, Nivarox hairspring, Incabloc shock resistance, Triovis fine adjustment.

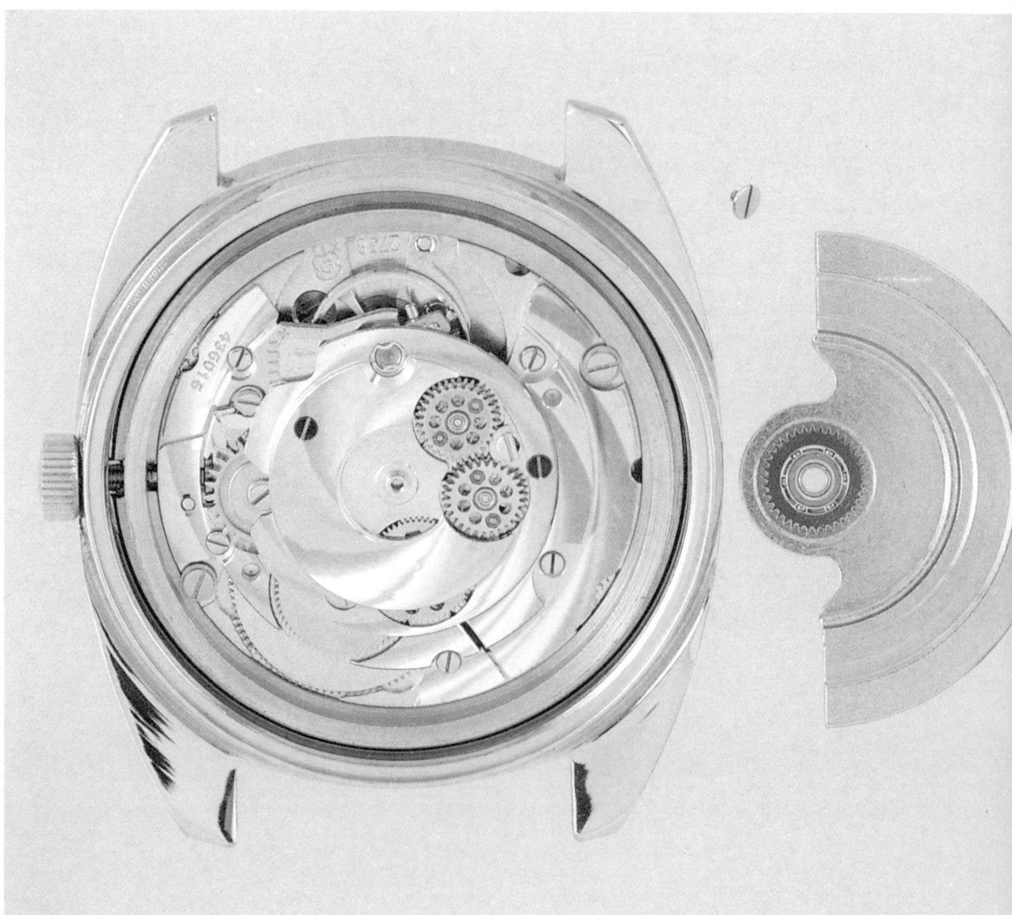

Eta Caliber (2824), "Ulysse Nardin"

Men's automatic as fast-swinging chronometer, Ulysse Nardin (Caliber NB11QU), with Eta raw movement (similar to Caliber 2824).

Stainless steel case with screw bottom, winding crown with Ulysse Nardin emblem, stopping device.

Rotor with ball bearings and heavy metal ring, winding in both directions via click-wheel switcher (half-concealed at 2:00), uncoupling of the hand-winding drive.

Screwless Glucydur balance, self-compensating flat hairspring, Incabloc shock resistance, fine regulator adjustment by eccentric screw.

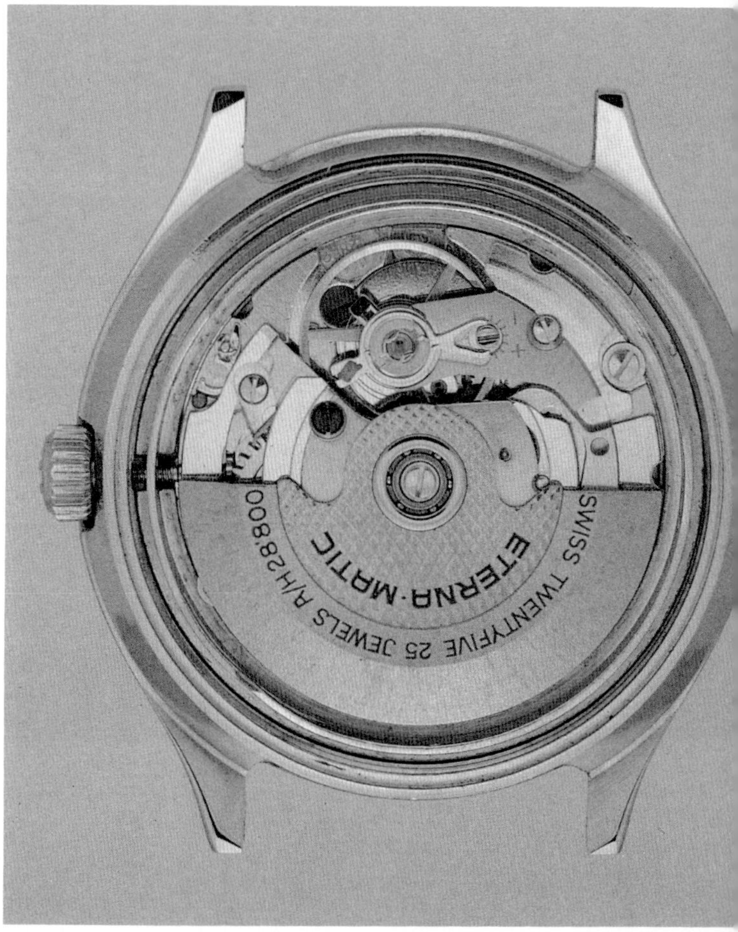

Eta Caliber 2872

Men's automatic by Eterna, circa 1975, with fast-swinging Eta Caliber 2872. Stainless steel case with screw bottom, winding crown with ball symbol, rotor with ball bearings and heavy metal ring, winding in both directions via patented click-wheel switcher, uncoupling of the hand-winding drive.
Screwless nickel balance, Nivarox hairspring, Incabloc shock resistance, fine regulator adjustment by eccentric screw.

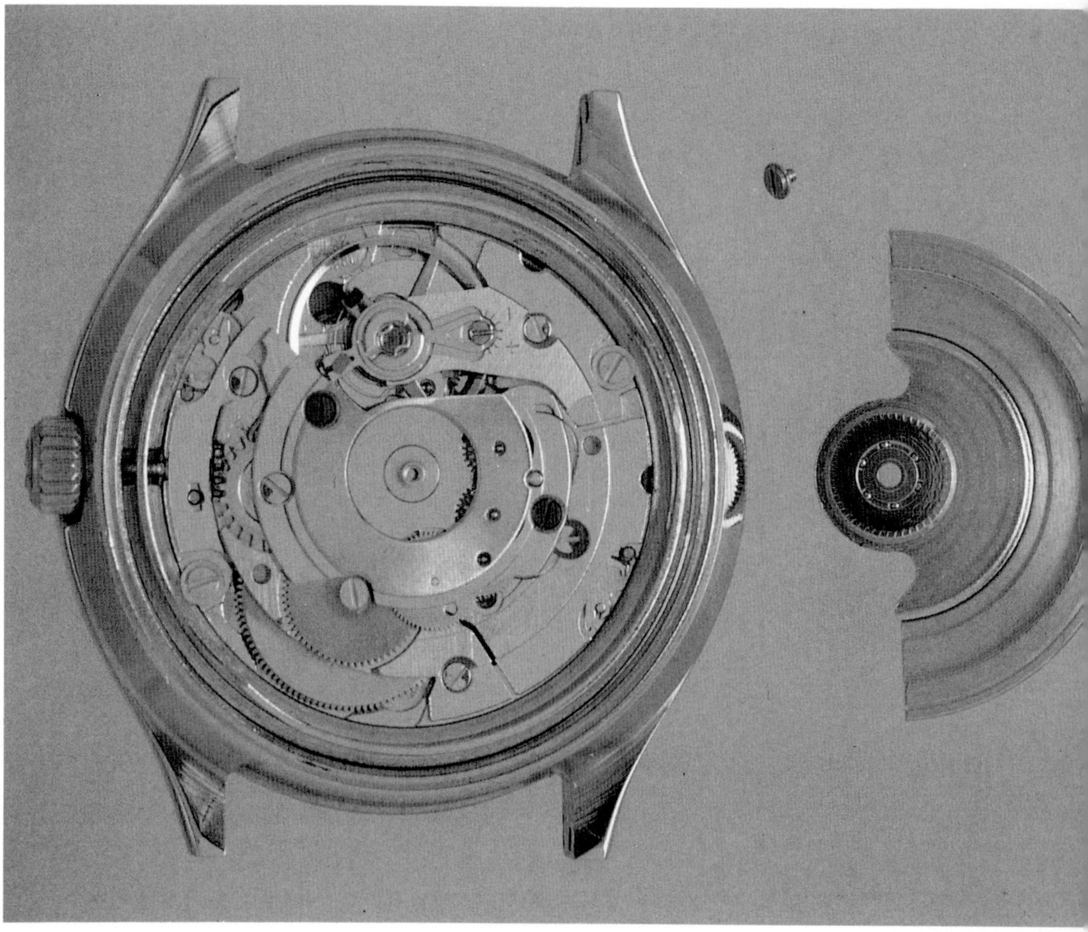

Caliber List: ETA

Caliber	Diameter Lignes	mm	Height mm	Other	BPH	Jewels	Running Reserve	Notes	Year	Photo
1216	9 1/4	20,5	5,35	SCD	18.000	17			1950	xx
1217	10		"	SCD	"	"		Basic Caliber 1216		
1218	8 1/4		"	SCD	"	"		"		
1256	11 1/2	25,6	5,9	SCD	18.000	17	40		1950	xx
1257	"	"		SCD	"	"	"			
1258	"	"		SCD, QG	"	"	"			
1259	"	"		QG	"	"	"			
1285	"	"		SCD	"	"	"			
1286	"	"		SCD	"	"	"			
1306	9 3/4	21,7	5,6	SCD	18.000					
1307	10 1/2	23,3	"	SCD	"					
1308	8 3/4	19,4	"	SCD	"					
1309	9 1/4	20,5	"	SCD	"					
1310	"	"	"	SCD	"					
1315	"	"		SCD	"					
1316	"	"		SCD	"					
1317	10 1/2	23,3		SCD	"					
1318	8 3/4	19,4		SCD	"					
1320	9 1/4	20,5		SCD	"					
1322	8 3/4	19,4		SCD	"					
2365	7 3/4	17,2	5,3	SCD	18.000	17			1952	
2366	"	"	"		"					
2367	8 1/4	18,3	"	SCD	"	"				
2368	"	"	"		"					
2369	7 3/4	17,2	5,6	SCD, QG, RDR	"	"				
2375	11 1/2	25,6	5,9	SCD	18.000				1953	
2377	13			SCD, QG	"					
2378	11 1/2	25,6		SCD, QG	"					
2379	"	"		SCD	"					
2380	"			SCD	"					
2381	"			SCD, QG	"					
2425	6 3/4	15,3	5,2		18.000				1956	
2428	"	"	"		21.600					
2446	7 3/4	17,3	5,3	SCD	18.000	21				
2447	"	17,5	5,6	SCD, QG	"	"				
2450	10 1/2	23,5	5,2	SCD	18.000		40			
2451	11 1/2	25,6	"	SCD	"		"		1956	
2452	"	"	"	SCD, QG, RDR	"		"			
2453	"	"	"	SCD	"		"			
2454 /55 /57	"	"	"	SCD, QG	"		"			
2456	10 1/2			SCD, QG	"					
2458	"			SCD	"					
2459	11 1/2	25,6		SCD	"					
2460	"	"		SCD, QG	"					
2461	"	26,0	5,2	SCD	"	25				

Caliber List: ETA

Caliber	Diameter Lignes	Diameter mm	Height mm	Other	BPH	Jewels	Running Reserve	Notes	Year	Photo
2462	11 1/2	26,0	5,2	SCD, QG	18.000	25				
2463	"			SCD, QG	"					
2464	10 1/2			SCD, QG	"					
2465	11 1/2	25,6		SCD	"					
2471	"	"		SCD	"					
2472 /74 /75	"	"	5,2	SCD, QG, RDR	"	21	40			xx
2483	"	26,0	5,2	SCD	"	25				
2484	"	"	"	SCD, QG	"	"				
2520 /R	11 1/2	25,6	4,5	SCD	18.000				1959	
2522 /R	"	"	"	SCD, QG	"			See Roamer Caliber 474	1959	(xx)
2524 /25	"	"	"	SCD	"					
2526 /27	"	"	"	SCD, QG	"					
2528	"	"	"	SCD	"					
2529	"	"	"	SCD, QG	"					
2532 /R	"	"	"	SCD, QG	"					
2538 /R	"	"		SCD, QG, JG	"					
2539 /R	13			SCD, QG, JG	"					
2550 /R	7 3/4	17,5	4,5	SCD	21.600	23		See Roamer 472	1966	(xx)
2551 /55	"	"	"	SCD, QG, RDR	"	21				
2554	"	"	4,5	SCD	"	"				
2570 /80	6 3/4			SCD	"					
2620 /R	11 1/2	25,6	4,6	SCD	21.600					
2622 /R	"	"	"	SCD, QG, RDR	"					
2623 /R	13	29,0	"	SCD, QG, RDR	"					
2630 /32 /37	11 1/2	25,6		SCD, QG	"					
2638 /R	"	"	5,05	SCD, QG, JG, RDR	"					
2639 /R	13	29,0	"	SCD, QG, JG, RDR	"					
2650	7 3/4	17,2	4,8	SCD	21.600					
2651	"	"	"	SCD, QG	"					
2658	"	"	5,35	SCD, QG, JG	"					
2670	7 3/4	17,5	4,8	SCD	28.800	21/25				xx
2671	"	"	"	SCD, QG	"	"				
2677	"	"	"	SCD, QG	"				1971	
2678	10 1/4	23,3	5,35	SCD, QG, JG	"				ca. 73	
2722 /R	11 1/2	25,6		SCD	28.800					
2724 /R	"	"		SCD	"					
2728	"	"		SCD	"					
2732 /R	"	"		SCD	36.000					
2734 /R	"	"		SCD	"					
2738	"	"		SCD, QG, JG, RDR, ST	"					xx
2770	11 1/2	25,6	5,2	SCD	21.600			19.7 Million made	1969	
2771	"	"		SCD	28.800					
2772 /73	"	"	5,2	SCD, QG, RDR	21.600	25		See Longines L 633.1	ca. 72	(xx)
2774	"	"		SCD, QG, RDR, ST	28.800					

Caliber List: ETA

Caliber	Diameter Lignes	mm	Height mm	Other	BPH	Jewels	Running Reserve	Notes	Year	Photo
2776	11 1/2	25,6	5,2	SCD	21.600	25				
2778	"	"	5,9	SCD, QG, JG, RDR	"					
2779	"	"		SCD, QG, JG, RDR	"					
2782	"	"	5,2	SCD, QG, JG	"					
2783	"	"	"	SCD, QG	"	25			ca. 74	
2784	"	"	"	SCD, QG	28.800	"				
2788	"	"	5,9	SCD, QG	21.600					
2789	"	"		SCD, QG, JG, RDR	"					
2790	"	"		SCD, QG, JG	28.800					
2820	11 1/2	25,6	4,6							
2824	"	"	"	SCD, QG, ST	28.800	17/25	40	Similar type: See Ulysse-Nardin NBIIQU	1972	xx
2834	13	29,0	5,05	SCD, QG, JG	"	25				
2870	11 1/2	25,6	5,2					Gabarit	1975	
2872	"	"		SCD, QG, RDR	28.800	25				xx
2890	12 1/2		3,6		28.800			Flat line, see Eterna 12890	1976	(xx)
2892	"		"		"					

Caliber (2824)
Ulysse-Nardin, D = 25.6 mm (11.5 lignes), H = 4.6 mm, 25 jewels
Basically similar to Caliber 2738.

Caliber 2872
Eta-Ebauches (Gabarit), D = 25.6 mm (11.5 lignes), 21 jewels
AUTOMATIC MODULE screwed to the basic movement as a complete unit, central rotor with ball bearings and heavy metal ring, winding in both directions via springless twin click wheels, two reduction wheels on the locking wheel. Crown wheel with moveable mount to uncouple hand-winding drive.
BASIC MOVEMENT: three-spoke screwless Glucydur balance, Nivarox hairspring, 28,800 BPH, Incabloc shock resistance, Etachron regulating system (moveable hairspring block carrier and fine adjustment), direct center seconds, Eloxal GL barrel, instantly changing date indication with quick set calendar.

Magazine advertisement (from JSH, 1948)

Caliber Development

The firm was founded by Urs Schild and Dr. Joseph Girard at Grenchen in 1856 and named "Eterna-Werke" in 1906. In 1932 the firm was divided into the "Eterna S.A." (making precision movements) and "ETA S.A." (making raw movements).

The first Eterna automatic (Caliber 834), made in 1938, had a hammer winding in one direction while swinging back and forth around the relatively small but very well crafted basic movement. A similar design appears in the ladies' automatic (Caliber 1033) introduced in 1942. Eterna's last hammer automatic, Caliber 1158, was developed in 1946. Here the automatic unit could be unscrewed from the basic movement completely, making servicing easier.

Though the earlier automatic calibers had been conventional in form, Eterna introduced several new features simultaneously in its first rotor automatic, the ladies' Caliber 1198, in 1948: the relatively small central rotor had a particularly low-friction and rugged mounting on its miniature ball bearings. The five balls of the bearing, each with a diameter of 0.65 mm, have formed the firm's trade mark since then. At the same time, it was the first automatic to carry the name of "Eterna-Matic", which all of the firm's subsequent automatic watches have borne on their dials and movements.

Another new feature was the use of a springless double click wheel (patented by Eterna), which marked the beginning of the most varied developments in the realm of direction changers. With these technical improvements, Eterna was able for the first time to make the winding capability of a ladies' caliber equal to that of a larger men's caliber and thus achieve a relatively flat movement construction.

The first man-size Eterna-Matic of 1950 (Caliber 1248) was designed similarly.

The replacement of the double click wheel by two click wheels side by side allowed a further decrease of movement height in the next developments. These twin click wheels were improved further through the use of beryllium and improved click operation, and since that time they have constituted a characteristic mark of the automatic calibers made by Eterna and ETA.

For the manufacturer's hundredth anniversary in 1956, there appeared the "Centenaire" (Caliber 1427 and later 1429), a men's flat automatic with a lens-formed movement cross-section achieved by making the movement's height decrease in steps toward the outside, thus allowing it to be installed in particularly elegant cases. This movement shape was made possible by new space-saving designs, such as that with an indirectly driven minute hand, as in Caliber 1411 of 1954, the direct center seconds and the use of beryllium wheels only 0.09 mm thick.

To improve running regularity, the screw balance was replaced by a smooth ring of Glucydur, thus achieving chronometer quality. In the anniversary year too, the smallest ladies' automatic of its time (Caliber 1419), called "Golden Heart" because of its 23-karat gold rotor, was introduced. Its ring balance, with its increased speed of 21,600 BPH, could be made relatively large and guaranteed remarkable precision for a caliber of this size.

With the "Eterna-Matic 3000" of 1962, the engineers of Eterna's development department created what was then the most slim caliber with center seconds and date indication (Caliber 1453) in series production. While the previous automatic designs had the automatic module on top of the basic movement, the latter was compressed here into the center and covered by the large but flat ball bearing of the central rotor. The automatic mechanism was moved further out and recessed in the basic movement. The date mechanism was also integrated into the plane of the plate without increasing the height of the movement.

The first rotor automatic, "Eterna-Matic", as a ladies' model of 1948, with rotor ball bearings and patented Eta-Eterna click-wheel switcher.

In the ensuing years, Eterna installed only ETA raw movements in its watches, for economic reasons. A listing of these movements by number will not be included.

All in all, it can be seen that the manufacture of the "Eterna-Matic" was not only a forerunner of the mass-production of ETA raw movements, but also played a major part in the development of the modern automatic watch with its introduction of, among other things, the miniature ball bearing and the springless click wheel.

Caliber Notes

Caliber 834
D = 27.45 mm (12 lignes), H = 5.3 mm
AUTOMATIC MODULE: hammer with ruby bearings, winding in one direction via click-carrier rocker on the crown wheel.
BASIC MOVEMENT: two-spoke screw balance, Nivarox hairspring, 18,000 BPH, Eterna-E shock resistance, seconds bit, bearing jewels with chatons.

Caliber 1034 H
D = 21.8 mm (9.75 lignes)
See Caliber 834, but with Eterna-H shock resistance and indirect center seconds.

Caliber 1159 H
D = 29.4 mm (13 lignes)
AUTOMATIC MODULE screwed to the basic movement as a complete unit, hammer with ruby bearings, limited by sprung sheet-metal buffers, winding in one direction via click-carrier rocker on the crown wheel.
BASIC MOVEMENT: two-spoke Glucydur screw balance, Nivarox hairspring, 18,000 BPH, Eterna-H shock resistance, direct center seconds.

Caliber 1237 T

D = 27.6 mm (12.5 lignes), H = 5.9 mm
AUTOMATIC MODULE screwed to the basic movement as a complete unit, brass central rotor with ball bearing, winding in both directions via patented springless double click wheels, two reduction wheels on the upper crown wheel, double crown wheel to uncouple the hand-winding mechanism.
BASIC MOVEMENT: two-spoke Glucydur screw balance, Nivarox hairspring, 18,000 BPH hour, Eterna-U shock resistance, direct center seconds, Eterniflex mainspring, slowly changing date indication.

Caliber 1248 T
D = 25.6 mm (11.5 lignes), H = 5.9 mm, movement as Caliber 1237 T.

Caliber 1419
D = 16.2 mm (7.75 lignes), H = 4.9 mm
AUTOMATIC MODULE screwed to the basic movement as a complete unit, 23-karat gold central rotor with ball bearing, winding in both directions via springless twin click wheels, two reduction wheels on the locking wheels, crown-wheel rocker to uncouple hand-winding mechanism.
BASIC MOVEMENT: two-spoke screwless Glucydur balance, Nivarox hairspring, 21,600 BPH, Eterna-U shock resistance, moveable hairspring block carrier, no second hand.

Eterna Caliber 834
The first Eterna automatic, made in 1938. Stainless steel case with screw bottom. Silver dial with luminous numerals and blued and luminous hands.
Hammer winding in one direction via click-carrier rocker made of special bronze.
Glucydur screw balance, Nivarox flat hairspring, Eterna-E shock resistance (case with slight signs of wear, "blooming" dial).

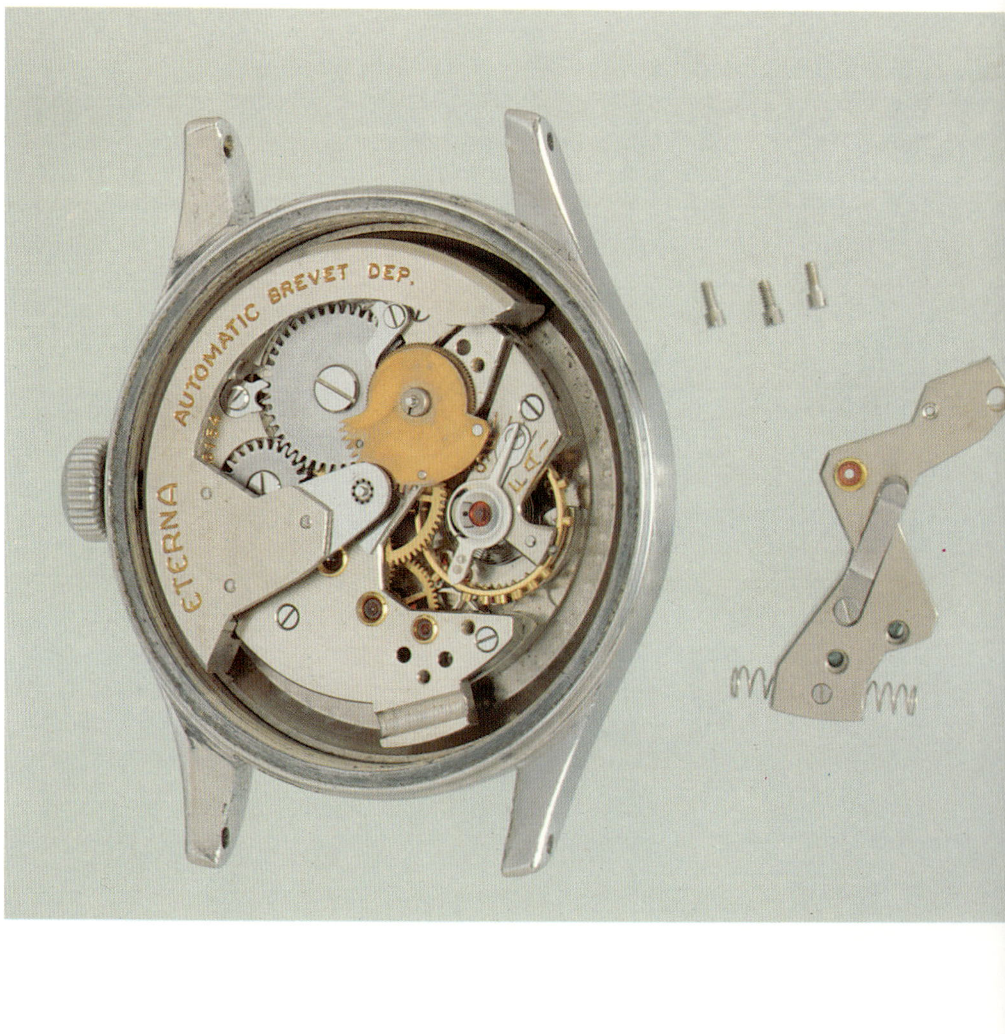

Eterna Caliber 1034 H
Ladies' automatic made by Eterna in 1943, with Caliber 1034 H.
Stainless Steel case with screw bottom.
Hammer winding in one direction via click-carrier rocker (bronze-colored segment).
Glucydur screw balance, Nivarox hairspring, Eterna-H shock resistance (dial restored, case with signs of wear).

Eterna Caliber 1159 H

Men's automatic made by Eterna in 1946 (Caliber 1159 H).
14-karat gold case with press-in bottom.
Hammer movement limited by sheet-metal buffer spring, winding in one direction via click-carrier rocker (bronze-colored segment), automatic module removable as a complete unit.
Glucydur screw balance, Nivarox hairspring, Eterna-H shock resistance (dial slightly flecked, winding crown not original).

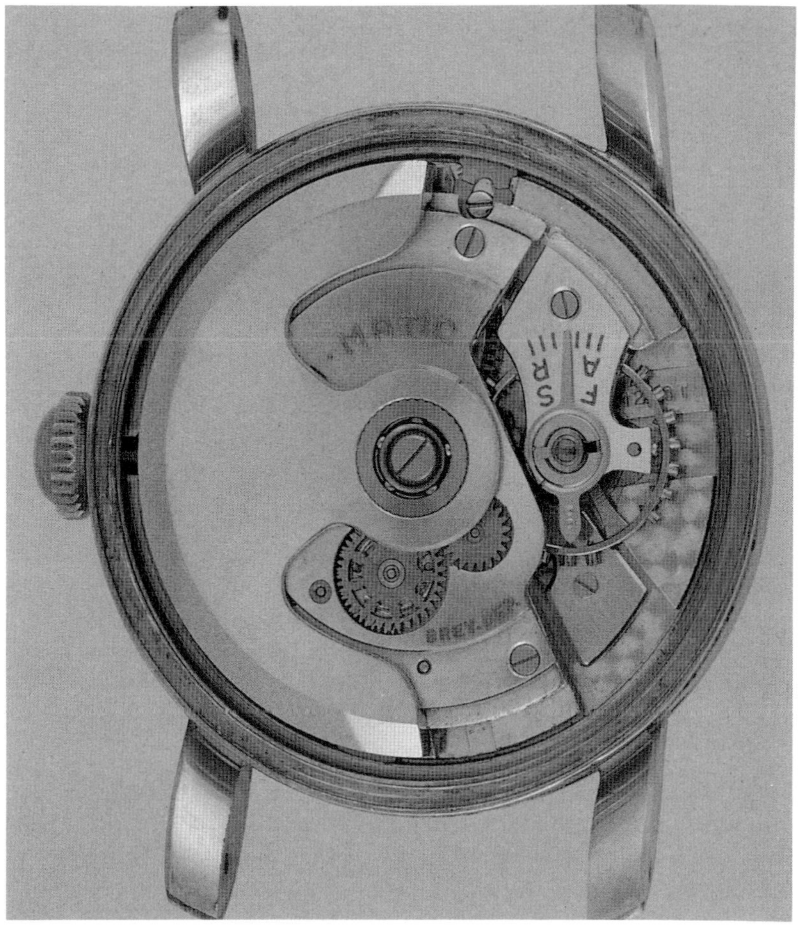

Eterna Caliber 1248 T

First men's rotor automatic, "Eterna-Matic", made by Eterna circa 1950, with Caliber 1248 T.
Stainless steel case with gold bezel and screw bottom.
Brass rotor with ball bearings, winding in both directions via patented click-wheel switcher (at 6:00), uncoupling of the hand-winding drive.
Glucydur screw balance, Nivarox hair-spring, Eterna-U shock resistance (dial inlays and hands corroded).

Eterna Caliber 1237 T

Men's automatic made by Eterna in 1951, with Caliber 1237 T.
Stainless steel case with gold bezel and screw bottom, winding crown with five balls. Movement corresponds to Caliber 1248 T (see page 109).

Eterna Caliber 1419, "Golden Heart"
Smallest ladies' automatic of its time, made by Eterna in 1956, with "Golden Heart" Caliber 1419.
Ball-shaped bowl case with 18-karat gold bracelet, structured gold dial.
23-karat gold rotor with ball bearings, winding in both directions via patented click-wheel switcher (at 1:00), uncoupling of the hand-winding drive.
Large screwless Glucydur balance, Nivarox hairspring, Eterna-U shock resistance (winding crown not original).

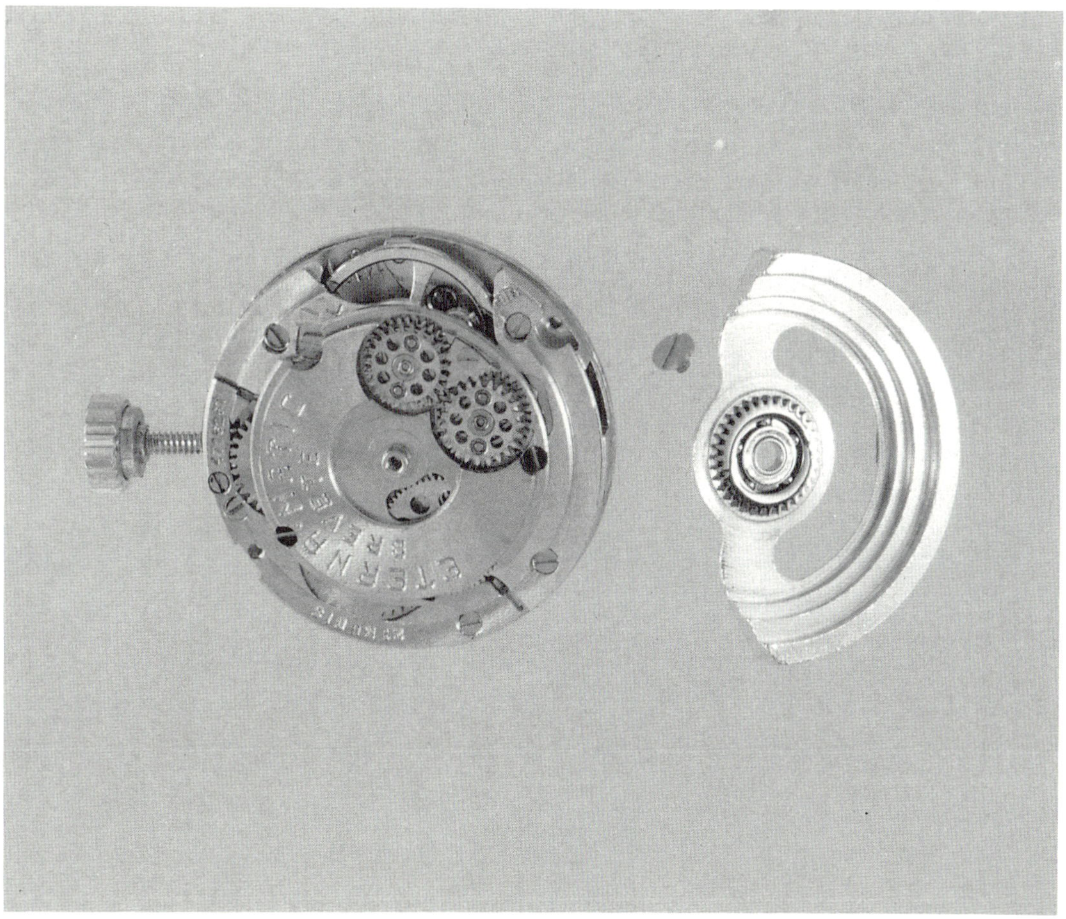

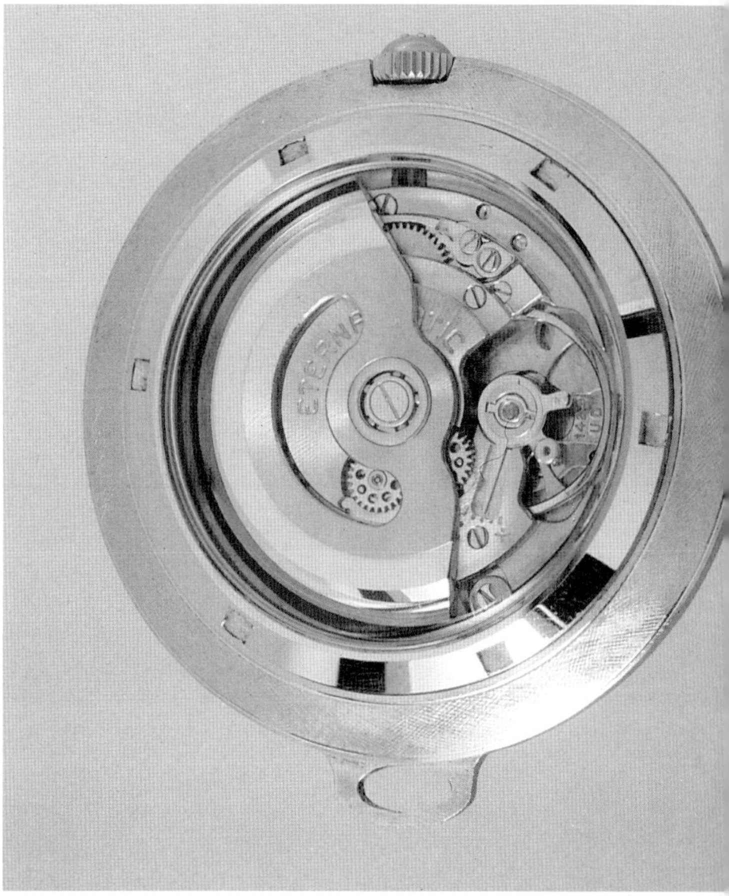

Eterna Caliber 1424 UD
Men's automatic "Golfer" model made by Eterna circa 1956, with Caliber 1424 UD.
Large stainless steel case with glassed screw bottom, winding crown with call symbol. Monometallic rotor with ball bearings, winding in both directions via click-wheel switcher in characteristic Eta-Eterna open twin fashion (at 2:00), uncoupling of the hand-winding drive.
Screwless Glucydur balance, Nivarox hairspring, Eterna-U shock resistance.

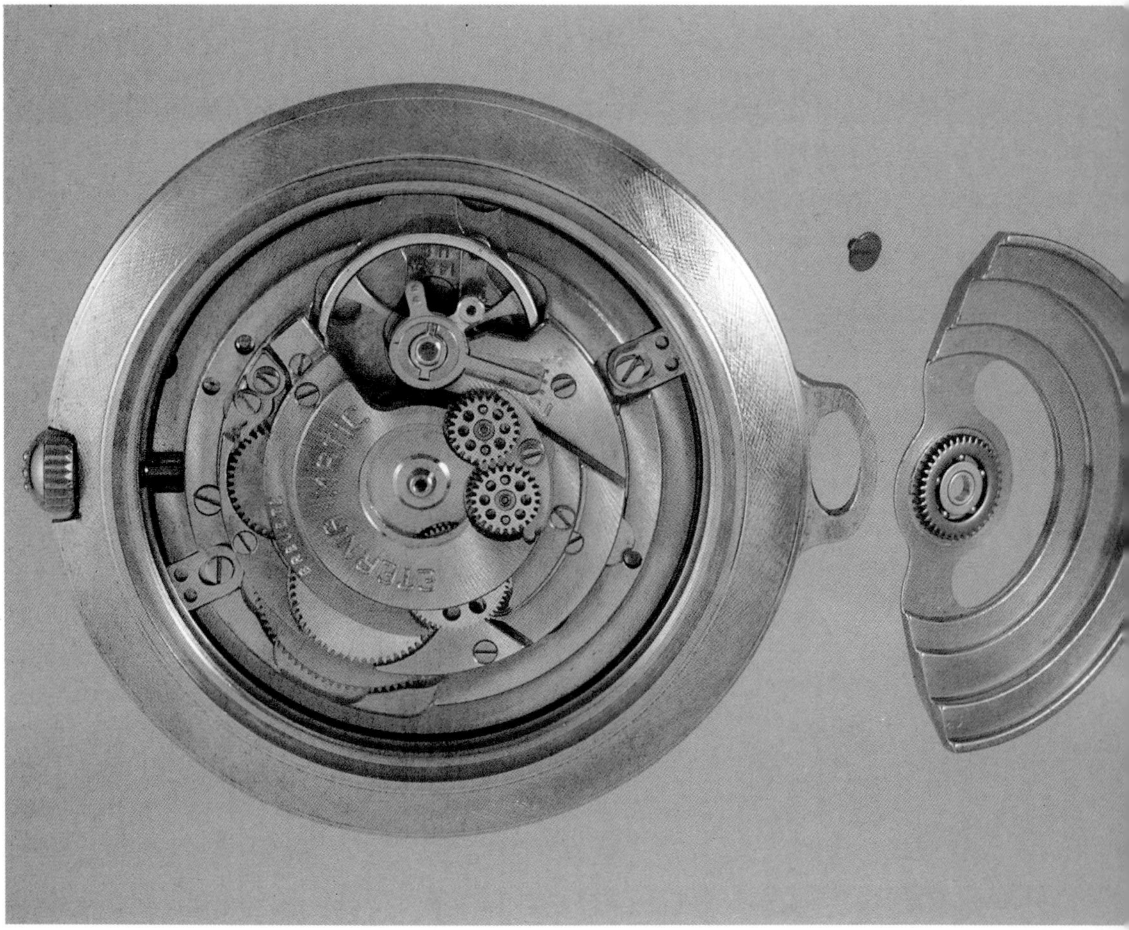

Eterna Caliber 1426 U

Men's automatic made by Eterna circa 1956, with Caliber 1426 U. Stainless steel case with screw bottom, winding crown with ball symbol. Movement largely identical to Caliber 1424 UD (see page 110), but here with a Glucydur screw balance.

Eterna Caliber 1429 U

Men's automatic, "Centenaire" chronometer model, made by Eterna circa 1956, for the 100th anniversary of the firm, with Caliber 1429 U.

Stainless steel case with screw bottom and gold bezel, winding crown with ball symbol, gold dial inlays and hands.

Rotor with ball bearings and heavy metal ring, winding in both directions via patented click-wheel switcher (at 2:00), uncoupling of the hand-winding drive.

Screwless Glucydur balance, Nivarox hairspring, Eterna-U shock resistance.

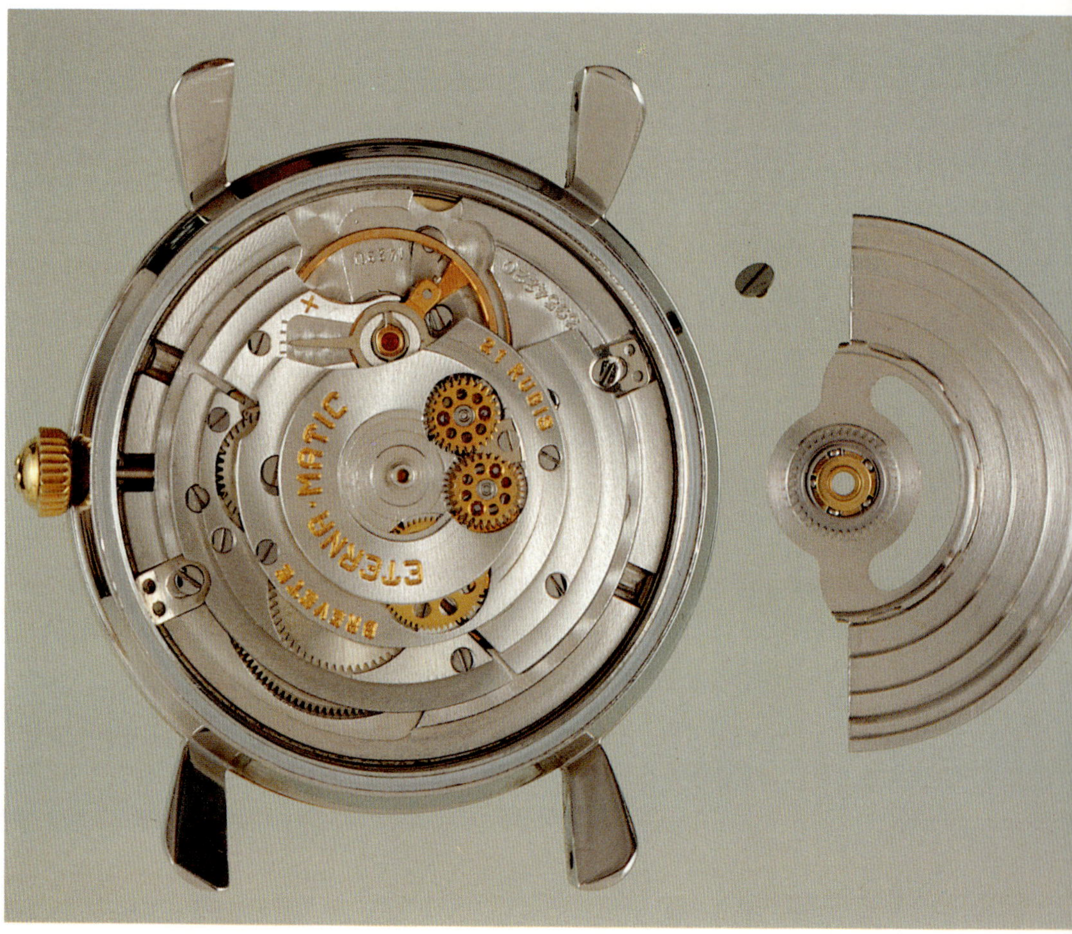

Eterna Caliber 1439 U

Men's automatic, "Centenaire Golfer" model, made by Eterna in 1959, with Caliber 1439 U. Large 18-karat gold case with glazed screw bottom, winding crown with ball symbol, plus dial inlays and hands of gold.

Movement identical to Caliber 1429 U (see page 114), but with fine regulator adjustment by eccentric screw.

Eterna Caliber 1466 U

Men's flat automatic, Model "3000", made by Eterna circa 1963, with Caliber 1466 U.
Stainless steel case with screw bottom and gold bezel, winding crown with ball symbol, gold dial inlays. Rotor with large ball bearings and heavy metal ring, winding in both directions via patented click-wheel switcher, uncoupling of hand-winding drive. Automatic module removable as a complete unit.
Screwless Glucydur balance, Nivarox hairspring, Eterna-U shock resistance, fine regulator adjustment by eccentric screw.

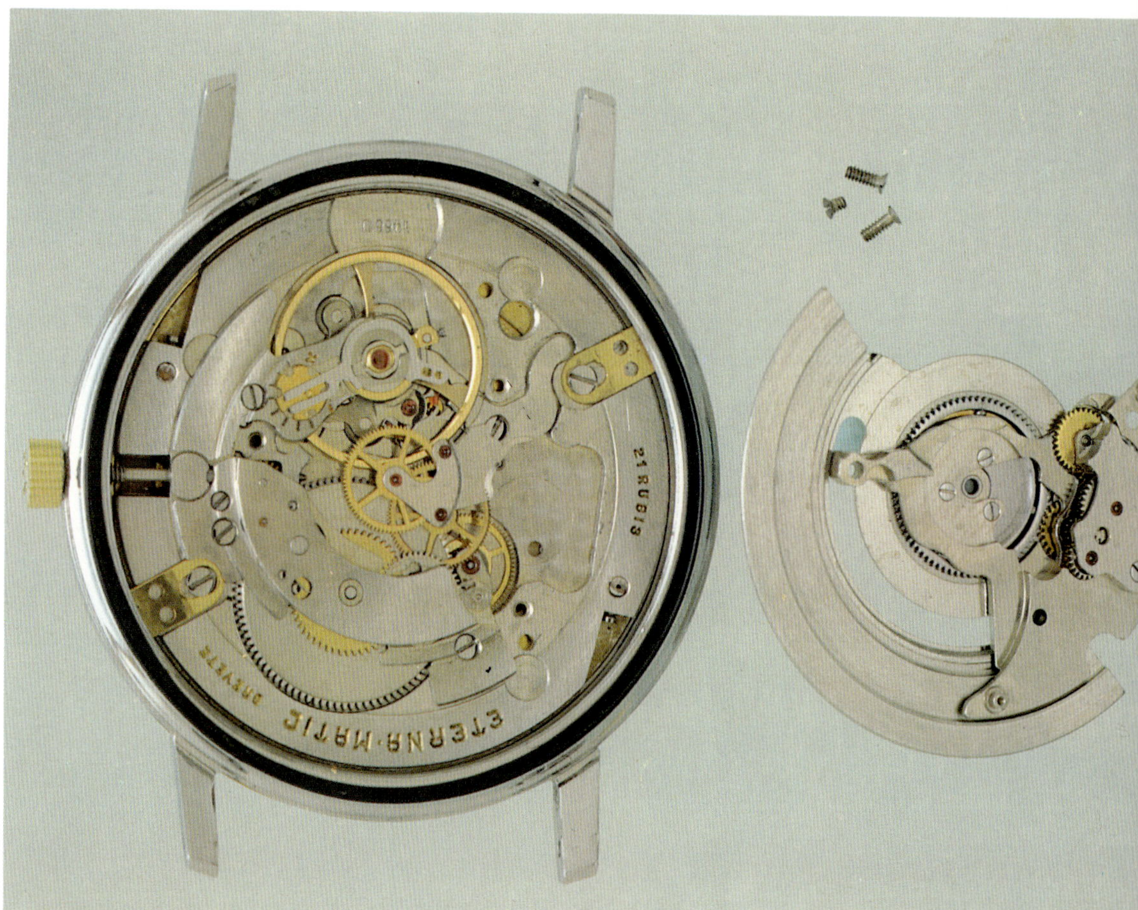

Eterna Caliber 1480 K

Men's automatic, Model "2000, Centenaire 71", made by Eterna in 1971, with Caliber 1480 K.
18-karat gold case with screw bottom, winding crown with ball symbol. Gold dial inlays and hands.
Movement largely identical to Caliber 1429 U (see page 114), but with Kif-Ultraflex shock resistance.

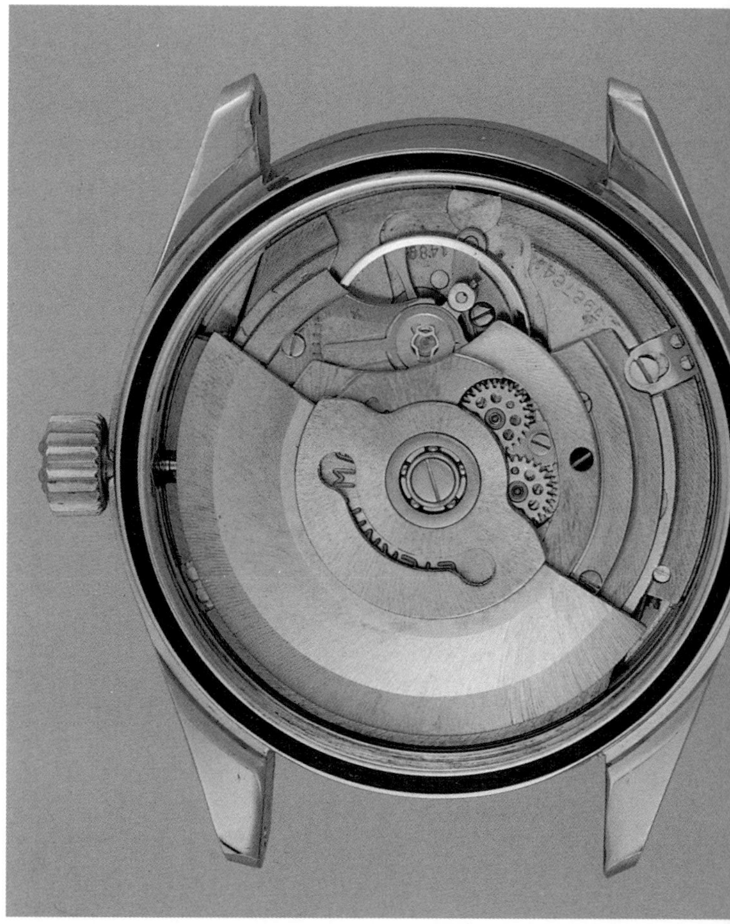

Eterna Caliber 1489

Men's automatic, "Kontiki 10" model, made by Eterna circa 1968, with Caliber 1489.

Stainless steel case with screw bottom, winding crown with ball symbol. Movement largely identical to Caliber 1480 (see page 117).

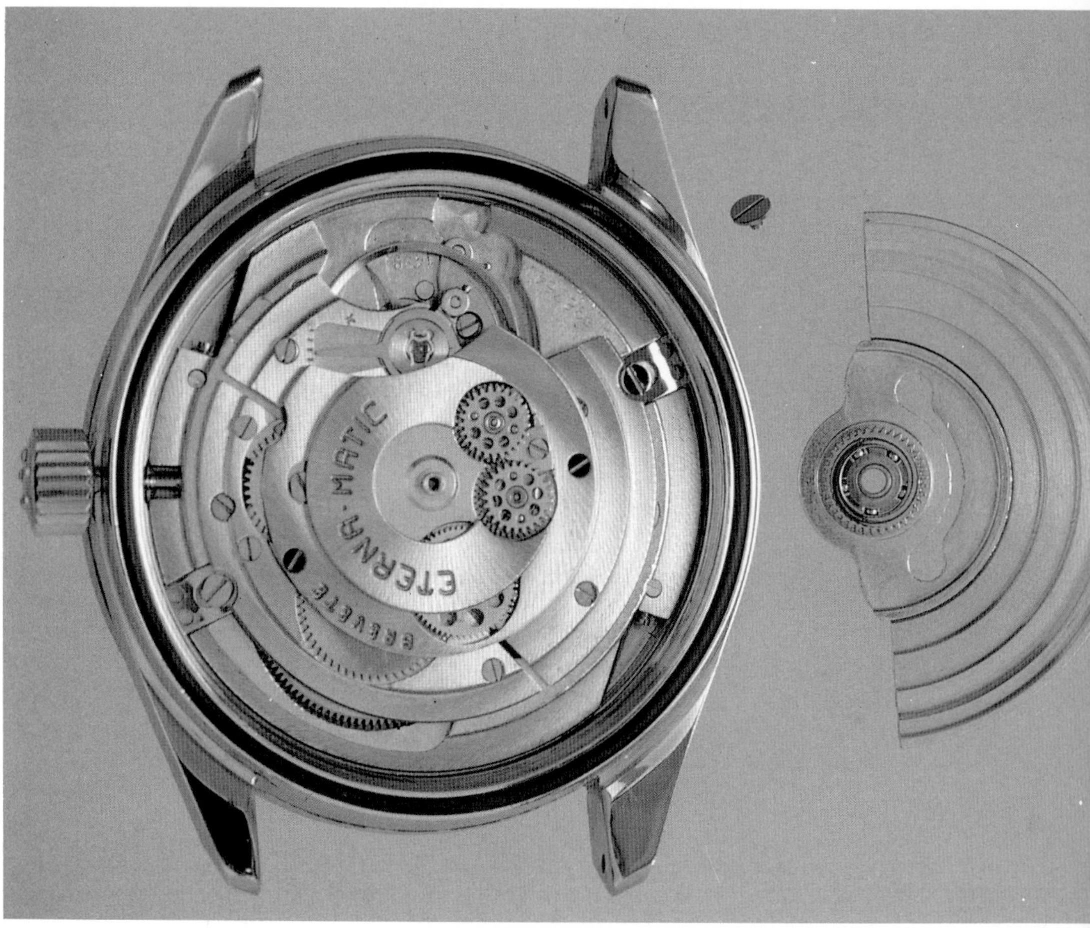

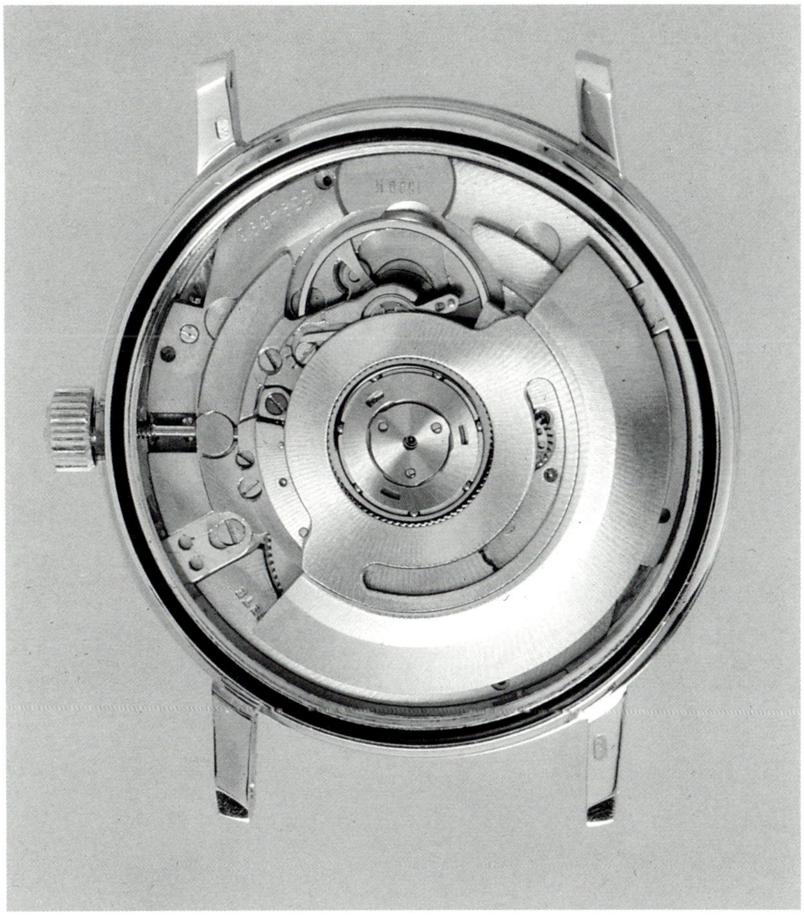

Eterna Caliber 1500 K

Flat men's automatic, Model "3000", made by Eterna circa 1970, with Caliber 1500 K.
Eighteen Karat gold case with screw bottom, winding crown with ball symbol. Gold dial inlays and hands. Movement largely identical to Caliber 1466 U (see page 116), but with 21,600 BPH, Kif-Ultraflex shock resistance and fine regulator adjustment by eccentric screw.

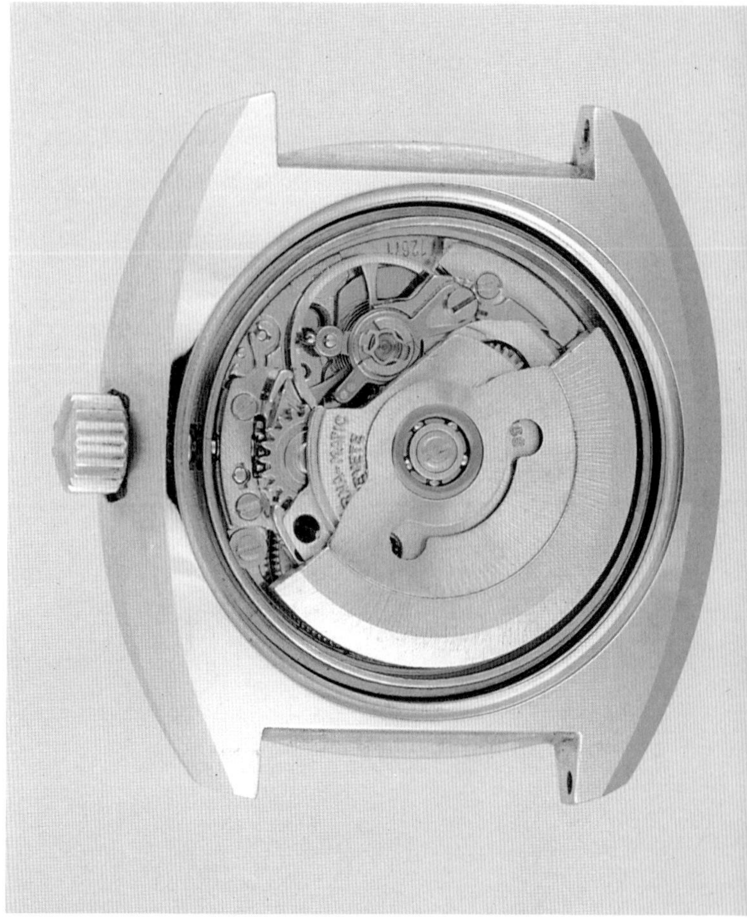

Eterna Caliber 12671

Ladies' automatic, "Sahida" model, made by Eterna circa 1971, with Caliber 12671 (Eta raw movement 2671). Barrel-shaped stainless steel case with screw bottom, winding crown with ball symbol.

Rotor with ball bearings and heavy metal ring, winding in both directions via patented click-wheel switcher (at 2-3:00), uncoupling of the hand-winding drive. Screwless Glucydur balance, Nivarox hairspring, Kif-Ultraflex shock resistance, fine regulator adjustment by eccentric screw.

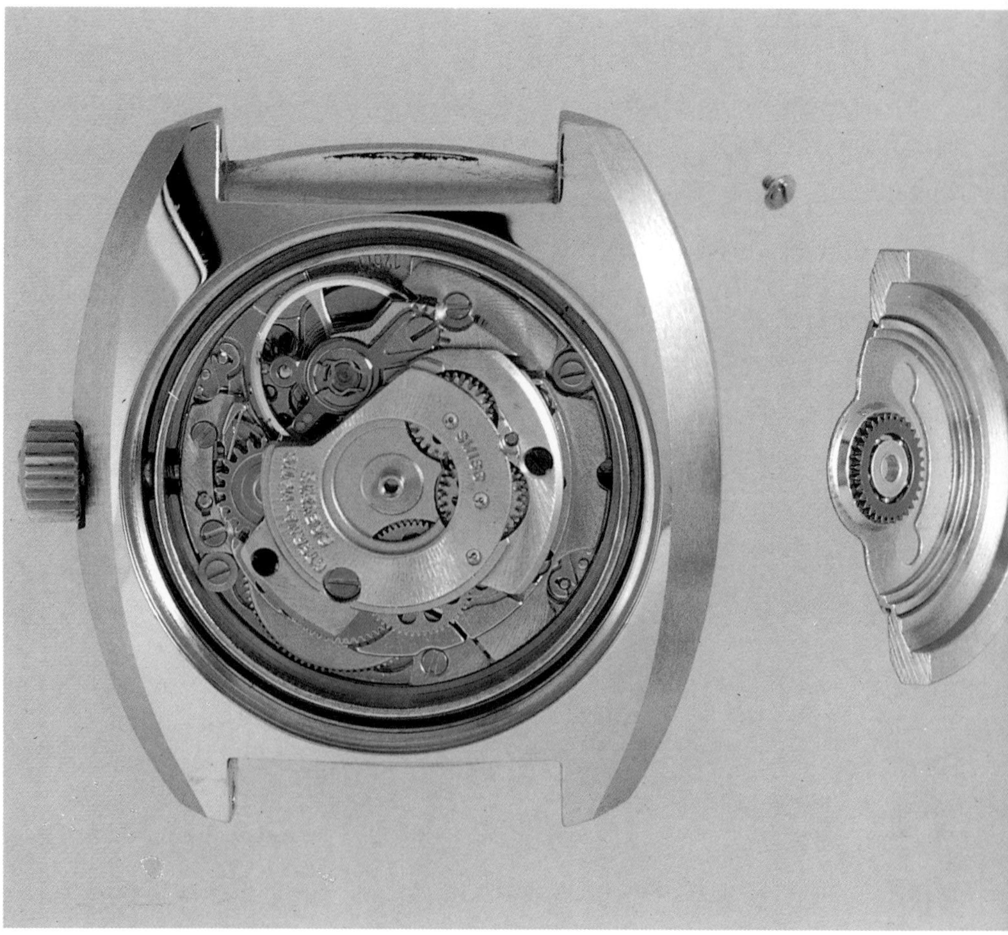

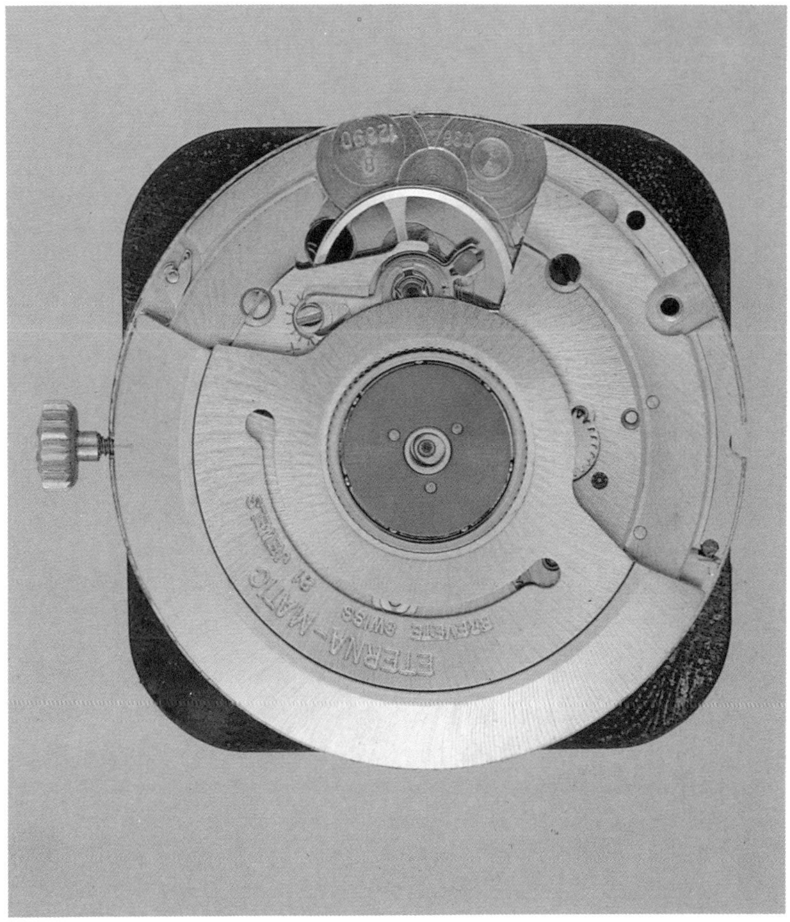

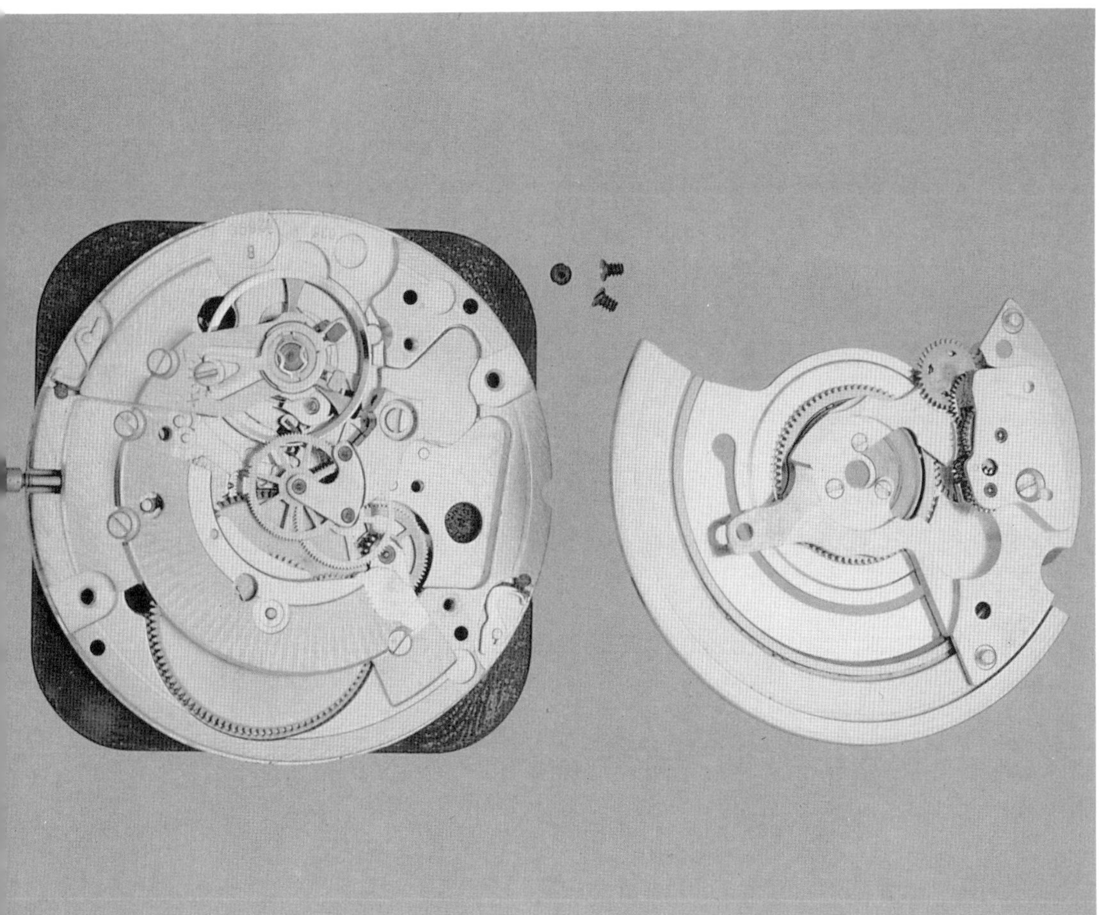

Eterna Caliber 12890

Flat men's automatic, Model "3003", made by Eterna in 1976, with fast-swinging Caliber 12890 (Eta Flat-line Caliber 2890).

Barrel-shaped stainless steel case with bowl-shaped press-in bottom, winding crown with ball symbol. Movement structure corresponds largely with Eterna Caliber 1466 U (see page 116).

Caliber 1424 UD
D = 29 mm (13 lignes), H = 5.6 mm, 21 jewels
AUTOMATIC MODULE screwed to the basic movement as a complete unit, brass central rotor with ball bearing, winding in both directions via patented springless twin click wheels, two reduction wheels on the locking wheel, crown-wheel rocker to uncouple hand-winding mechanism.
BASIC MOVEMENT: three-spoke screwless Glucydur balance, Nivarox hairspring, 18,000 BPH, Eterna-U shock resistance, moveable hairspring block carrier, direct center seconds, Eloxal GL barrel, date indication with quick set calendar.

Caliber 1426 U
D = 30.6 mm (13.5 lignes), H = 5.6 mm
See Caliber 1424 U.

Caliber 1429 U
D = 29.0 mm (13 lignes), H = 4.5 mm, 21 jewels
AUTOMATIC MODULE screwed to the basic movement as a complete unit, central rotor with ball bearing and heavy metal ring, winding in both directions via springless twin click wheels, two reduction wheels on the locking wheel, crown-wheel rocker to uncouple the hand-winding mechanism.
BASIC MOVEMENT: two-spoke screwless Glucydur balance, Nivarox hairspring, 18,000 BPH, Eterna-U shock resistance, moveable hairspring block carrier, direct center seconds, wheels of beryllium bronze.

Caliber 1439 U
D = 29.0 mm (13 lignes), H = 4.5 mm, 21 jewels
Movement like Caliber 1429 U, additional date indication, and quick set calendar.

Caliber 1466 U
D = 29.0 mm (13 lignes), H = 3.6 mm, 21 jewels
AUTOMATIC MODULE as a complete unit with its drive recessed in the basic movement, central rotor with heavy metal ring turning on a large adjustable ball bearing (with seven balls), winding in both directions via springless double click wheel, three reduction wheels and two intermediate wheels on the locking wheel, rocker link between crown and locking wheels to uncouple winding mechanism.
BASIC MOVEMENT: three-spoke screwless Glucydur balance, Nivarox hairspring, 18,000 BPH, Eterna-U shock resistance, moveable hairspring block carrier, fine adjustment via eccentric screw, direct center seconds, micro-geared beryllium wheels, date indication with instant change and quick set calendar.

Caliber 1480 K
D = 27.6 mm (12.5 lignes), H = 4.5 mm, 21 jewels
AUTOMATIC MODULE: See Caliber 1429 U.
BASIC MOVEMENT: three-spoke screwless Glucydur balance, Nivarox hairspring, 21,600 BPH, Kif-Ultraflex shock resistance, moveable hairspring block carrier, fine adjustment by eccentric screw, direct center seconds, date indication, and quick set calendar.

Caliber 1489
D = 29.3 mm (13 lignes)
See Caliber 1480 K

Caliber 1500 K
D = 29.3 mm (13 lignes), H = 3.6 mm, 21 jewels
AUTOMATIC MODULE: See Caliber 1466 U.
BASIC MOVEMENT: three-spoke screwless Glucydur balance, Nivarox hairspring, 21,600 BPH, Kif-Ultraflex shock resistance, moveable hairspring block carrier, fine adjustment by eccentric screw, direct center seconds, Eloxal GL barrel, date indication with instant change and quick set calendar.

Caliber 12671
Eta Ebauche
D = 17.2 mm (7.75 lignes), H = 4.8 mm
AUTOMATIC MODULE screwed to basic movement as a complete unit, central rotor with ball bearings and heavy metal ring, winding in both directions via springless twin click wheels, two reduction wheels on the locking wheel, movably mounted crown wheel to uncouple hand-winding mechanism.
BASIC MOVEMENT: three-spoke screwless Glucydur balance, Nivarox hairspring, 28,800 BPH, Kif-Ultraflex shock resistance, moveable hairspring block carrier, fine adjustment by eccentric screw, direct center seconds, second stop, date indication with instant change and quick set calendar.

Caliber 12980
Eta Ebauche (Caliber 2980), D = 27.2 mm (12.5 lignes), H = 3.6 mm, 21 jewels
AUTOMATIC MODULE similar to Caliber 1466 but with modified rotor ball bearings.
BASIC MOVEMENT: three-spoke screwless Glucydur balance, Nivarox hairspring, 28,800 BPH, Incabloc shock resistance, Etachron regulation system, no second hand, Eloxal GL barrel.

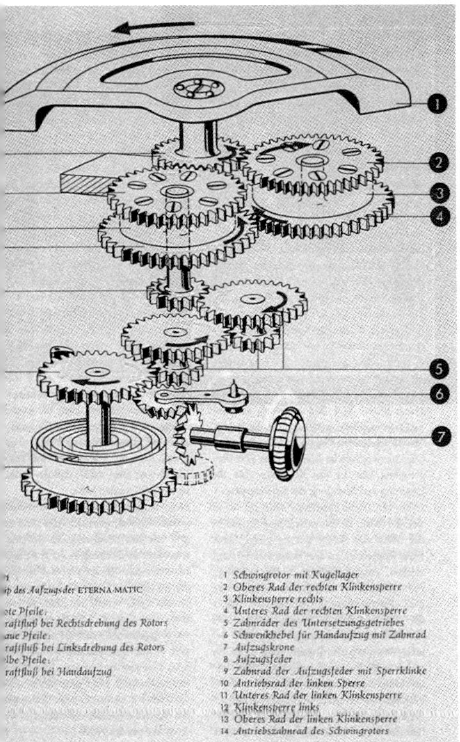

The functioning principles of an Eterna-Matic are shown in this drawing (twin click-wheel switchers 2-4 and 10-13).

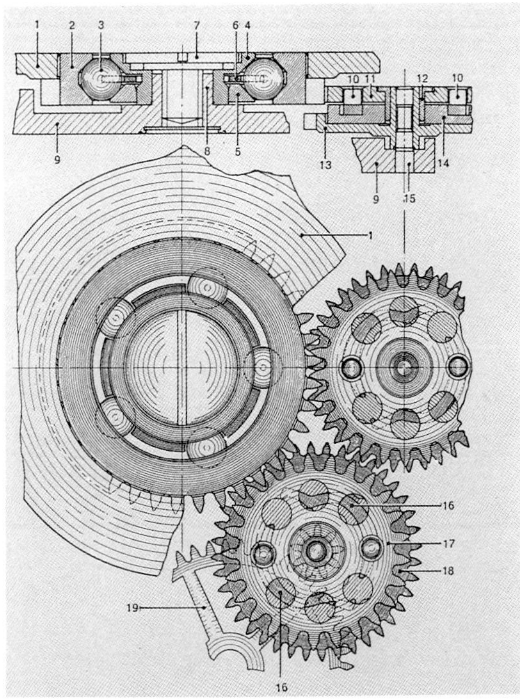

The coupling of the rotor pinion to the twin click-wheel switcher: the rotor pinion simultaneously contacts both upper click wheels, while the two lower ones are in direct contact with each other, so that one of them must always turn passively along with the other.

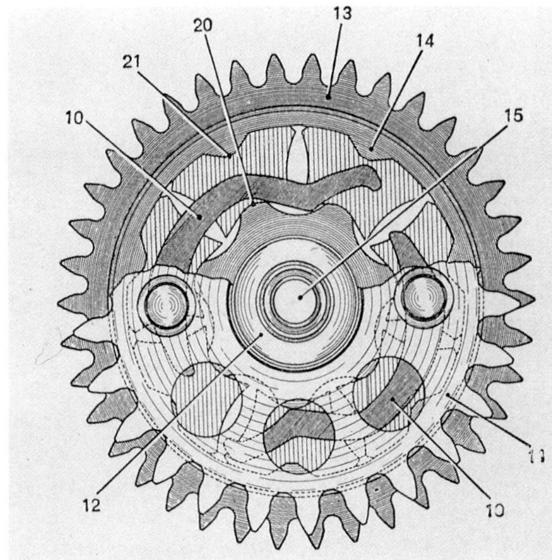

Cutout click wheel (Eta-Eterna patent): in its working position, one of the clicks (10) of the upper click wheel (11), extended by one of the inner cams (20), contacts one of the outer cams (14) of the lower click wheel (13) and carries it along.
When winding goes in the opposite direction, the clicks (10) slide between the outer and inner cams and run free.

Caliber List: Eterna

Caliber	Diameter Lignes	mm	Height mm	Other	BPH	Jewels	Running Reserve	Notes	Year	Photo
833	12 1/2			SCI	18.000				1941	
834	"		5,3	s	"				1938	xx
835	"			s	"					
1033	9 3/4	21,8	4,5		18.000				1942	
1034 /H	"	"		SCI	"				1943	xx
1076	13	29,4		s	18.000			Like Caliber 834		
1077	"	"		SCI	"			Like Caliber 833		
1158	12 1/4	27,4		SCD	18.000				1946	
1159 /H	13	29,4		SCD	"				1946	xx
1194	7 3/4	17,2	5,35	SCD	18.000		40		1949	
1195	8 1/4	18,3	"	SCD	"		"			
1198	9 1/4	20,5	"	SCD	"		"	1st Eterna- Matic	1948	
1199	10	22,1	"	SCD	"		"			
1237 /T	12 1/2	27,6	5,9	SCD, QG	18.000				1951	xx
1239	13	29,0	"	SCD, QG	"					
1243	"	"	"	SCD, QG	"					
1246	13 1/2		"	SCD, QG	"				ca. 53	
1247	12 1/2	27,6	"	SCD	"					
1248 /T	11 1/2	25,6	"	SCD	"				1950	xx
1249	13	29,0	"	SCD	"					
1253	"	"	"	SCD	"					
1254	13 1/2		"	SCD	"					
1319	9 1/4	20,5	5,6	SCD	18.000					
1321	10 1/2	23,3	"	SCD	"					
1323	9 1/4	20,5	"	SCD	"					
1324	10 1/2	23,3	"	SCD	"					
1401	7 1/4	16,2	5,3	SCD	18.000		40		1953	
1402	7 3/4	17,2	"	SCD	"		"		1953	
1403	8 1/4	18,3	"		"		"			
1411	11 1/2	25,6	5,6	SCD	18.000		44		1954	
1412 /UD	12 1/2	27,6	"	SCD	"		"			
1413	"	"	"	SCD	"		"			
1414	13	29,0	"	SCD	"		"			
1415	"	"	"	SCD	"		"			
1416	13 1/2	30,6	"	SCD	"		"			
1419	6 3/4	15,3	5,0		21.600			Golden Heart	1956	xx
1420	7 3/4	16,2	4,9		21.600				1956	
1421	11 1/2	25,6	5,6	SCD, QG	18.000			Eterna Matic Dato	1957	
1422	12 1/2	27,6	"	SCD, QG	"			"	1957	
1423	"	"	"	SCD, QG	"			"		
1424 /UD	13	29,0	"	SCD, QG, RDR	"			Eterna Matic Golfer		xx
1425	"	"	"	SCD, QG	"					
1426	13 1/2	30,6	"	SCD, QG, RDR	"			Eterna Matic Dato		xx

Caliber List: Eterna

Caliber	Diameter Lignes	Diameter mm	Height mm	Other	BPH	Jewels	Running Reserve	Notes	Year	Photo
1427	11 1/2	25,6	4,5	SCD	18.000	21		Centenaire	1956	
1428	12 1/2	27,6	"	SCD	"	"		"	1956	
1429 U	13	29,0	"	SCD	"	"		"	ca. 61	xx
1436	10 1/2			SCD	18.000					
1438	12 1/2		4,5	SCD, QG	"					
1439 /U	13	29,0	"	SCD, QG/RDR	"	21		Centenaire Golfer	1959	xx
1441	7 3/4			QG	18.000				1959	
1444	8 1/4			SCD, QG	"					
1445	7 3/4		4,0		21.600			Sahida	1962	
1446	8 1/4		"		"					
1453	13 1/2		3,6	SCD, QG	18.000		50	Eterna Matic 3000 Dato	1962	
1455	12 1/2		"	SCD	"		"	Eterna Matic 3000		
1456	13		"	SCD	"		"	"		
1457	13			SCD, QG, JG	"		"			
1459	7 3/4			SCD	21.600					
1463	6 3/4			SCD	"					
1465	12 1/2		3,6	SCD, QG	18.000					
1466 U	13	29,0	"	SCD, QG, RDR	"	21	50	Eterna Matic 3000 Dato	1963	xx
1469	7 3/4			SCD, QG	21.600					
1470	12 1/2			SCD	21.600					
1471	13			SCD	"					
1477	11 1/2			SCD	21.600					
1478	12 1/2			SCD	"					
1479 /K	13			SCD	"			Eterna Matic 1000		
1480 /K	12 1/2	27,6	4,5	SCD, QG/RDR	"	21		Eterna Matic 2000 Centenaire	1971	xx
1481	13			SCD, QG	"	"				
1487	11 1/2			SCD, QG	"	"				
1488	12 1/2			SCD, QG	"	"				
1489	13	29,3		SCD, QG, RDR	"	"		Kontiki 10	ca. 68	xx
1500 /K	13	29,3	3,6	SCD, QG/RDR	21.600	21		Eterna Matic 3000		xx
1501	"	"		SCD, QG, JG	"	"				
2872	11 1/2	25,6		SCD, QG, RDR, ST	28.800	25		Ebauche 2872	ca. 75	
12671	7 3/4	17,2	4,8	SCD, QG	28.800			Ebauche 2671	ca. 71	xx
12890	12 1/2	27,2	3,6		28.800		50	Ebauche 2890	1976	xx

Caliber Development

This firm, which later moved its headquarters to Geneva, was founded by Fritz Favre-Leuba at Le Locle in 1815.

The firm has become known in the profession as a manufacturer of special wristwatches such as the 1963 "Bivouac" with altimeter and the "Bathy 50", a diver's decompression watch. The first automatic watch that it developed was introduced in 1956 as Caliber 103. The automatic module was screwed to the basic movement as a complete unit and equipped with a switching rocker by which the rotor could wind in both directions. Along with the standard version with center seconds, the movement could be had with date indication (Caliber 104) and with up-down indicator.

Collaboration with the CHP (Communautèe Horlogère de Précision) resulted in a caliber that was produced in 1961 by the AS raw-movement factory for the firms of Doxa, Eberhard, Girard-Perregaux and Zodiac as well as for Favre-Leuba. On the basis of this FL Caliber 1149 there appeared, at the end of the 1960s, a fast-beat watch using the Clinergic 21 escapement made by FAR (Fabriques d'Assortiments Réunies).

This movement was a rotor automatic type that wound in both directions via patented springless AS click wheels.

The firm's second development was the automatic Caliber 269 Duomatic of 1968, based on the hand-wound Caliber 251 of 1962. This was an elegant flat design with two barrels which were meant to guarantee uniform transmission of power to the movement and, with their smaller dimensions, allow better space utilization. The simple automatic mechanism, mounted under a small horseshoe-shaped plate, wound in only one direction via the ball-bearing rotor and the sprung pusher pinion.

Caliber Notes

Caliber 1153
AS Ebauche (CHP)
D = 25.6 mm (11.5 lignes), H = 5.05 mm, 21 jewels
AUTOMATIC MODULE partly integrated into the basic movement, central rotor with heavy metal ring, winding in both directions via springless twin click wheels, one reduction wheel on the locking wheel.
BASIC MOVEMENT: three-spoke screwless Glucydur balance, Nivarox hairspring, 21,600 BPH, Incabloc shock resistance, moveable hairspring block carrier, patented fine adjustment, direct center seconds, date indication with instant change.

Caliber 269
D = 25.6 mm (11.5 lignes), H = 4.05 mm
AUTOMATIC MODULE mounted on the basic movement under a centrally located plate, central heavy-metal rotor with ball bearing, winding in one direction via sprung pusher pinion, three reduction wheels, the first with ruby click bearing, parallel on the locking wheels of the two barrels.
BASIC MOVEMENT (Caliber 259): three-spoke screwless Glucydur balance mounted under the balance bridge, Nivarox hairspring, 18,000 BPH, moveable hairspring block carrier, Incabloc shock resistance, direct center seconds, two barrels working via an intermediate wheel to activate the central minute wheel, locking wheels riveted to the barrel bosses, date indication with instant change.

Caliber List: Favre-Leuba

Caliber	Diameter Lignes	mm	Height mm	Other	BPH	Jewels	Running Reserve	Notes	Year	Photo
103	11 1/2	25,6	6,05	SCD/RM	18.000	21	43		1956	
104	"	"	"	SCD, QG	"	"	"	Daymatic	1956	
269	11 1/2	25,6	4,05	SCD, QG	18.000		48	Twinmatic/Duomatic	1968	xx
1149	11 1/2	25,6	4,75	SCD	21.600		40		1961	
1152	"	"	5,05	SCD, QG	"		"	Daymatic		
1153	"	"	"	SCD, QG	"	21	"	"		xx
1162	"	"	"	SCD, QG	36.000		34		ca. 70	
1164	"	"	"	SCD, QG, JG	"		"			
1165	"	"	"	SCD, QG	"		"			

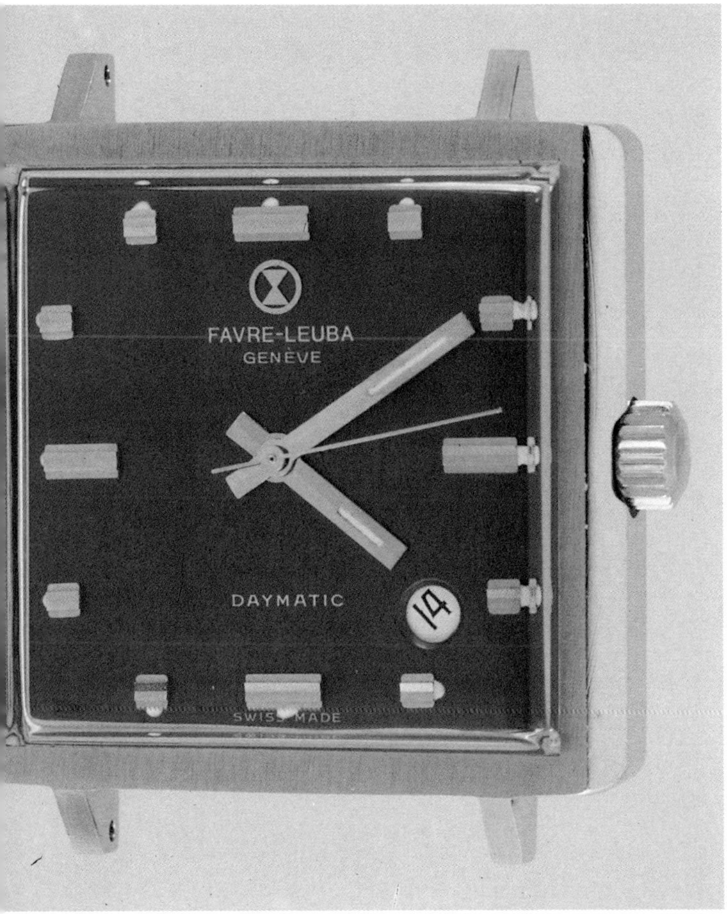

Favre-Leuba Caliber 1153

Men's automatic, "Daymatic" model, by Favre-Leuba circa 1965, with Caliber 1153 (CHP raw movement by AS).
Square stainless steel case with screwed bowl bottom, winding crown with firm's emblem, sectioned winding stem.
Rotor with heavy metal ring, winding in both directions via click-wheel switcher (at 2 and 4:00).
Screwless Glucydur balance, Nivarox hairspring, Incabloc shock resistance, patented fine regulator adjustment.

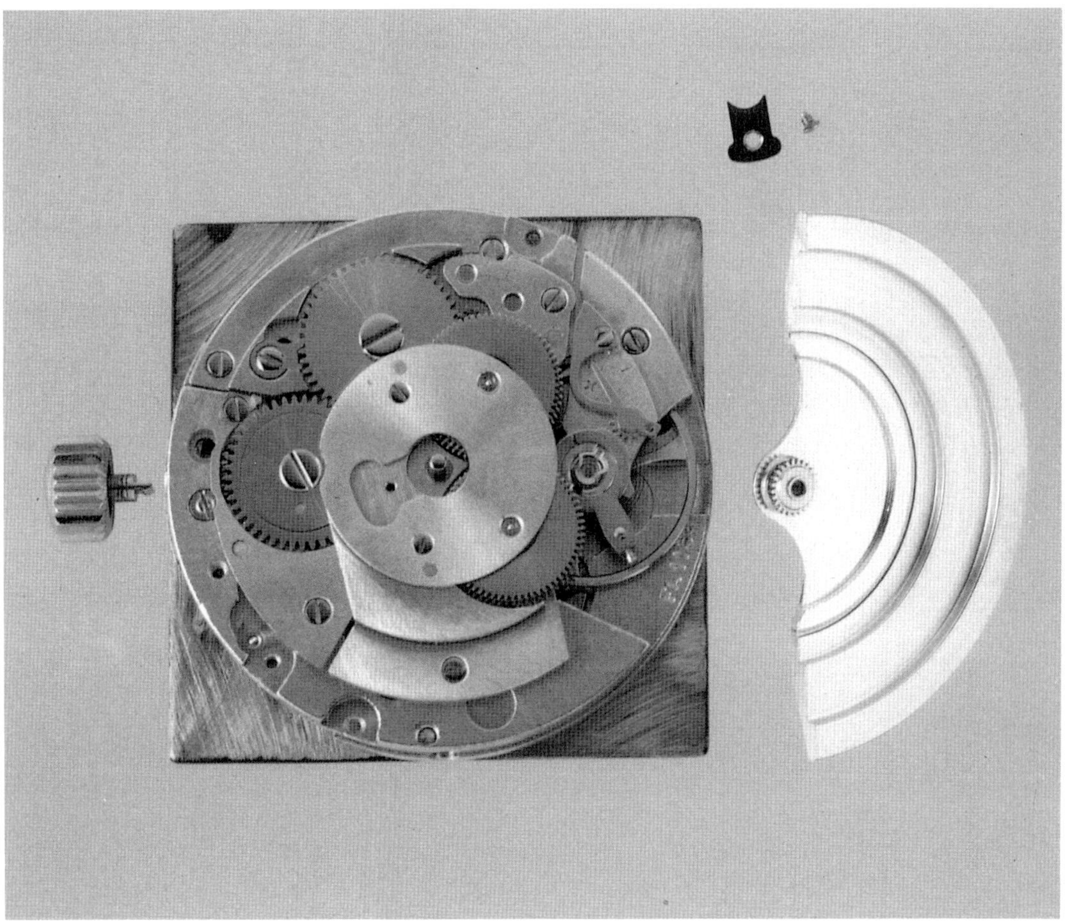

Favre-Leuba Caliber 269
Men's automatic made by Favre-Leuba in 1968, with "Duomatic" Caliber 269 with double barrels. 18-karat gold case with enameled bezel, winding crown with firm's emblem.
Heavy metal rotor with ball bearings, winding in one direction via pusher pinion (at 7:00 under the center), winding both barrels parallel (at 1 and 11:00). Screwless Glucydur balance mounted under a bridge, Nivarox hairspring, Incabloc shock resistance.

Caliber Development

In 1918 a raw movement factory was founded in Lengnau by H. Mägli, O. Rüfenacht and A. Tschudin. This firm was bought by Ebauches S.A. in 1928, and in 1940 the development and production of Felsa and the firm of A. Michel (A.M.) were included. The first automatic watch produced by Felsa consisted of a small basic movement and a hammer automatic, working in one direction, set atop the movement as a complete module (ca. 1938). As early as 1942 Felsa introduced, in the "Bidynator" Caliber 410, a concept that pointed the way to automatic development in general. It was the first rotor automatic that wound in both directions and thus used the energy of the swinging mass optimally, hence its name. The basic movement had the largest dimensions that would fit in the case. Its barrel and balance thus were big enough to guarantee a higher running reserve and better running performance. The rotor, turning above the large basic movement, had had its performance enhanced by a heavy metal inlay and could also be kept relatively flat on the rim. A locking-wheel coupling working in two directions uncoupled the hand-winding mechanism and thus minimized winding losses. During hand-winding, the wheels of the reduction drive would turn particularly fast and thus be protected by the uncoupling.

The 690 ff caliber family, introduced in 1947, represented the second Bidynator generation and did not differ markedly from the original model. But the locking-wheel coupling, which often needed to be repaired, was replaced by one that worked in only one direction during hand-winding. The barrel was a patented Felsa design in which the locking wheel and barrel arbor formed a unit and the barrel cover could be eliminated. In the ensuing years, this movement was equipped with date (Caliber 692), moon-phase calendar (Caliber 693) and running reserve, or up-down indicators ("Permutator" Caliber 699), and installed in watches with a wide variety of Swiss brands. The Caliber 760 was also equipped with this Bidynator movement, and had not only up and down indication on a small dial by the 9, but also a date aperture by the 3.

The "Inversator" Caliber 1565 can be regarded as a special automatic design. Here the rotor was blocked when the watch was fully wound, the guiding stud fell away, and the space in the barrel that became empty could be used to attain a greater running reserve. Like the first self-winding watch of A. L. Perrelet, this design had a pin lever resting in one of four notches in the rotor ring, and withdrawing when the mainspring had wound down somewhat. Blocking and releasing were controlled via the differential mechanism to indicate the running reserve, activating a lever-release mechanism.

In the mid-1960s, practical considerations resulted in the new design of a caliber in the modular manner. The basis was a simple movement to which an automatic module and a date mechanism could be added. The disadvantage of this design was the unfavorable height of the movement, as can be seen in the 4000 caliber family. Here, as also in the 4460 caliber family developed by Felsa in 1968, the lever switcher used in the Bidynator had been replaced by a switching rocker or by springless twin click wheels.

Caliber List: Felsa

Caliber	Diameter Lignes	mm	Height mm	Other	BPH	Jewels	Running Reserve	Notes	Year	Photo
279	11 1/4			s	18.000				ca. 38	xx
282	"			SC	"					
380 (AM)	8 3/4			s	18.000				1944	
381 (AM)	"			s	"					
382 (AM)	"			s	"					
410	11 1/2			s	18.000			1st Bidynator	1942	
415	"			SCD	"				1944	xx
690	11 1/2	26,0	5,81	SCD	18.000	17		2nd Bidynator	1947	
691	"	"	"	SCD	"	"				
692 /N	"	"	6,56	SCD, QG	"	"			1948	
693	"	"	7,01	SCD, QA, JG, PH	"	"				
694	"	"	6,56	SCD, QG, RDR	"	"				
695	"	"	5,81	s	"	"				
696	"	"		SCD	"	"				
697	"	"		SCD, QA, JG	"	"				
698	"	"		SCD	"	"				
699	"	"		SCD, RMG	"	"		Permutator		xx

Caliber List: Felsa

Caliber	Diameter		Height mm	Other	BPH	Jewels	Running Reserve	Notes	Year	Photo
	Lignes	mm								
700 /N	9 1/4	21,0	5,43	SCD	18.000					
701 /N	"	"	"	SCD	"					
702	"	"		SCD, QG	"					
703 /N	"	"		SCD	"					
705 /N	"	"	5,43	s	"					
706	"	"		SCD	"					
708	10 1/2				18.000					
709	"			SCD	"					
710	11 1/2	26,0		SCD	18.000					
711	"	"		SCD	"					
712	"	"		SCD, QG	"					
713	"	"		SCD, RMA	"					
714	15 1/4			SCD	"					
714-2	"			SCD, QG	"					
715	11 1/2			QG	18.000					
760	11 1/2			SCD, QG, RMA	18.000	30	44	Permutator	1954	
790	"			SCD	"					
792	"			SCD, QG	"					
799	"			SCD, RMG	"			Permutator		
1560	11 1/2			SCD	18.000					
1561	"			SCD	"					
1565	"			SCD, RMA	"	30	48	Inversator	1954	xx
1572	"			SCD, QG	"					
4000 /N	11 1/2	26,0	5,55	SCD	18.000		43			xx
4002	"	"	"	SCD, QG	"		"			
4004	"	"	"	SCD, QG	"					
4005 /N	11 1/2	26,0			18.000					
4006 /N	"	"	6,2	SCD, QG, JG	"				1966	
4007 /N	"	"	5,55	SCD, QG	"					
4008	"	"		SCD, QG, JG	"					
4009	"	"	6,25	SCD, QG, JG, RDR	"					
4031	13			SCD, QG	18.000					
4033	"			SCD, QG, JG	"					
4050	11 1/2	26,0		SCD	18.000					
4051	"	"		SCD, QG	"					
4460	11 1/2	26,0		SCD	21.600	17				
4462	"	"	5,2	SCD, QG	"	"			ca. 68	xx
4464	"	"	"	SCD, QG	"					
4466	"	"	5,9	SCD, QG, JG	"					
4468	"	"		SCD, QG	"					
4560	11 1/2	26,0		SCD	18.000					
4562	"	"		SCD, QG	"					
4568	"	"		SCD, QG	"					

Felsa Caliber 279

Men's automatic by "Dom Watch", circa 1940, with the first Felsa raw movement, Caliber 279.

Stainless steel case with screw bottom.

Hammer swinging mass, winding in one direction via Harwood-type automatic drive.

Nickel screw balance, Nivarox hairspring, Incabloc shock resistance.

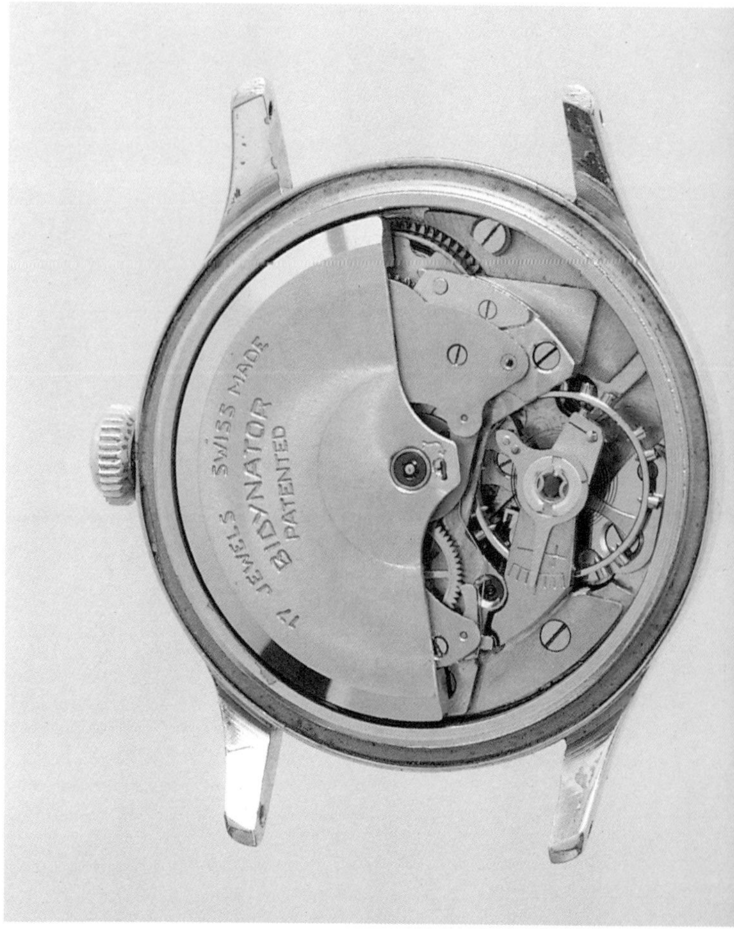

Felsa Caliber 415

Men's automatic of the "Arcadia" brand, made in 1944 with Felsa's first "Bidynator" Caliber 415, the first mass-produced rotor automatic with winding in both directions, produced since 1942. Gilded case with stainless steel screw bottom.

Rotor with ruby bearings and heavy metal inlay, winding in both directions via switching lever (at 8:00, near the rotor axle), complete reciprocal uncoupling of automatic and hand-winding drives.

Glucydur screw bearing, Nivarox hairspring, Incabloc shock resistance (case with signs of wear, winding crown replaced).

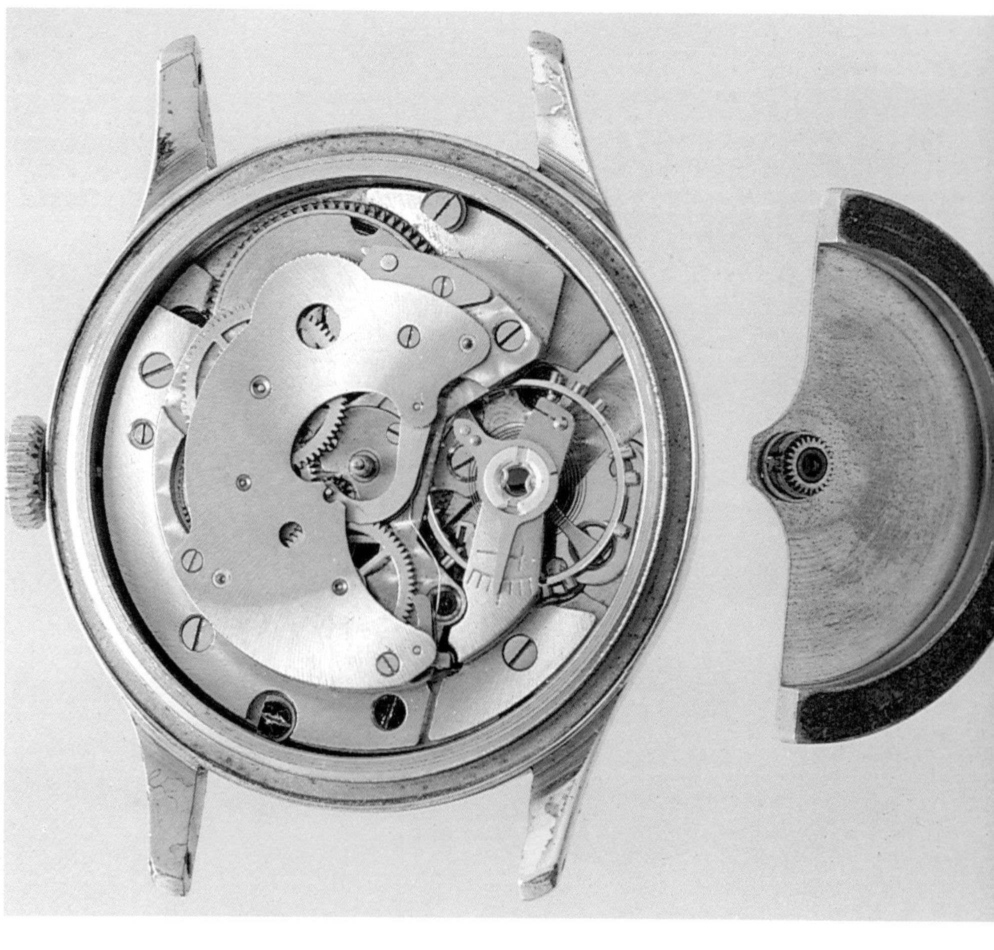

Felsa Caliber 669

Men's automatic of the "Orator" brand, circa 1951, with "Permutator" Caliber 699 by Felsa (second-generation "Bidynator" caliber). Gilded case with stainless steel screw bottom.

Up and down indicator in dial cutout over the 6. Rotor with ruby bearings, heavy metal inlay and fast unbolting, winding in both directions via a shifting lever (at 11:00 above the rotor axle), uncoupling of hand-winding drive.

Glucydur screw balance, Nivarox hairspring, Incabloc shock resistance. (The complete watch was reassembled out of original parts in the 1980s.)

Felsa Caliber 1565

Automatic Caliber 1565, "Inversator", made by Felsa in 1954.

The basic caliber is a second-generation Bidynator movement (see Caliber 699, page 133], equipped with a fourth central hand to indicate the running reserve and a patented rotor block. The latter provides protection against overwinding by letting a lever drop into one of the notches on the rotor ring. Blocking and freeing of the rotor are controlled by a lever drive on the dial side (at 12:30) from the differential mechanism.

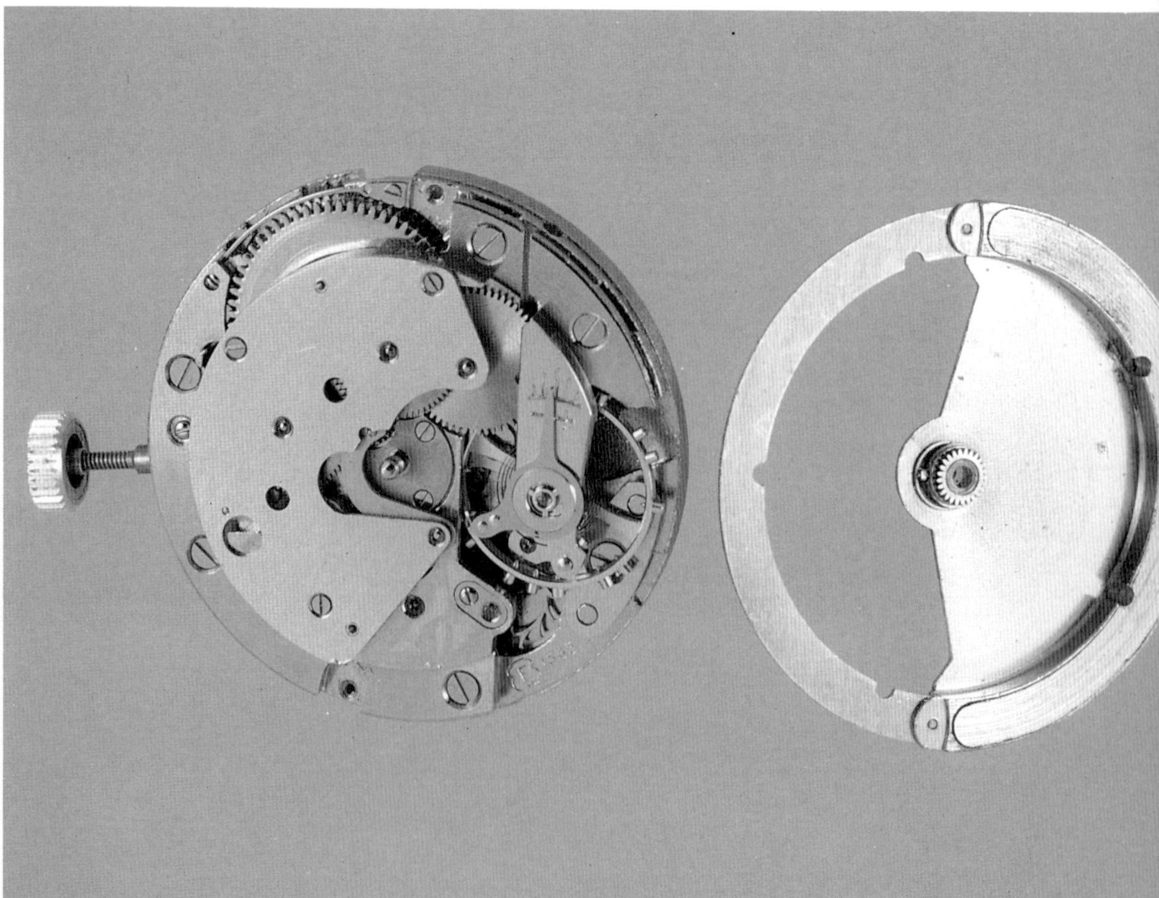

Felsa Caliber 4000 N

Men's automatic made by Minerva, circa 1960, with Felsa Caliber 4000 N. Gilded case with heavy metal inlay and quick unbolting, winding in both directions via switching rocker. Glucydur screw balance, Nivarox hairspring, Incabloc shock resistance.

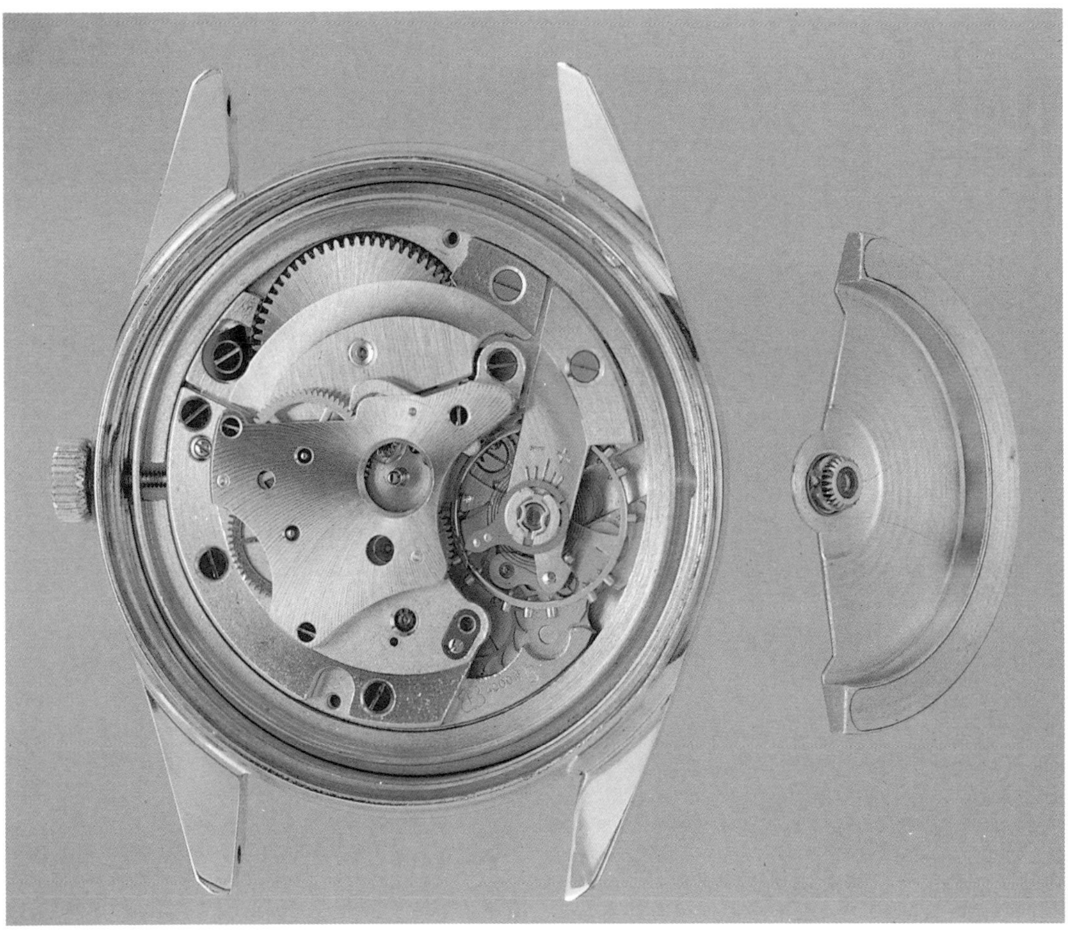

Felsa Caliber 4462
Men's automatic by "Paul Garnier", circa 1968, with Felsa Caliber 4462.
Gilded case with stainless steel screw bottom.
Rotor with heavy metal ring, winding in both directions via click-wheel switcher (at 5 and 6:00).
Screwless nickel balance, Nivarox hairspring, Incabloc shock resistance.

Caliber Notes

Caliber 279
D = 11.25 lignes, 17 jewels
AUTOMATIC MODULE screwed to the basic movement as a complete unit, hammer swinging mass winding in one direction via winding click on hammer and central pinion, two reduction wheels (the first with spring click) on the locking wheel.
BASIC MOVEMENT: two-shank nickel screw balance, Nivarox hairspring, 18,000 BPH, Incabloc shock resistance, eccentric seconds hand.

Caliber 415
D = 26.0 mm (11.5 lignes), 17 jewels
AUTOMATIC MODULE set on the basic movement, central rotor with ruby bearings and peripheral heavy-metal inlay, winding in both directions via switching lever, two reduction wheels on the locking wheel, which functions as a coupling for mutual uncoupling of automatic and hand-winding mechanisms.
BASIC MOVEMENT: two-shank Glucydur screw balance, Nivarox hairspring, 18,000 BPH, Incabloc shock resistance, direct center seconds via double intermediate wheel.

Caliber 699
D = 26.0 mm (11.5 lignes), 17 jewels
AUTOMATIC MODULE screwed to basic movement, central rotor with ruby bearings and peripheral heavy-metal inlay, winding in both directions via switching lever, one steel tension wheel with spring click, two reduction wheels (the first made of steel, the second as coupling wheel) on the locking wheel, reduction drive uncoupled during hand-winding.
BASIC MOVEMENT: two-shank nickel screw balance, Nivarox hairspring, 18,000 BPH, Incabloc shock resistance, direct center seconds via double intermediate wheel, patented barrel, running reserve indication via differential drive.

Caliber 1565
D = 26.0 mm (11.5 lignes), 30 jewels
Movement structure as Caliber 699, but with a half-ring screwed to the rotor with five notches to engage the blocking lever when fully wound. Blocking drive, controlled by the differential mechanism, under the dial, and running reserve indication by a fourth central hand.

Caliber 4000 N
D = 26.0 mm (11.5 lignes), H = 5.55 mm, 25 jewels

Exploded view of the second-generation "Bidynator" caliber: 1143—rotor; 1485/1—switching lever; 1480—tension wheel; 1481—first reduction wheel; 1489—second reduction wheel as click coupling to protect the reduction drive during hand-winding.

AUTOMATIC MODULE screwed to the basic movement, central rotor with ruby bearings and peripheral heavy-metal inlay, winding in both directions via switching rocker, two reduction wheels on the crown wheel.
BASIC MOVEMENT: two-shank Glucydur screw balance, Nivarox hairspring, 18,000 BPH, Incabloc shock resistance, escape wheel mounted between cap jewels, direct center seconds via double intermediate wheel.

Caliber 4462
D = 26.0 mm (11.5 lignes), H = 5.2 mm, 17 jewels
AUTOMATIC MODULE screwed to basic movement as a complete unit, central rotor with heavy metal ring, winding in both directions via springless twin click wheels, two reduction wheels on the crown wheel.
BASIC MOVEMENT: three-shank screwless nickel balance, Nivarox hairspring, 21,600 BPH, Incabloc shock resistance, moveable hairspring block carrier, direct center seconds, date indication.

Illustrated parts catalog for FHF Caliber 65.
View of the automatic drive on the basic movement after removal of the half-ring rotor and upper plate: 1485: switching lever; 1499—switching lever pinion (contacting the toothed arc of the half-ring rotor).

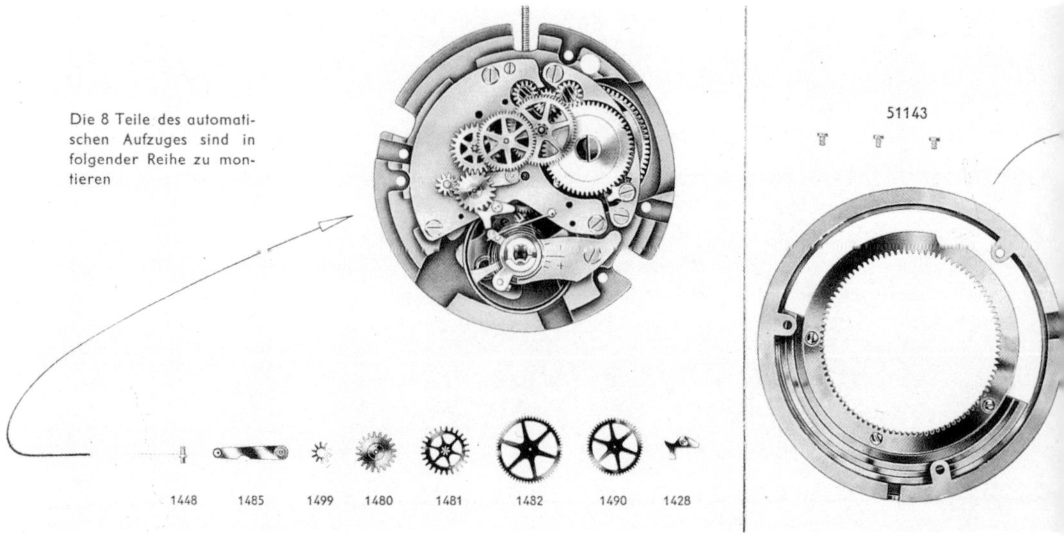

Caliber Development

The Fabrique d'Horlogerie de Fontainemelon (FHF) was founded at Fontainemelon in 1793 by two pairs of brothers, I. and S. Bengerel and J. and F. Humbert-Droz, and is thus the oldest factory producing raw movements. In 1926 the firm merged with other raw-movement producers, A, Schild and Ad. Michel of Grenchen, to form the basis of the Ebauches S.A.

FHF was the last raw-movement manufacturer to produce its own automatic caliber, doing so only in 1959. This was the "Fontomatic" Caliber 65 which offered a new feature, an arbor-less half-ring rotor mounted on ball bearings only peripherally, which surprised the firm's established competitors. The reduction drive of the swinging mass, which wound in both directions, was located in space in the interior area, thus allowing a relatively flat movement form sloping off toward the rim.

The Etablisseurs who utilized this much-honored movement in their watch models soon began to hear criticism, though, about the loud rolling sound of the big ball bearings and the watch's sensitivity to shocks and difficulty to repair. The subsequent automatic developments were then designed strictly on the basis of tested and economical practices; this applies to the Caliber 90-4 of 1968 as well as the 403 caliber family a few years later.

Caliber Notes

Caliber 65
D = 30.0 mm (13.25 lignes), H = 4.5 mm, 25 jewels
AUTOMATIC MODULE screwed to the basic movement, half-ring rotor of heavy metal with six peripheral ball bearings, ring geared on the inside, reduction drive located in free interior space, winding in both directions via a simple switcher lever, one tension wheel with spring click, two reduction wheels on the locking wheel.
BASIC MOVEMENT: two-shank Glucydur screw balance, Nivarox hairspring, 18,000 BPH, Incabloc shock resistance, moveable hairspring block carrier, direct center seconds.

Caliber 90-5
D = 25.6 mm (11.5 lignes), H = 5.3 mm, 25 jewels
AUTOMATIC MODULE screwed to the basic movement, central rotor with heavy metal ring, winding in both directions via twin click wheels with springless anchor clicks, two reduction wheels on the locking wheel.
BASIC MOVEMENT: three-shank screwless nickel balance, Nivarox hairspring, 21,600 BPH, Incabloc shock resistance, moveable hairspring block carrier (Fontofix), direct center seconds.

Caliber List: FHF

Caliber	Diameter Lignes	mm	Height mm	Other	BPH	Jewels	Running Reserve	Notes	Year	Photo
65	13 1/4	30,0	4,5	SCD	18.000	25	48	Frontomatic see	1959	xx
65-4	"	"		SCD, QG	"		"	Gruen 666 ss		(xx)
90-4	11 1/2	25,6	5,3	SCD, QG	21.600				1968	
90-5	"	"	"	SCD, QG	"	25				xx
403	11 1/2	25,6	5,4							
411	"	"								
412	"	"		SCD, QG, JG					1975	
905	11 1/2	26,5		SCD, QG, RDR	21.600		44	Basic Caliber 90-5	ca. 71	
908	"	"		SCD, QG, JG, RDR	"		"	"		
909	"	"		SCD, QG, JG	"			"		

FHF Caliber 65
Men's automatic, "Futurama" model made by Suter in 1959, with Caliber SU 65 (Fontomatic 65 raw movement made by FHF).
Gilded case with stainless steel screw bottom, winding crown with Suter emblem.
Arborless half-ring rotor of heavy metal mounted on ball bearings, with large toothed arc, winding in both directions via switching lever (at 6:00).
Glucydur screw balance, Nivarox hairspring, Incabloc shock resistance.

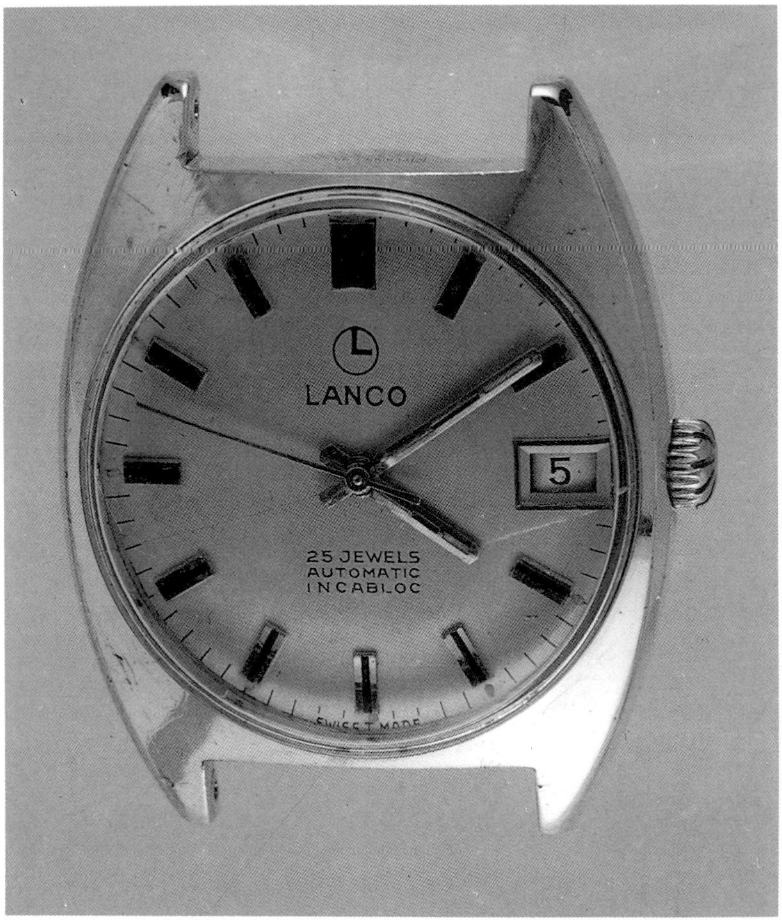
LANCO
5
25 JEWELS
AUTOMATIC
INCABLOC
SWISS MADE

Caliber Development

C. O. Girard and C. Robert founded the firm of "Girard-Perregaux" at La Chaux-de-Fonds in 1856 and purchased the firm of Bautte in 1906, which had existed since 1791.

GP attained a special degree of renown with its pocket chronometers that used a special three-bridge design. Only in 1956 did the firm offer an automatic wristwatch, known as "Gyromatic", utilizing a basic movement by the raw-movement manufacturer Peseux (that firm's Caliber 330 or 335, GP Caliber 18 or 20) and completed by an automatic module developed by GP itself. The two directions of the rotor were utilized with the help of two so-called Gyrotrones, originally using springless click wheels (Calibers 19 and 21) and as of 1957 patented free-running components in the manner of a clamp lock, each with seven steel or ruby rollers (Calibers 21, 19 and 22).

With the 17 jewels of the basic movement, seven ruby rollers for each of the two Gyrotones, and a total of eight bearing jewels for the reduction drive and rotor, there was an unusual total of 39 functioning jewels that served the manufacturers as a feature well worth advertising.

The second gyromatic generation came on the market in 1961 as Caliber 31. It included a raw movement by AS, which was also produced for the Doxa, Favre Leuba, Eberhard and Zodiac firms within the framework of the Communautée Horlogère de Précision (CHP). GP replaced the typical AS click wheels with their Gyrotrones and used this caliber in 1965 as the basis of the firm's first fast beat automatic watch, Caliber 31.7 or 32.7. An important component was the special escapement made by the Ebauches S.A. branch "Fabriques d'Assortiments Réunies" (FAR) in Le Locle.

In this special factory, A. Simon-Vermont developed the "Clinergic 21" escapement, using a 21-toothed escape wheel, for the extremely fast beat combination of balance and hairspring (36,000 BPH). The teeth of the escape wheel were treated with a dry lubricant of molybdenum bisulfate. At the fast speeds involved, oil on the teeth would simply be thrown off!

The need for greater energy to power a fast beat watch required a stronger mainspring, so a smaller running reserve had to be accepted, given the same barrel size. The rotor bearing jewels, no longer needed thanks to the installation of miniature ball bearings, were included in the movement with nothing to do, so that the impressive number of 39 jewels could still be advertised to the public.

This caliber was modified by GP to make the first "HF" chronometer (Caliber 32 A), which received a certificate from the Swiss Observatory at Neuenburg in 1966.

For ladies' automatic watches and the automatic models before 1956, GP purchased complete movements from AS and Eta.

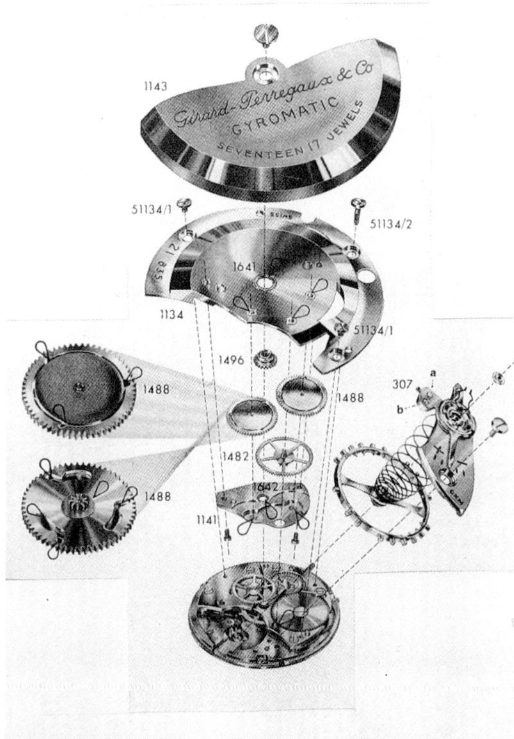

Exploded view of Gyromatic Caliber 21 by Girard-Perregaux (1488—first-generation Gyrotrone).

Caliber Notes

Caliber 21
D = 25.6 mm (11.5 lignes), H = 5.64 mm, 17 jewels
AUTOMATIC MODULE screwed to the basic movement as a complete unit, central rotor with ruby bearings and heavy metal inlay, winding in both directions via twin free-running couplings, each with seven steel rollers (Gyrotone), one reduction wheel on the locking wheel.
BASIC MOVEMENT: two-shank Glucydur screw balance, Nivarox hairspring, 18,000 BPH, Incabloc shock resistance, moveable hairspring block carrier.

Caliber 32 A
Ebauches AS (CHP)
D = 25.6 mm (11.5 lignes), H = 5.05 mm, 37 functioning jewels
AUTOMATIC MODULE partly integrated into the basic movement, heavy metal central rotor with ball bearings, winding in both directions via twin free-running couplings, each with seven ruby rollers (Gyrotron), one reduction wheel on the locking wheel.
BASIC MOVEMENT: three-shank screwless Glucydur balance, Isoval flat hairspring (Konstante 1.9), 36,000 BPH, moveable hairspring block carrier, fine adjustment by regulator screw, Clinergic 21 anchor escapement, direct center seconds, date indication with quick change.

Girard-Perregaux Caliber 21

Men's automatic made by Girard-Perregaux in 1956, with the firm's first self-developed "Gyromatic" Caliber 21. Stainless steel case with screw bottom. Ruby-mounted rotor with lead inlay, winding in both directions via free-running switchers (patented Gyrotrone with 7 steel rollers each, at 11 and 2:00). Glucydur screw balance, Nivarox hairspring, Incabloc shock resistance (winding crown not original, dial inlays and hands corroded).

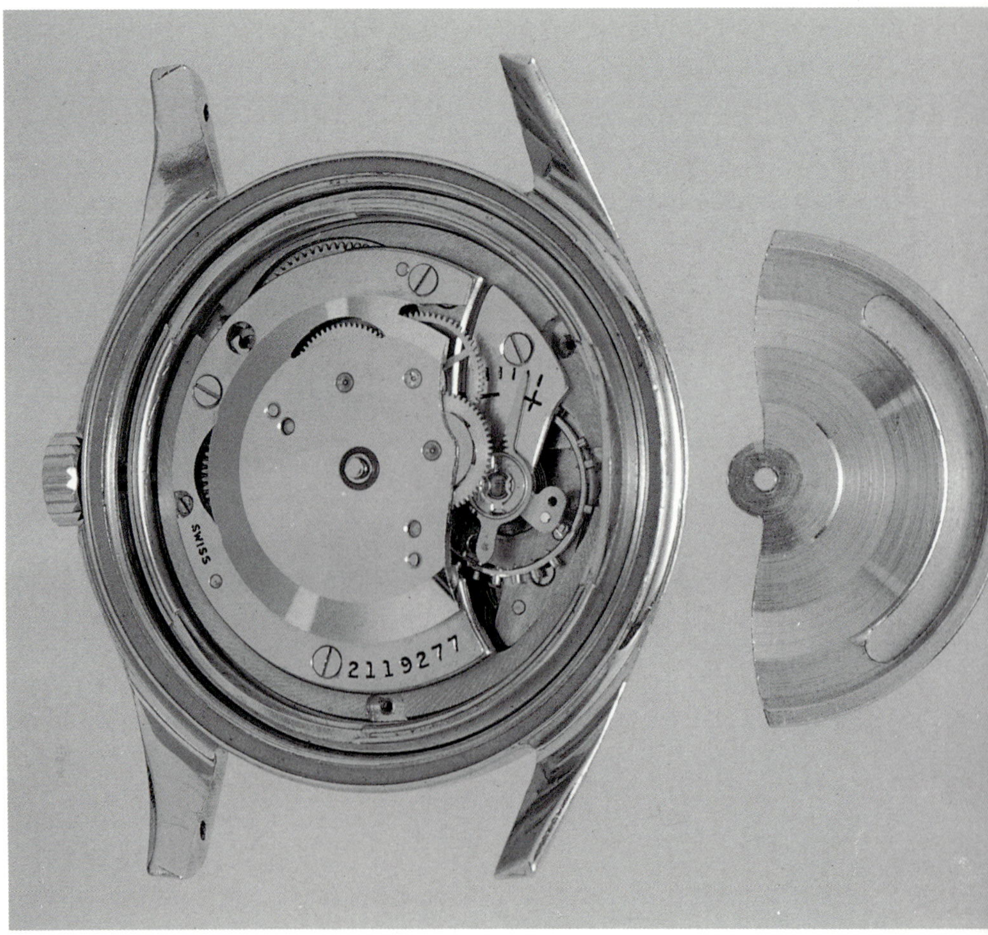

Girard-Perregaux Caliber 32 A, "HF-Chronometer"

Men's automatic made by Girard-Perregaux in 1966, with the firm's first fast-swinging chronometer ("HF") Caliber 32 A.

Eighteen Karat gold case with screw bottom, winding crown with GP emblem, gold dial inlays and hands.

Rotor with heavy metal ring and ball bearings, winding in both directions via patented Gyrotron switcher (twin free-running units, each with 7 ruby rollers). Screwless Glucydur balance, self-compensating flat hairspring, Incabloc shock resistance, fine regulation by a kind of pulley drive with side screw. Two jewels at the foot of the bridge (at 12:00) to reach the number of 39 jewels engraved on the rotor.

Caliber List: Girard-Perregaux

Caliber	Diameter Lignes	mm	Height mm	Other	BPH	Jewels	Running Reserve	Notes	Year	Photo
01	11 1/2			SCD						
00	11 1/2			SCD						
080	13	30,0	7,45	SCD, QG, REV	28.800	17		Ebauche J. LeCoultre	ca. 72	(xx)
19	11 1/2	25,6	5,5	s	18.000	17	36	Gyromatic	1956	
21	"	"	5,64	SCI	"	"	40	"	1956	xx
21.0	"	"	"	SCI	"		"	"		
21.19	"	"	"	SCI	"	39	"	"	1957	
21.29	"	"	"	SCI	"		"	"		
22	"	"	6,29	SCI, QG	"	39	"	"	1957	
22.09	"	"	"	SCI, QG	"	17	"	"		
22.19	"	"	"	SCI, QG	"		"	"		
22.29	"	"	"	SCI, QG	"	17	"	"		
31	11 1/2	25,6	4,75	SCD	21.600	17-39	40	Gyromatic (II)	1961	
31.09	"	"	"	SCD	"	"	"	"		
31.19	"	"	"	SCD	"	"	"	"		
31.29	"	"	"	SCD	"	"	"	"		
31.3	"	"	"	SCD	"	"	"	"		
31.4	"	"	"	SCD	"	"	"	"		
31.5	"	"	"	SCD	"	"	"	"		
31.6	"	"	"	SCD	"	"	"	"		
31.7	"	"	"	SCD	36.000	"	34	"	1965	
31.8	"	"	"	SCD	21.600	"	40	"		
31.9	"	"	"	SCD	"	"	"	"		
31.A	"	"	"	SCD	"	"	"	"		
32	"	"	5,05	SCD, QG	"	"	"	"		
32.09 / 32.6	"	"	"	SCD, QG	"	"	"	"		
32.7	"	"	"	SCD, QG	36.000	"	34	"	1965	
32.8 / 32.9	"	"	"	SCD, QG	21.600	"	40	"		
32 A	"	"	"	SCD, QG	31.000	37	34	" , HF-Chronometer	1966	xx
41.0	11 1/2	26,5		SCD	21.600					
42.0	"	"	5,2	SCD	36.000				1966	
42.2	"	"		QG	21.600					
42.4	"	"			36.000					
44				s, QA, JG, MG		17		AS-Ebauche		
47	11 1/2			SCD, QG		17		Eta- Ebauche 1256		
62.0	6 3/4				21.600			Eta-Ebauche		
74	7	16	5,3		18.000			AS-Ebauche 1477		
81.0	7 3/4			SCD	21.600			Eta-Ebauche 2550		
82.0	"			SCD, QG	"			Eta-Ebauche 2551		
94	7			SCD	21.600			AS-Ebauche 1654		
94.09	"			SCD	"	17/25				
94.19	"			SCD	"					

Caliber Development

Dietrich Gruen founded the firm at Biel in 1874 and built up a large sales organization in the USA in 1923. At the end of the 1940s, the firm developed its first hammer automatic, Caliber 475. The winding mechanism was a typical design but could be kept very flat despite the balance with its self-compensating Breguet hairspring. With its rounded square bottom plate, it was particularly suitable for installation in square cases.

A rotor automatic, Caliber 480 with a small seconds hand and Caliber 480 SS with center seconds, was introduced by the firm at the beginning of the 1950s. With its relatively small size and rounded square shape, it could also be used in the larger ladies' watches. The rotor was made primarily of heavy metal and wound in both directions via a gliding pinion. To improve its performance, the drive was fitted with a mutual uncoupling apparatus between the automatic and hand-winding drives.

The firm's designers developed a special device to protect against overwinding in their new automatic caliber 560 SS, which came on the market with a lot of advertising in 1959 under the name of "Power-Glide". The device was a patented self-regulating friction coupling outside the barrel, and was meant to replace the sliding stud of the barrel with its well-known weaknesses. This drive was very expensive to build and required additional height, so the watch looked unusually hefty for its time. Unfortunately, signs of premature wear appeared, resulting in the coupling releasing long before the maximum running reserve had been attained, so the watch often stopped during the night. This problem was solved a short time later with the introduction of the Eloxal GL barrel, developed by Dr. Straumann. After this interesting but unsuccessful development, Gruen generally used movements from outside, such as Caliber 666 SS, a ring-rotor Fontomatic type built by the raw-movement producer FHF.

Caliber Notes

Caliber 475
D = 24.3 x 24.3 mm (10.75 x 10.75 lignes), 17 jewels
AUTOMATIC MODULE screwed to the basic movement, hammer automatic winding in one direction via a click-carrier rocker with a combined winding and reduction wheel on the crown wheel.
BASIC MOVEMENT: two-shank nickel screw balance, self-compensating Breguet hairspring, 18,000 BPH, Incabloc shock resistance, eccentric seconds hand.

Caliber 480 SS
D = 23.4 x 20.1 mm (10.5 x 9.25 lignes), H = 5.8 mm, 17 jewels
AUTOMATIC MODULE screwed to the basic movement as a complete unit, central heavy-metal rotor, winding in both directions via a sliding pinion, one tension wheel with spring click, two reduction wheels on the upper locking wheel. Complete mutual uncoupling of automatic or hand-winding drives via locking-wheel coupling.
BASIC MOVEMENT: two-spoke nickel screw balance, self-compensating Breguet hairspring, 18,000 BPH, Incabloc shock resistance, escape wheel mounted between cap jewels.

Magazine advertisement (from ÖU, 1960)

Caliber 560 SS
D = 25.6 mm (11.5 lignes), H = 6.0 mm, 23 jewels
AUTOMATIC MODULE screwed to the basic movement, central rotor with heavy metal ring, winding in both directions via a gliding pinion, one tension wheels with spring click, two reduction wheels over the self-regulating "Power Glide" friction coupling on the locking wheel. Automatic drive uncoupled during hand-winding by a second reduction wheel with gliding mount.
BASIC MOVEMENT: two-spoke nickel screw balance, Nivarox hairspring, 18,000 BPH, Incabloc shock resistance, direct center seconds.

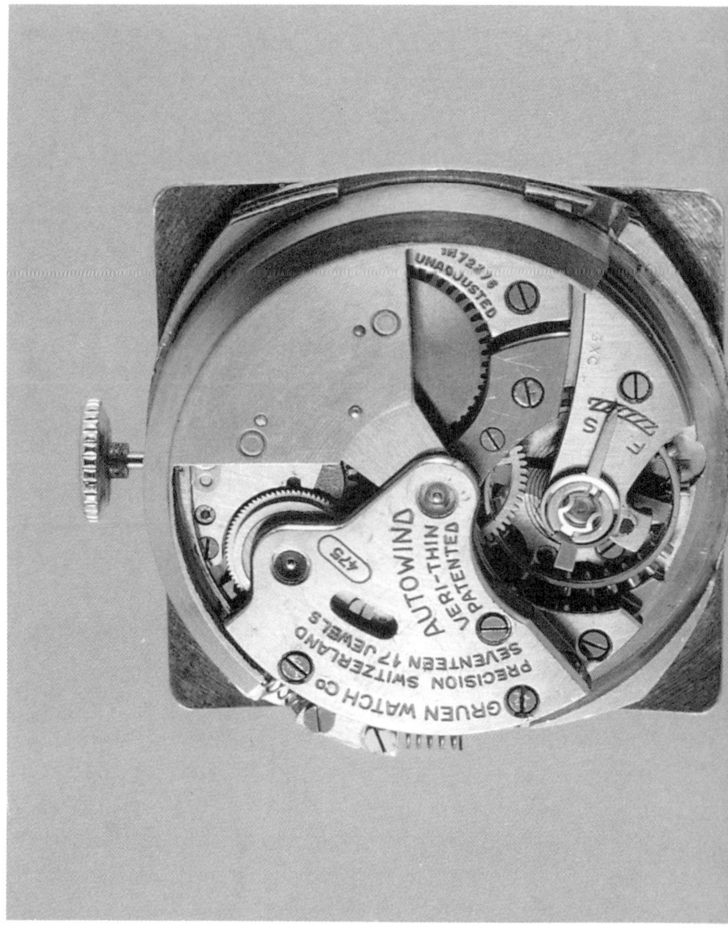

Gruen Caliber 475

Men's automatic made by Gruen circa 1950, with "Autowind" Caliber 475. Square 10-karat "gold-filled" case. Hammer winding in one direction via click-carrier rocker. Nickel screw balance, self-compensating Breguet hairspring, Incabloc shock resistance (dial "grayed", hands corroded).

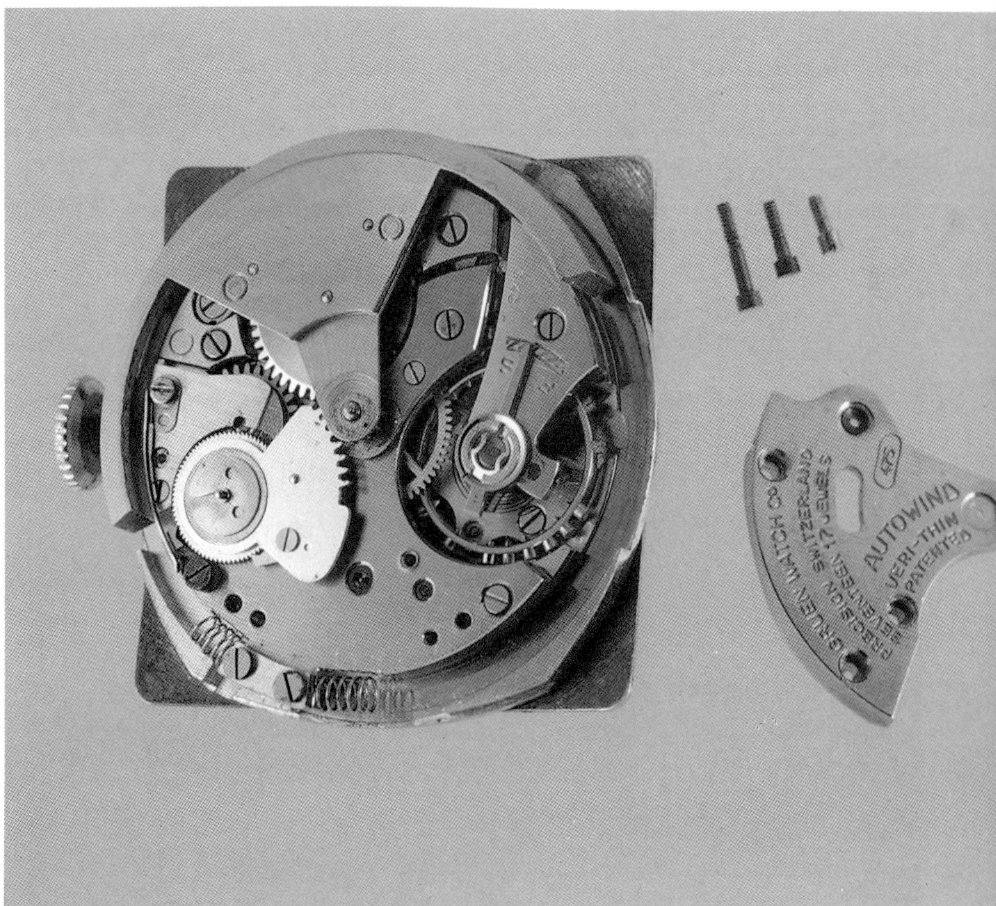

Gruen Caliber 480 SS

First rotor automatic made by Gruen, circa 1953, in medium size, with Caliber 480 SS.
Stainless steel case with screw bottom.
Rotor with heavy metal ring, winding in both directions via a gliding-pinion switcher (at 7:00 next to the rotor axle), automatic module removable as a complete unit.
Nickel screw balance, self-compensating Breguet hairspring, Incabloc shock resistance (movement corroded).

Gruen Caliber 560 RSS

Men's automatic made by Gruen in 1959, "Power-Glide" model, with "Autowind" Caliber 560 RSS.
Stainless steel case with screw bottom.
Rotor with heavy metal ring, winding in both directions via gliding-pinion switcher (at 1:00, above the rotor axle), uncoupling during hand-winding. Self-regulating "Power-Glide" friction coupling on the barrel (at 5:00) as overwinding protection.
Nickel screw balance, Nivarox hairspring, Incabloc shock resistance.

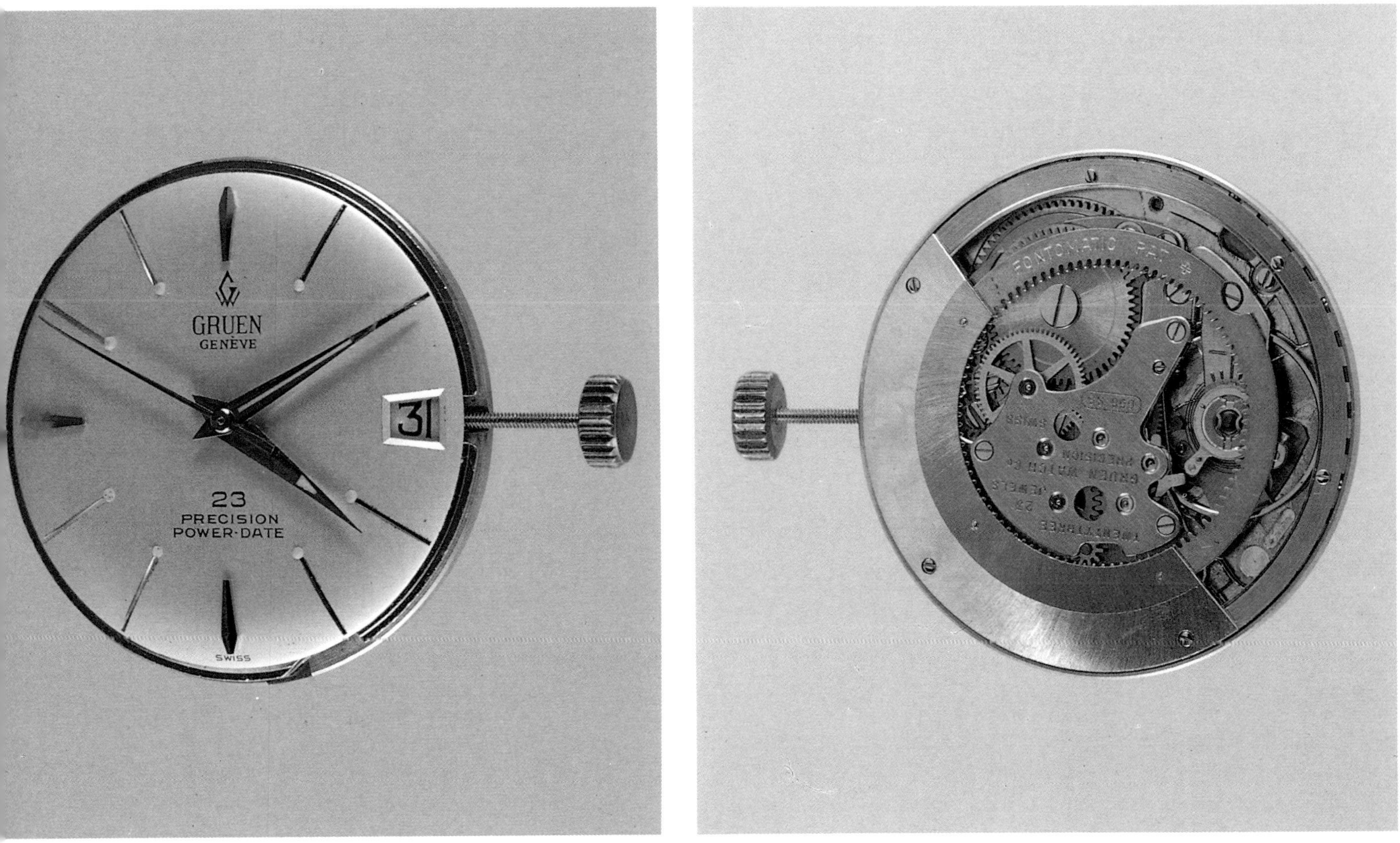

Gruen Caliber 666 SS movement
Movement of a men's automatic by Gruen (Caliber 666 SS) circa 1960, with "Fontomatic" Caliber 65.4 by the raw-movement manufacturer FHF (see page 139).

Caliber 666 SS
Ebauche from FHF
D = 30.0 mm (13 lignes), 25 jewels
AUTOMATIC MODULE: half-ring rotor of heavy metal with peripheral ball bearings and ring geared on the inside, located in movement area, winding in both directions via a simple lever switcher, one tension wheel with spring click, two reduction wheels on the locking wheel.
BASIC MOVEMENT: two-spoke screwless Glucydur balance, Nivarox hairspring, 18,000 BPH, Incabloc shock resistance, moveable hairspring block carrier, direct center seconds, date indication.

Caliber list: Gruen

Caliber	Diameter Lignes	mm	Height mm	Other	BPH	Jewels	Running Reserve	Notes	Year	Photo
362	6 3/4 x 7 3/4				18.000					
363	"				"					
460	11 1/2				18.000				1947	
460 SS	"			SCI	"					
462	"				"					
462 B	"				"					
462 SS	"			SCI	"					
470	11 1/2				18.000				1948	
470 SS	"			SCI	"					
475	10 3/4 x 10 3/4			s	18.000	17		Autowind	ca. 50	xx
480	9 3/4 x 11		5,8	s	18.000	17	36			
480 SS	"		"	SCI	"	"	"			xx
560	11 1/2	25,6	6,0	s	18.000	23	45	Power-Glide	1959	
560 RSS	"	"	"	SCD	"	"	"	"	1959	xx
560 CA	"	"		SCD, QG	"	"	"	"		
570 SS	12			SCD						
666 SS	13 1/4	30,0		SCD, QG	18.000	23	48	FHF-Ebauche	ca. 60	xx
710 CA	11 1/2			SC, QG						
710 SS/RSS	"			SC						
712 SS	"			SC						
712 CA	"			SC, QG						
730 SS	"			SC						
730 CA	"			SC, QG						

Caliber Development

The Montres Helvetia General Watch Co. S.A. firm was founded at Biel in 1895 and built a second factory in Reconvilier after World War II. The firm made a name for itself with, among other things, the development of a waterproof wristwatch in 1929 and a special shock-resistance system.

Its first automatic caliber, built in 1951, had a rotor winding in one direction. In order to decrease losses in automatic winding, the hand-winding mechanism could be uncoupled by a geared rocker. One result of this was the unusual turning of the crown to the left for hand-winding.

In the mid-1950s, Helvetia modified the automatic module while retaining the modernized basic movement. The rotor ran on miniature ball bearings and now wound in both directions. Special uncoupling devices between the automatic and hand-winding mechanisms could be eliminated, so that this watch could be wound by turning the crown to the right, and the movement was definitely flatter.

In the ensuing years, the firm used only movements from outside in its automatic models, such as the chronograph Caliber 7750 made by Valjoux (see Valjoux section).

Caliber Notes
Caliber 837
D = 26.3 mm (11.5 lignes), H = 6.13 mm, 21 jewels
AUTOMATIC MODULE screwed to the basic movement, monometallic central rotor, winding in one direction via one Breguet free-running device, two reduction wheels, the first with a spring click, on the double locking wheel, mutual uncoupling of the automatic or hand-winding drive (crown winding to the left) by a locking-wheel coupling and toothed rocker between the crown and locking wheels.
BASIC MOVEMENT: two-spoke Glucydur screw balance, Nivarox hairspring, 18,000 BPH, Incabloc shock resistance.

Caliber 845
D = 26.3 mm (11.5 lignes), 27 jewels
AUTOMATIC MODULE screwed to the basic movement under a half-plate. Central rotor with ball bearings and heavy metal inlay, winding in both directions via springless twin click wheels, one reduction wheel on the crown wheel.
BASIC MOVEMENT: two-spoke screwless Glucydur balance, Nivarox hairspring, 18,000 BPH, Incabloc shock resistance, moveable hairspring block carrier, escape wheel mounted with combined jewel housing (Duofix).

Caliber List: Helvetia

Caliber	Diameter Lignes	mm	Height mm	Other	BPH	Jewels	Running Reserve	Notes	Year	Photo
510	7 3/4			SCD	21.600					
515	"			SCD, QG	"					
836	11 1/2	26,3		s	18.000	21	40		1951	
837	"	"	6,13	SCI	"	"	"		1952	xx
838	"	"	"	SCI	"			Basic Caliber 837	ca. 55	
839	11 1/2			SCD				AS-Ebauche 1680		
842	"			SCD, QG				"	1956	
844	11 1/2	26,3		SCI, QG						
845	"	"		SCI	18.000	27				xx
846	11 1/2		4,5	SCD	18.000	25		AS-Ebauche 1673		
847	"			SCD, QG	"	"		AS-Ebauche 1674		
861	11 1/2			SCD, QG						
868	11 1/2			SCD						
869	"			SCD, QG						
1330	6 3/4				21.600					
8390	11 1/2			SCD						
8420	"			SCD, QG						

Helvetia Caliber 837

Men's automatic made by Helvetia in 1952, with Caliber 837.
18-karat gold case with screw bottom. Lead-weighted rotor with fast unbolting device (at 1:00), winding in one direction via Breguet coupling wheel, complete reciprocal uncoupling between automatic and hand-winding. Glucydur screw balance, Nivarox hairspring, Incabloc shock resistance.

Helvetia Caliber 845

Men's automatic made by Helvetia circa 1956, with Caliber 845.
Chrome plated case with stainless steel screw bottom. Rotor with ball bearings and heavy metal inlay, winding in both directions via click-wheel switcher (at 7:00).
Screwless Glucydur balance, Nivarox hairspring, Incabloc shock resistance, escape wheel with Duofix combination mount (hour hand and winding crown not original, case with clear signs of wear).

Magazine advertisement (from JSH, 1952)

Caliber Development

The American F. A. Jones and the Swiss J. H. Moser founded the "International Watch Company" (IWC) firm at Schaffhausen on the Rhine in 1868. In 1889 it was purchased by Johannes Rauschenbach, and it remained in his family's ownership until it was changed into a stock company in 1971. It was taken over by the German VDO company in 1978.

The firm introduced its first automatic development, Caliber 81, in 1950. It was produced only in small numbers and replaced in the same year by its successor, Caliber 85. Here the IWC engineers decided in favor of robust eccentric winding and combined this principle, introduced by Longines in 1945, with that of double-lever winding, which is also found in the hammer automatic made by the Lanco firm. In 1952 IWC improved the basic movement by equipping it with a self-compensating Breguet hairspring instead of the flat type, and by adding a regulator for fine adjustment. These measures, though, resulted in a thicker movement. This movement, Caliber 852, was produced in large quantities and as of 1953 was also available with date indication as Caliber 8521 in, among others, the well-known anti-magnetic Engineer model of 1954.

The Calibers 853 without and 8531 with date indication, introduced in 1958 and 1959, also differed only slightly from the previous movement (for example, in having equalizing screws on both spokes of the balance). This was also true of the Calibers 854 and 8541, produced as of 1964, of which only the running section was modernized.

In 1956 IWC developed a smaller movement, Caliber 44, patterned after the men's caliber, and ranking among the flattest ladies' automatics of the time without a seconds hand.

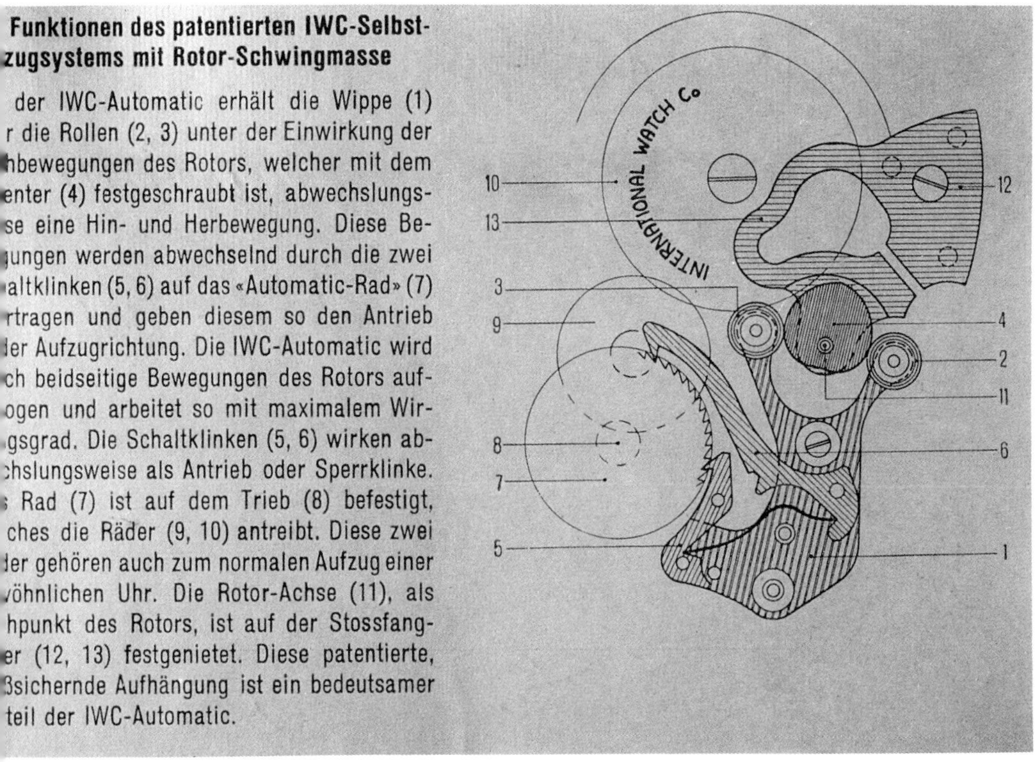

Patent description of IWC eccentric winding (from a brochure).

Magazine advertisement (from SUZ, 1959)

On the occasion of the firm's 100th anniversary, experiments were made on a completely new automatic development, Caliber 100, for the purpose of replacing the rather thick eccentric movement. Only a test series was built, and the firm purchased flat raw movements for its elegant modern watch models from Eta (for example, for the successor to the "Engineer", the "Ocean 2000" and "Ocean 500" divers' watches and the compass watch), Valjoux (for the "Da Vinci" and the "Grand Complication" of 1990), Piguet Fr. S.A. (for the rectangular "Novecento" model and ornamental watches), and Jaeger LeCoultre.

Caliber Notes

Caliber 85
D = 29.0 mm (12.75 lignes), H = 5.45 mm, 21 jewels
AUTOMATIC MODULE set atop the basic movement. Monometallic central rotor with ruby bearings on a sprung arbor, winding in both directions via a heart-shaped eccentric disc on the rotor, rocker lever on rollers with two sprung lever clicks, that alternately operate a sawtoothed winding wheel, on the crown wheel.
BASIC MOVEMENT: two-spoke Glucydur screw balance, self-compensating flat hairspring, 18,000 BPH, Incabloc shock resistance.

Caliber 44
D = 17.6 mm (7.75 lignes), H = 4.8 mm, 21 jewels
AUTOMATIC MODULE screwed to the basic movement, central rotor with ruby bearings and heavy metal ring on a sprung arbor, winding in both directions via a heart-shaped eccentric disc on the rotor, rocker lever with ruby rollers and two sprung lever clicks, that alternately operate a sawtoothed winding wheel, on the locking wheel.
BASIC MOVEMENT: two-spoke Glucydur screw balance, self-compensating flat hairspring, 19,800 BPH, Incabloc shock resistance, fine adjustment via toothed disc, no seconds hand.

IWC Caliber 85

Men's automatic made by IWC in 1951, with Caliber 85. Two-piece bowl case of Staybrite steel, divided winding stem, crown with fish symbol.

Monometallic rotor with ruby bearings on sprung arbor with fast unbolting, winding in both directions via patented eccentric switcher (lever clicks at 8 and 9:00, rocker-driving rollers at both sides of the rotor arbor, also of beryllium bronze).

Glucydur screw balance, self-compensating flat hairspring, Incabloc shock resistance (dial bubbled, hands corroded).

IWC Caliber 44

Ladies' automatic by IWC, post-1960, with Caliber 44, flattest ladies' caliber, first made in 1956.
Stainless steel case with screw bottom, winding crown with fish symbol.
Rotor with ruby bearings and heavy metal ring, running on sprung arbor with fast unbolting, winding in both directions via patented eccentric switcher (similar to Caliber 85, see page 156), rocker driven via two ruby rollers on rotor eccentric.
Glucydur screw balance with one equalizing screw on each spoke, self-compensating flat hairspring, Incabloc shock resistance, fine regulator adjustment via eccentric screw (glass with scratches, case and movement show signs of wear).

IWC Caliber 8541

Men's automatic made by IWC circa 1965, with Caliber 8541.

Eighteen Karat gold case with waterproof press-in bottom, gold winding crown with fish symbol, gold hands and dial inlays.

Movement largely identical to Caliber 85 (see page 156), but rotor has heavy metal ring, screwless Glucydur balance has three equalizing screws, self-compensating Breguet hairspring, fine regulator adjustment by eccentric screw.

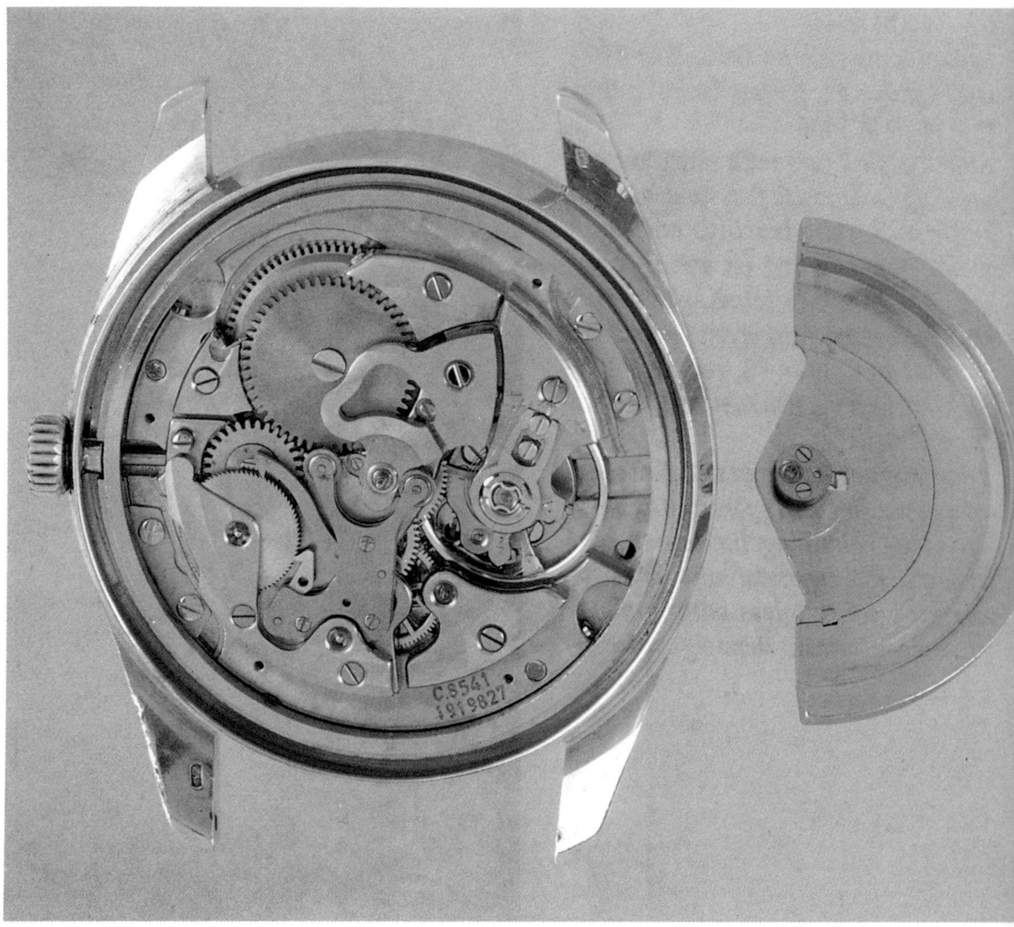

Caliber List: IWC

Caliber	Diameter Lignes	mm	Height mm	Other	BPH	Jewels	Running Reserve	Notes	Year	Photo
44	7 3/4	17,6	4,8		19.800	21	40		1956	xx
81	12 3/4	29,0	5,05	s	18.000				1950	
85	12 3/4	29,0	5,45	SCI	18.000	21	40		1950	xx
100	12 1/4	28,0	5,9	SC	25.200	27		Experimental Series	1965	
270 271	12 "	27,4 "	2,4 "		18.000 "	34 "		Piguet-Ebauche P70 "	1976 1976	
320	9	20,4	3,2		21.600	19	40	Piguet-Ebauche 951	1986	
375	12 1/4	28,0	3,75	SCD, QG, ST	28.800	22		Eta-Ebauche 2890	1978	
442	7 3/4	17,6	4,8		21.600	23			1972	
480	7 3/4	17,2	4,8	SCD, QG, ST	28.800	25		Eta-Ebauche 2671	1983	
790	13 1/2	30,0	7,9	s, QG, JG, RDR, CHR	28.800	25		Valjoux-Ebauche 7750	1978	
852 853 854	12 3/4 " "	29,0 " "	5,6 " "	SCI SCI SCI, ST	19.800 " "	21 " 21-25	46 " "		1952 1958 1964	
1001	12 1/2	28,0	3,95	SC, QG	25.200	27		Versuchs-Serie	1965	
2717	12	27,4	2,4		18.000	34		Piguet-Ebauche P70	1976	
3206	9	20,4	4,5	"Perpetual" Calendar	21.600	33		Piguet-Ebauche 951	1988	
3251 3252 3254	11 1/2 " "	26,0 " "	3,25 " "	 QG, RDR SCD, QG, RDR, ST	28.800 " "	30 32 33		JLC-Ebauche 888 " "	1976 1976 1976	
3752 3753 3754 3755 3757 3758 3759	12 1/2 11 1/2 " " 13 1/2 12 1/2 11 1/2	28,0 30,0 28,0 	3,75 " 4,15 " 5,0 5,0 3,75	SCD, QG, RDR, ST SCD, QG, RDR, ST SCD, QG, RDR, ST SCD, QG, RDR, ST "Perpetual" Calendar "Perpetual" Calendar SCD, QG, RDR, ST	28.800 " " " " " "	22 " " " 37 " 22		Eta-Ebauche 2890 " " " " " "	1978 1986 1987 	
7906	13 1/2	30,0	8,75	"Perpetual" Calendar	28.800	36		Valjoux-Ebauche 7750	1985	
8521 8531 8541	12 3/4 " 12 1/2	29,0 " 28,0	6,4 " 5,9	SCI, QG SCI, QG SCI, QG, ST	19.800 " "	21 " 21-25	 40 		1953 1959 1964	 xx

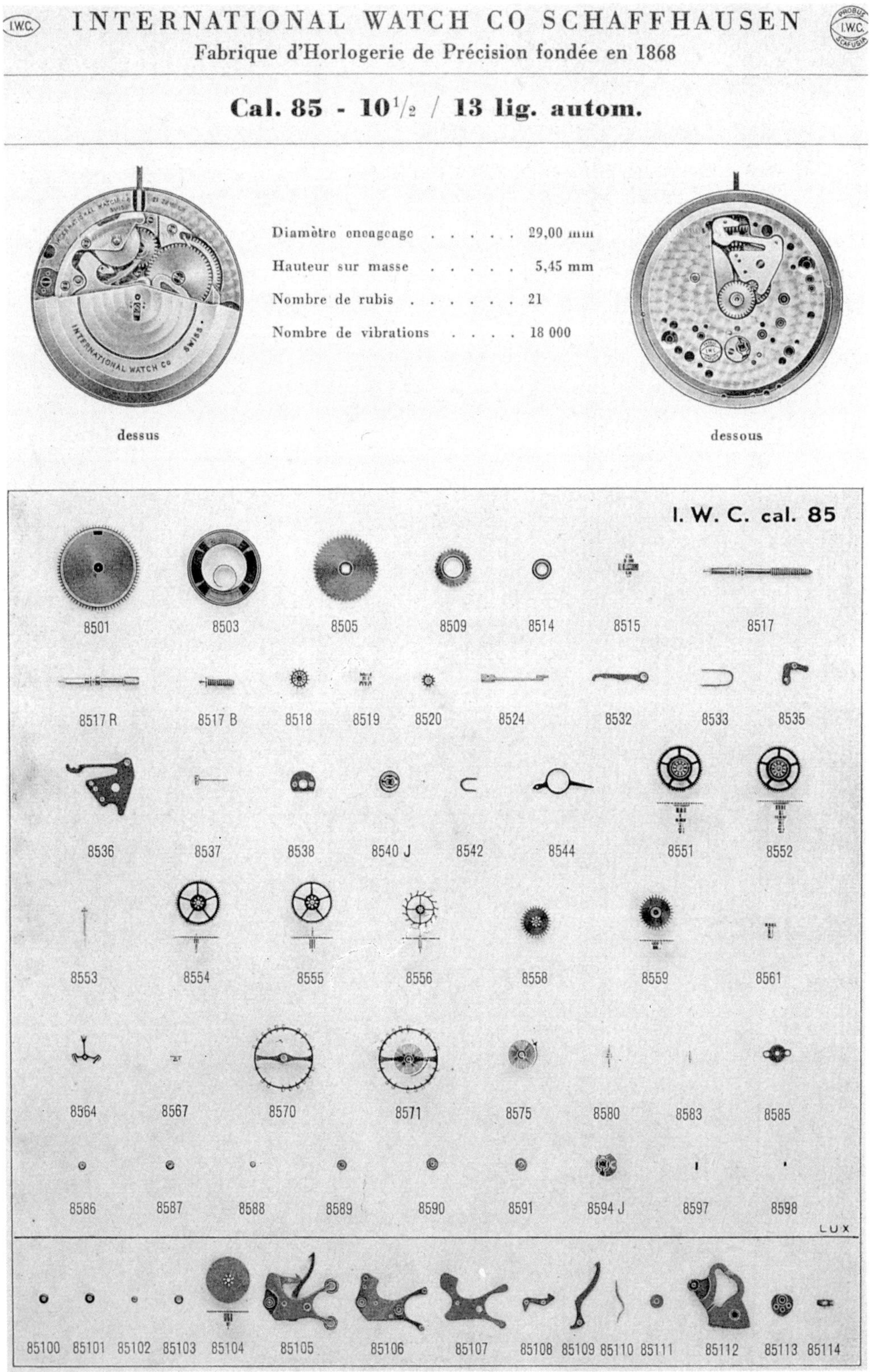

Illustrated parts catalog for IWC Caliber 85 (automatic parts at the bottom):
85 105—complete click-carrier rocker
85-112—elastically suspended rotor arbor
85 113—rotor-eccentric with ruby bearing

Caliber 8541

D = 28.0 mm (12.5 lignes), H = 5.9 mm, 21 jewels

AUTOMATIC MODULE screwed to the basic movement, central rotor with ruby bearings and heavy metal ring on sprung arbor, winding in both directions via a heart-shaped eccentric disc on the rotor, rocker lever with rollers and two sprung lever clicks, that alternately operate a sawtoothed winding wheel, on the crown wheel.

BASIC MOVEMENT: three-spoke screwless Glucydur balance with three equalizing screws, self-compensating Breguet hairspring, 19,800 BPH, Incabloc shock resistance, double fine adjustment by regulator and hairspring block carrier via eccentric screws, second stop, date indication.

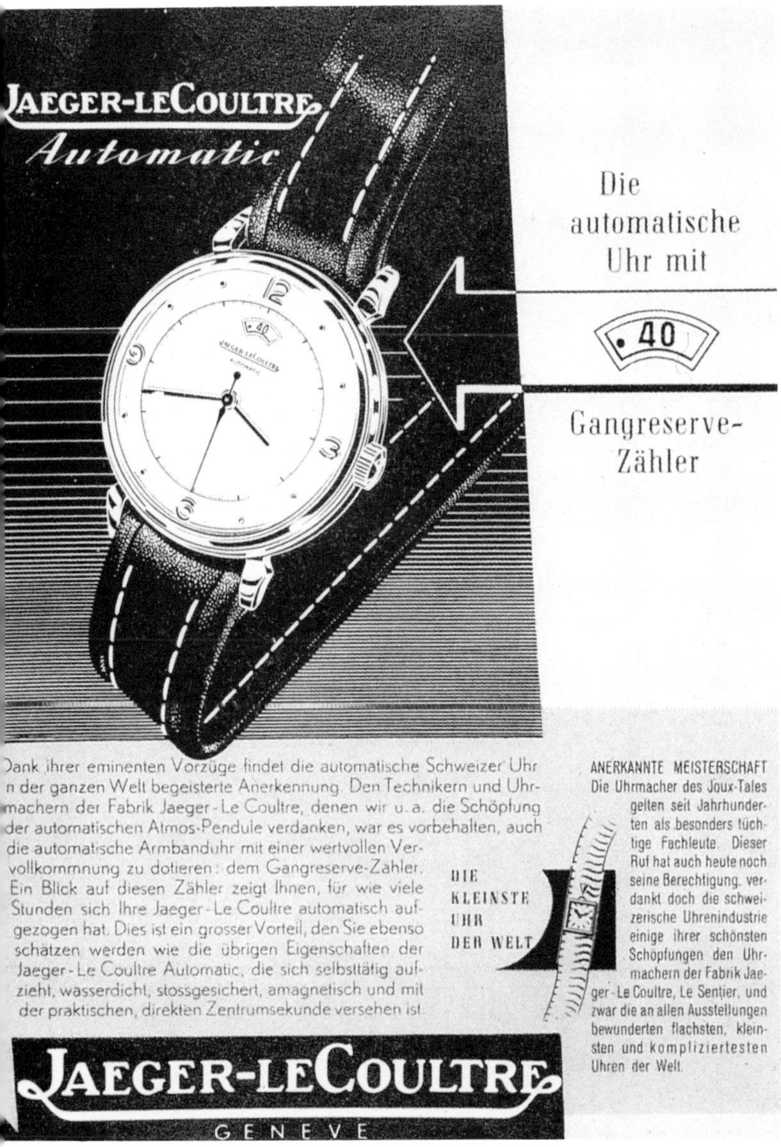

Magazine advertisement (from SUZ 1949)

Caliber Development

The firm was founded by Antoine Le Coultre at Le Sentier in the Vallée de Joux in 1833 to produce raw movements, and continued under the name of "Jaeger LeCoultre" after Edmond Jaeger joined the firm in 1937. Watches made by this firm were sold under the name of LeCoultre in the USA, as according to import laws the original Swiss movements could be imported but they had to be mounted in cases made in the USA.

The first JLC automatic was introduced in 1947 as Caliber 476. It had hammer winding in one direction in the usual manner, and after a differential drive was installed it became the first series-production automatic wristwatch with up and down indication; it was produced as of 1948 under the model name of "Powermatic" (Caliber 481).

Although the manufacturers developed their first rotor automatic (Caliber 493) with winding in both directions and a swinging mass on ball bearings in 1951, they decided to return to hammer winding in their next designs. There were several reasons for this decision: the hammer automatic had proved itself as a sturdy and flat design, additional features could be added without essentially increasing the thickness of the movement.

By using both directions of the hammer to wind the mainspring, the Caliber 497 "Futurematic", developed in 1952, could attain such great dependability that hand-winding seemed unnecessary and was eliminated. The empty space allowed the installation of a larger balance and barrel. The up-down indicator was meant to meet the wearer's needs, and at the same time it offered the possibility of regulating the stopping of the swinging mass. For the first time, the example of the Perrelet automatic pocket watch was followed to block the whole automatic mechanism at full winding and thus protect not only the balance from damage but also the drive from wear. A normal mainspring without a sliding stud could be installed in the barrel, with its driving power available after a few shakes with the help of a prewinding device. The watch could be set to the second by a flat bottom crown.

The "Memovox" (Caliber 815), introduced in 1956, was the first automatic alarm watch and was also equipped with hammer winding in one direction. The alarm could be hand-wound and set by a second crown.

Only in 1959 did the rotor finally begin to find favor at the house of Jaeger LeCoultre, although movements similar to Caliber 493, but in larger sizes, had been built previously for Audemars Piguet and Vacheron Constantin.

The rotor of this new 880 caliber family wound in both directions using a switching rocker. This sturdy equalizing principle has been found in all of this manufacturer's rotor automatics since then.

Caliber 888, still being built today, was already developed as a particularly flat design by 1967.

In 1969 JLC marketed the successor model to the "Memovox", a modern automatic alarm watch with a central rotor on ball bearings. It was the firm's only movement in which the direction was changed with the help of an eccentric switcher (Caliber 916).

Jaeger LeCoultre ranks among the most renowned Swiss manufacturers and also is a major supplier of raw movements to the firms of Audemars Piguet and Vacheron Constantin.

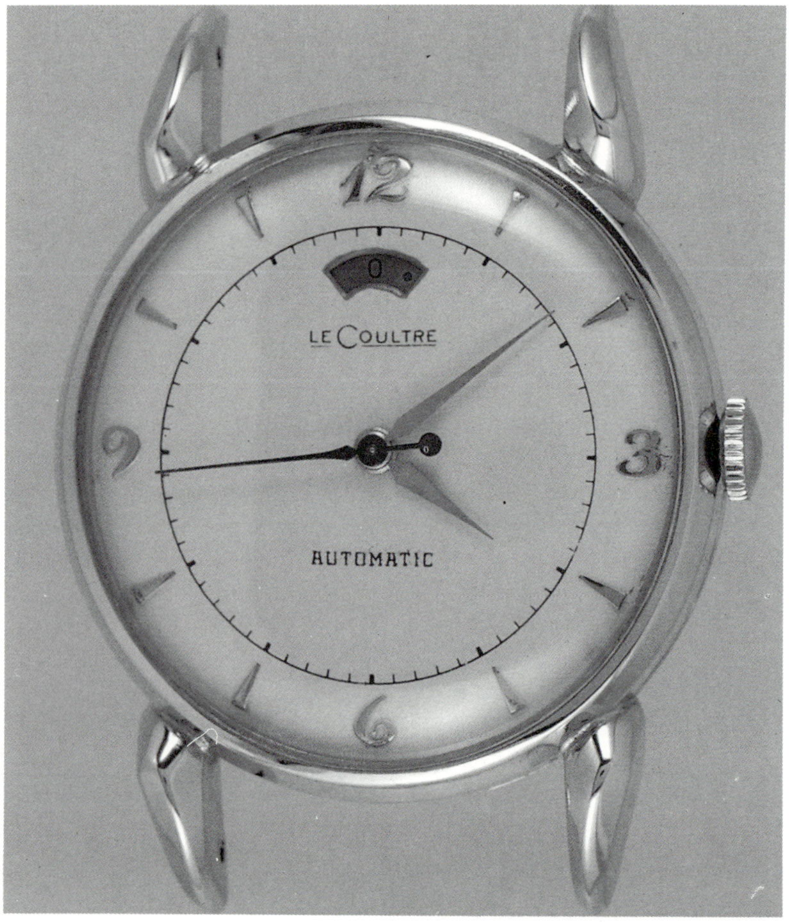

Jaeger LeCoultre Caliber 481
Men's automatic made by LeCoultee in 1948, with Caliber 481, known as "Powermatic", the first production model with up and down indicator (in the dial aperture under the 12).
Fourteen Karat gold case with press-in bottom.
Hammer winding in one direction via click-carrier rocker.
Glucydur screw balance, Nivarox hairspring, JLC shock resistance (dial is freshened up).

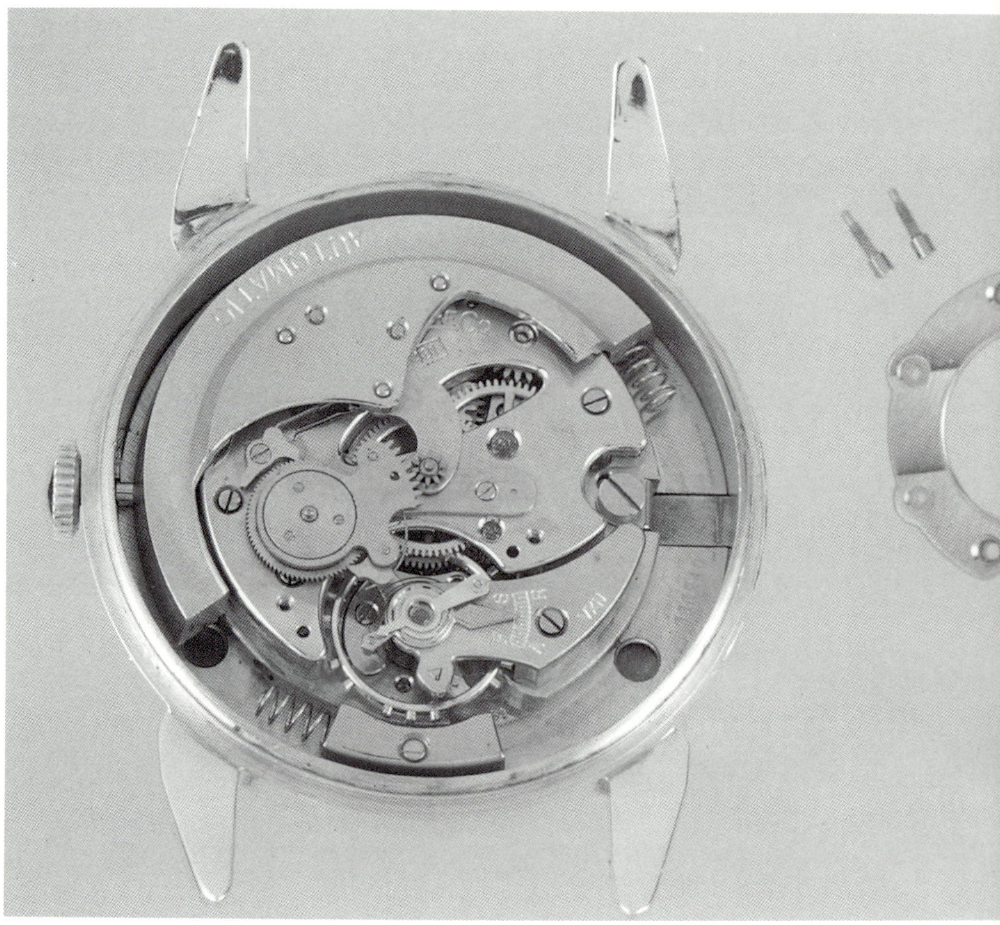

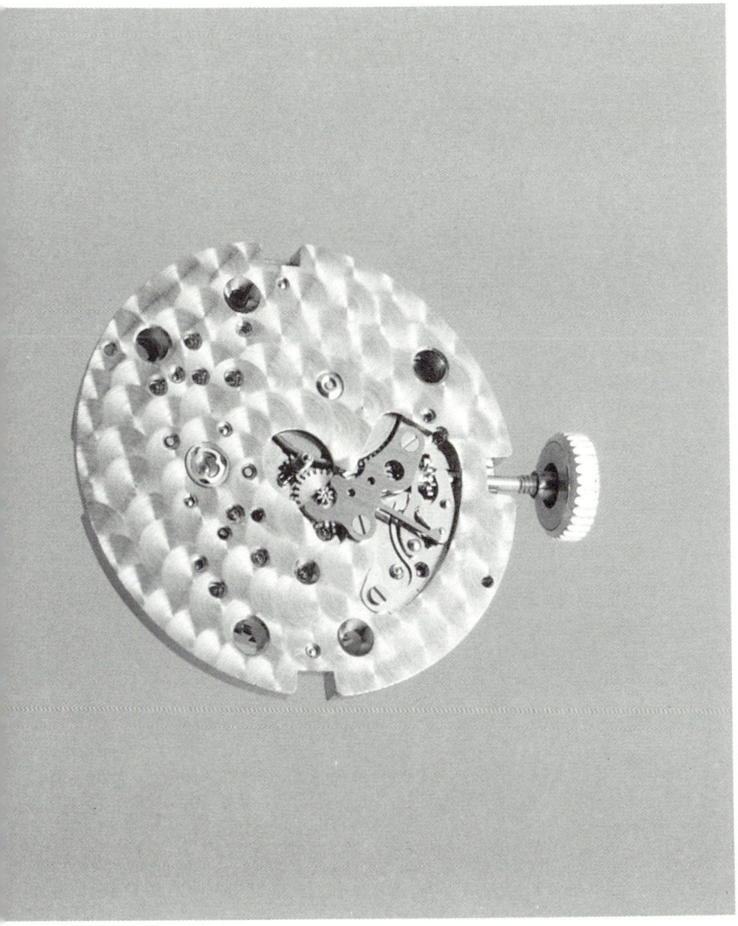

Jaeger LeCoultre Caliber 493

First rotor caliber made by Jaeger LeCoultre in 1951 (Caliber 493), chiefly for rectangular watch models.
Ball-supported rotor with heavy metal ring (4 balls), winding in both directions via switching rocker.
Glucydur screw balance, Nivarox hairspring, Kif-Protechoc shock resistance.

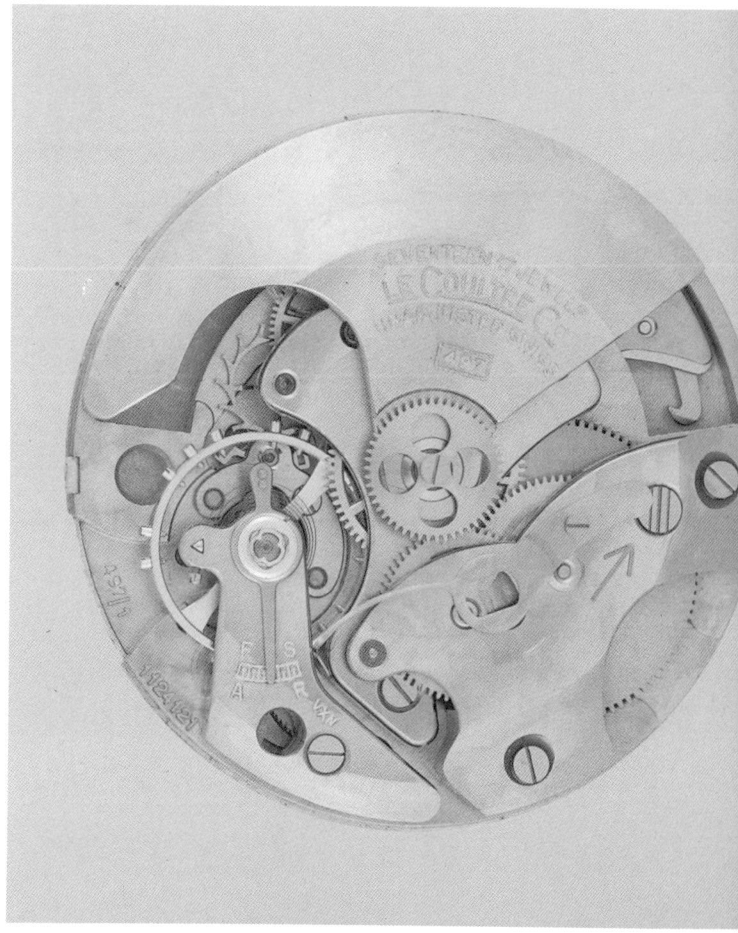

LeCoultre Caliber 497/1

Men's model made by LeCoultre circa 1955, the so-called fully automatic "Futurematic", with Jaeger LeCoultre Caliber 497 first made in 1951.
Fourteen Karat gold case (two-piece, with bottom bowl), flat crown on bottom only for exact-second hand-setting. Up and down indicator on the auxiliary dial near the 9 (marked sector indicates critically short reserve). Hammer mounted on both sides of the movement with small range of action, winding in both directions via switching rocker. Overwinding protection by regulated hammer blocking.
On underside of movement (lower right) the automatic drive, lever mechanism for running reserve indication, and blocking mechanism with catch lever (at 3:00) are shown. The hands are set by the central toothed wheel on the hammer (upper right).
Glucydur screw balance, self-compensating flat hairspring, Kif-Protechoc shock resistance.

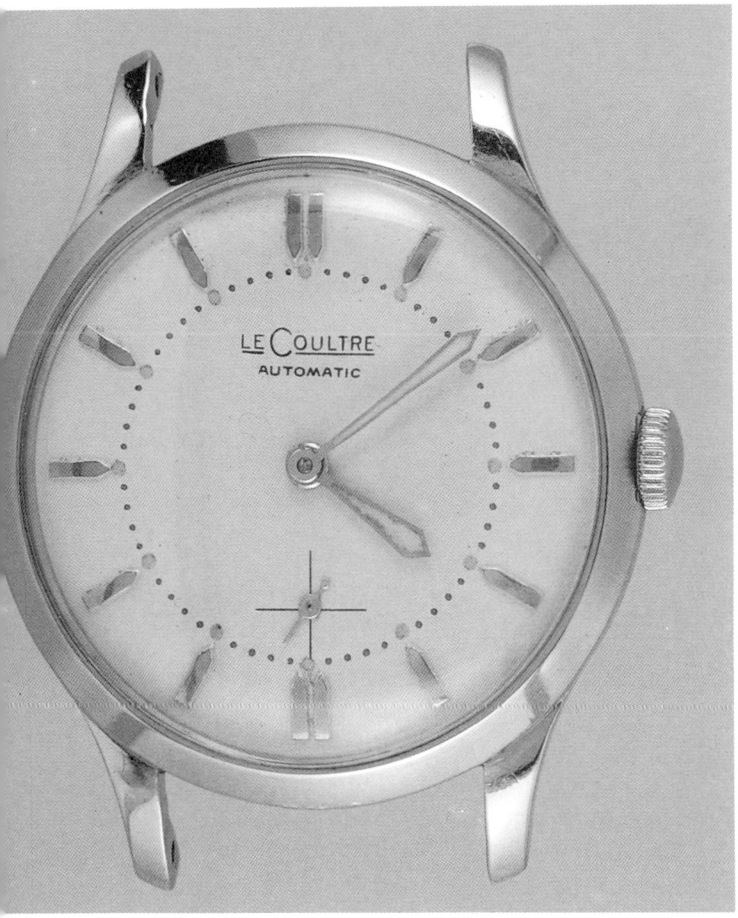

LeCoultre Caliber 812

Men's automatic made by LeCoultre circa 1955, with Jaeger LeCoultre Caliber 812.
Stainless steel case with screw bottom. Hammer winding in one direction via click-carrier rocker.
Glucydur screw balance, self-compensating flat hairspring, Kif-Flector shock resistance (dial is freshened up, second hand not original, case with signs of wear, surface or movement corroded).

Jaeger LeCoultre Caliber 817

Men's model made by Jaeger LeCoultre circa 1955 as so-called fully automatic watch, with Caliber 817.

Stainless steel case (three parts, with screwed bottom bowl), flat crown on rear only for exact-second hand-setting.

Up and down indicator in round dial aperture near the 9 (the size of a light-colored segment represents the available reserve).

Movement identical to Caliber 497/1 (see page 164). (Dial is freshened up, second arrow redrawn, case with signs of wear.)

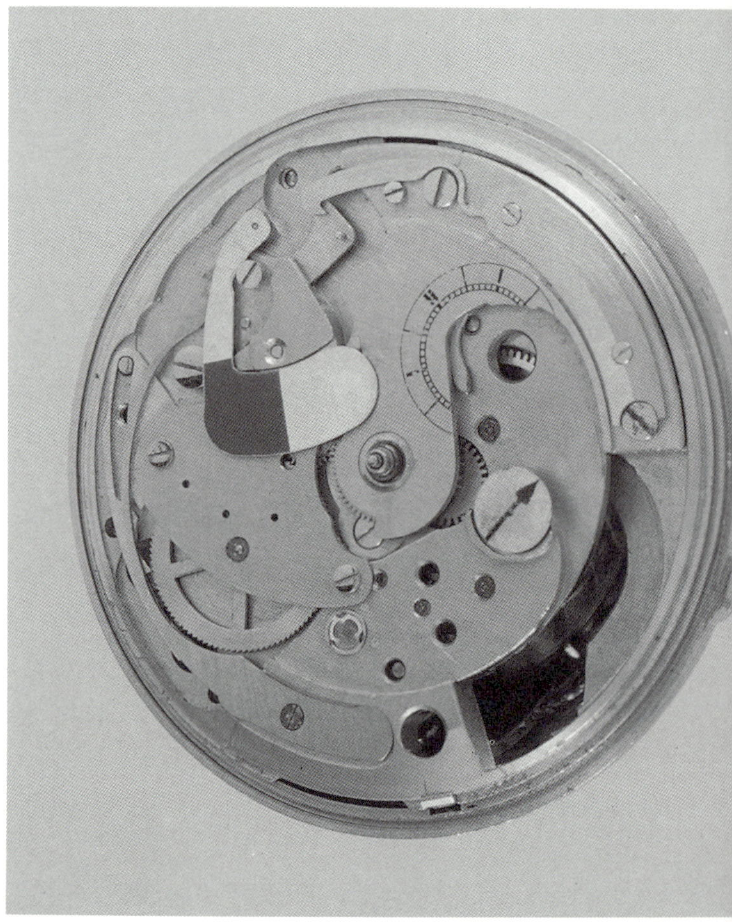

Jaeger LeCoultre Caliber 825

Men's model made by Jaeger LeCoultre in 1958, with Caliber 825, the first automatic alarm watch ever made (here with date), called "Memovox". Gilded case with stainless steel screw bottom, winding crowns with JLC emblem. The crown by the 2 winds the alarm mechanism and sets the alarm time on the central dial.

Hammer swinging mass, winding in one direction via click-carrier rocker. The alarm hammer (at 2:00) strikes against a rod on the case bottom that projects into the movement through the kidney-shaped cutout in the swinging-mass shank.

Glucydur screw balance, self-compensating flat hairspring, Kif-Flector shock resistance.

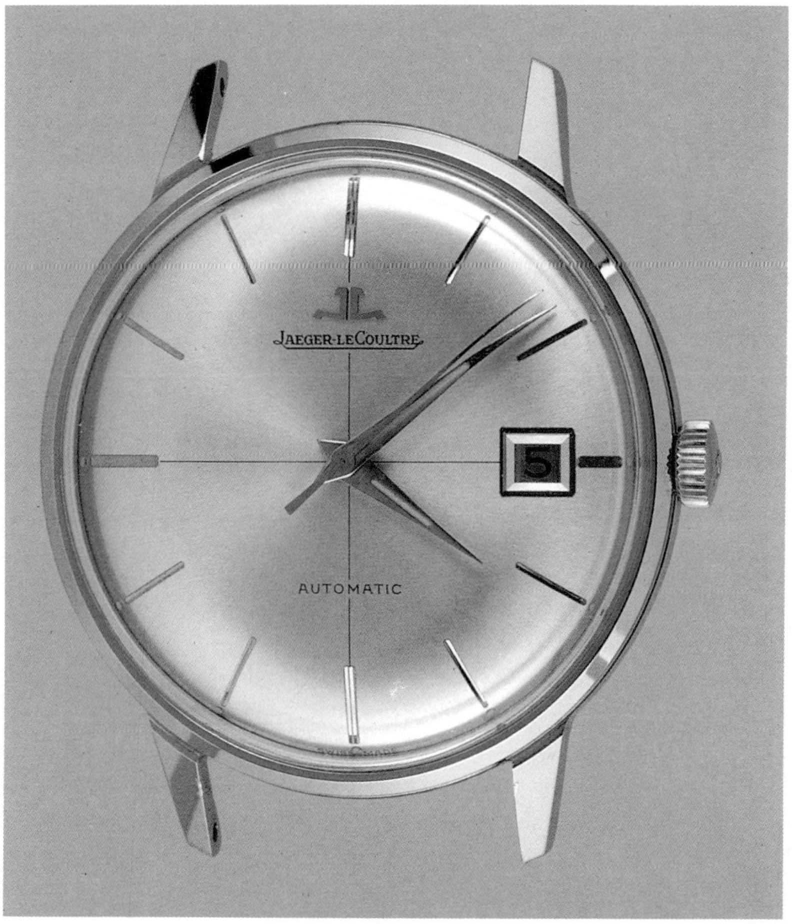

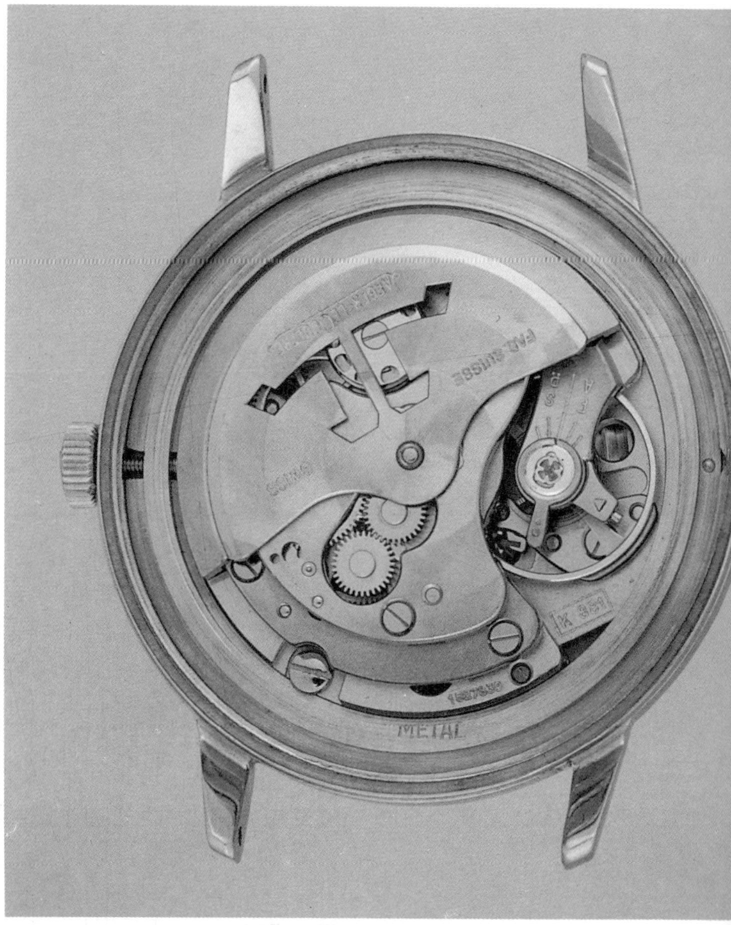

Jaeger LeCoultre Caliber 881

Men's automatic made by Jaeger LeCoultre in 1959, with Caliber 881. Stainless steel case with screw bottom, winding crown with firm's emblem. Rotor with heavy metal ring, winding in both directions via switching rocker (at 7:00), uncoupling during hand-winding via locking-wheel coupling (at 11:00).

Screwless Glucydur balance, self-compensating flat hairspring, Kif-Flector shock resistance.

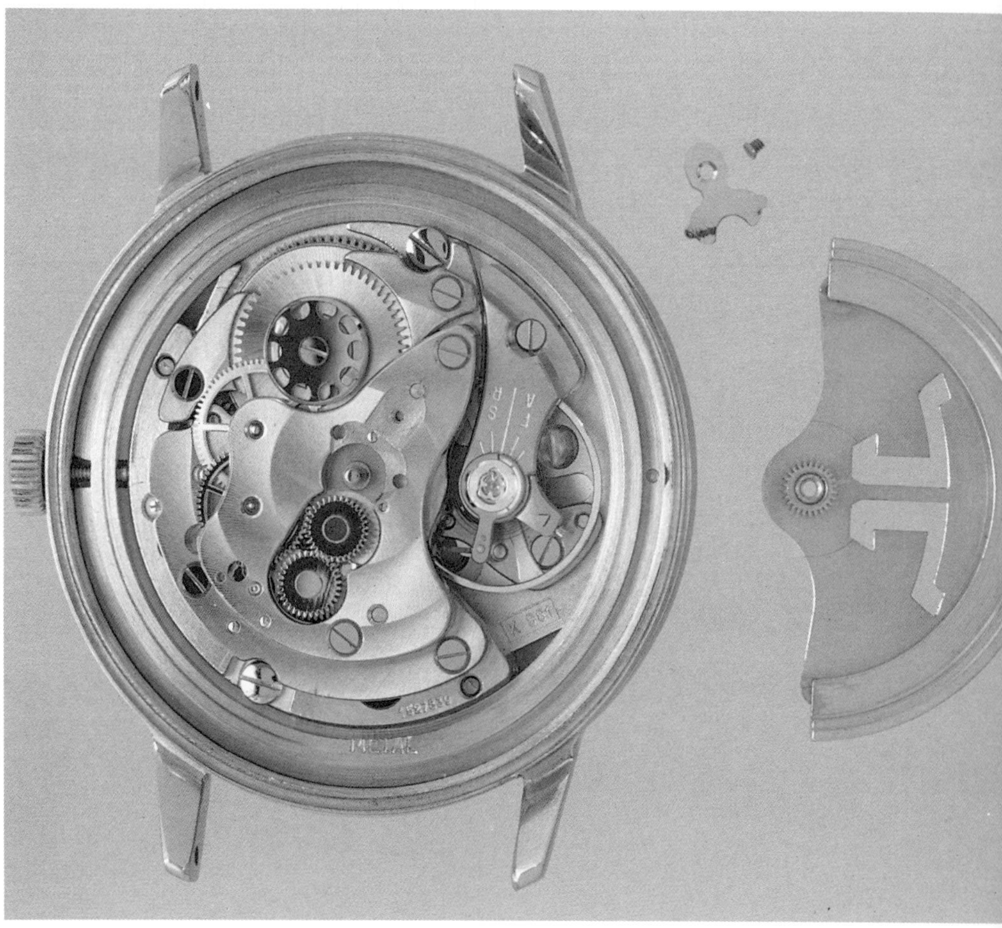

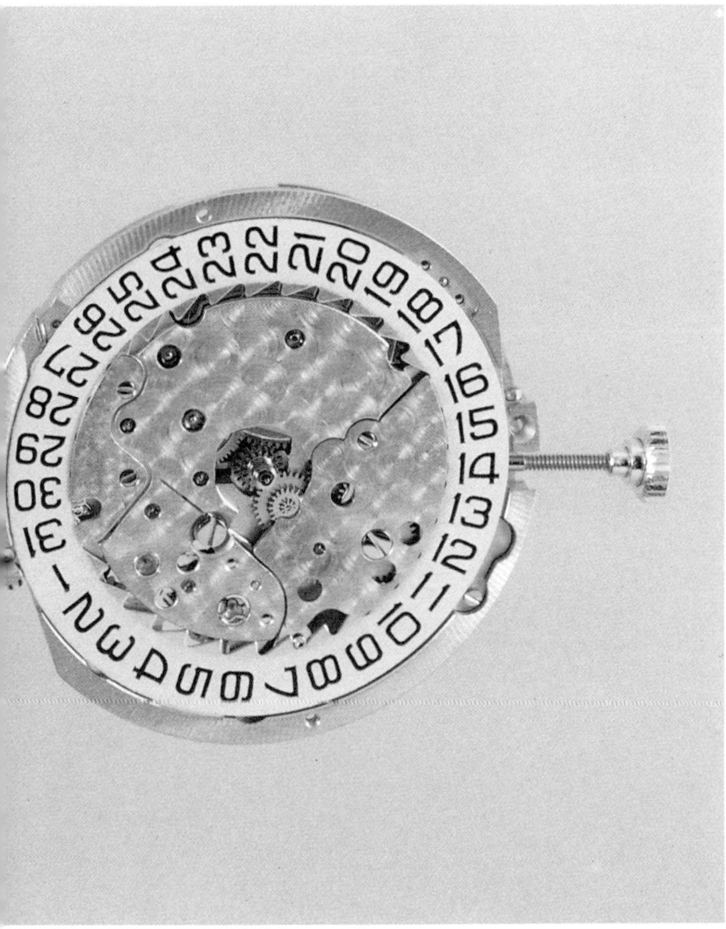

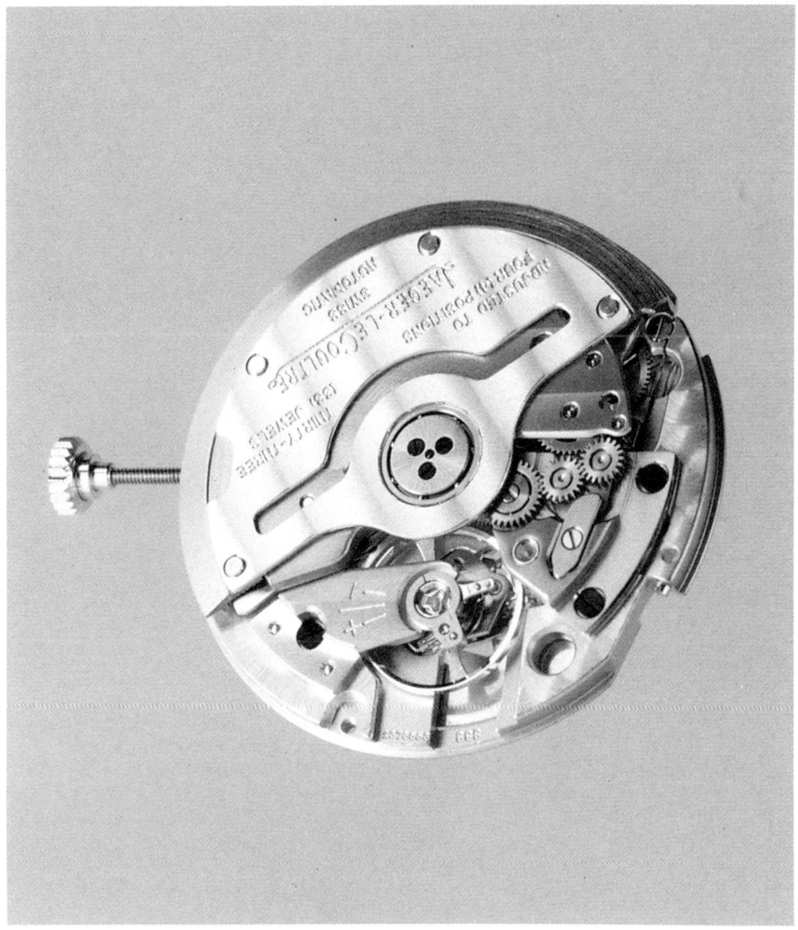

Jaeger LeCoultre Caliber 888 Movement

Automatic movement, Caliber 888 for men's watches, made by Jaeger LeCoultre as of 1967.

Rotor with ball bearings and heavy metal ring, winding in both directions via switching rotor (at 3:00), uncoupling during hand-winding.

Screwless Glucydur balance, self-compensating flat hairspring, Kif-Ultraflex shock resistance, Spirofin fine adjustment, second stop.

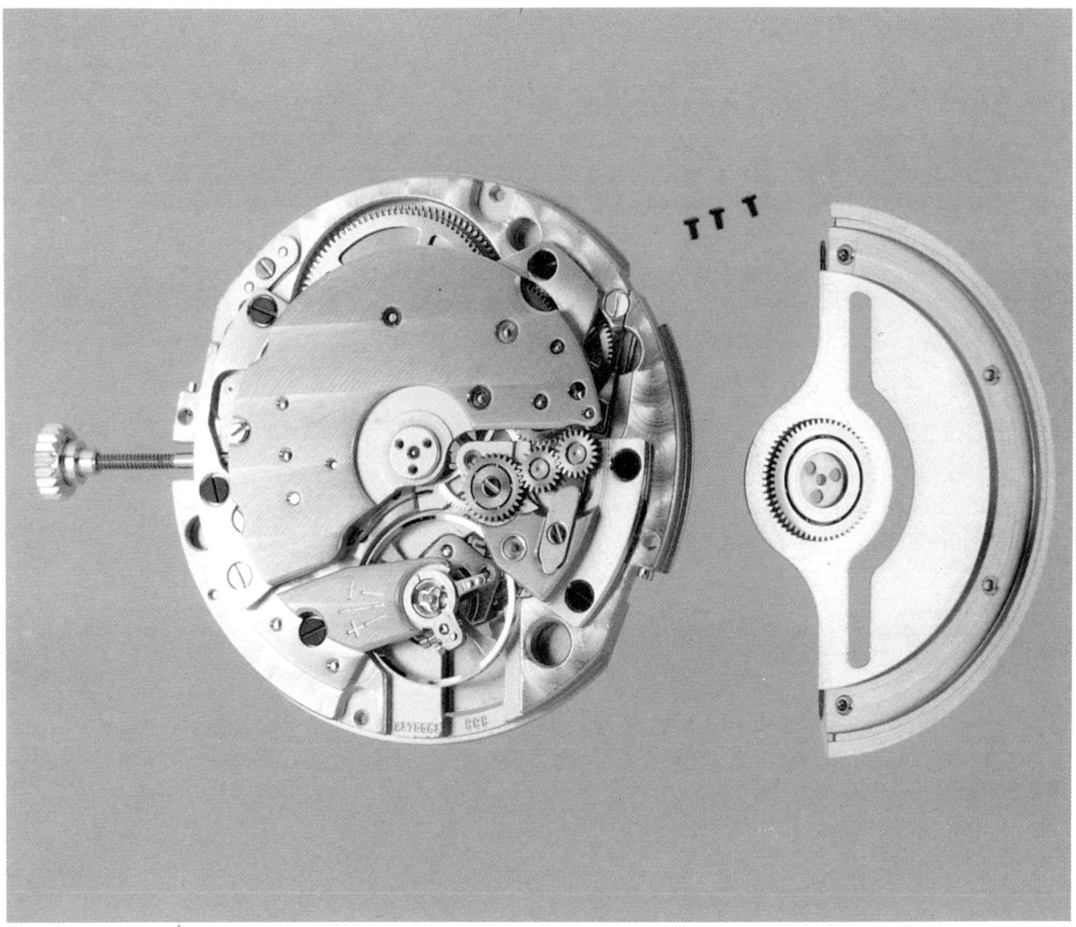

Jaeger LeCoultre Caliber 916

Automatic alarm watch made by Jaeger LeCoultre in 1969, with Caliber 916.
Stainless steel case with screw bottom, winding crowns with JLC emblem (at the 2 for the alarm, at the 4 for the movement).
Rotor with ball bearings and heavy metal ring, winding in both directions via eccentric switcher (at 5:00), uncoupling during hand-winding, alarm mechanism wound only by hand, alarm hammer strikes against the plate (at 1:00).
Screwless Glucydur balance, self-compensating flat hairspring, Kif-Elastor shock resistance, Spirofin fine adjustment (glass defective).

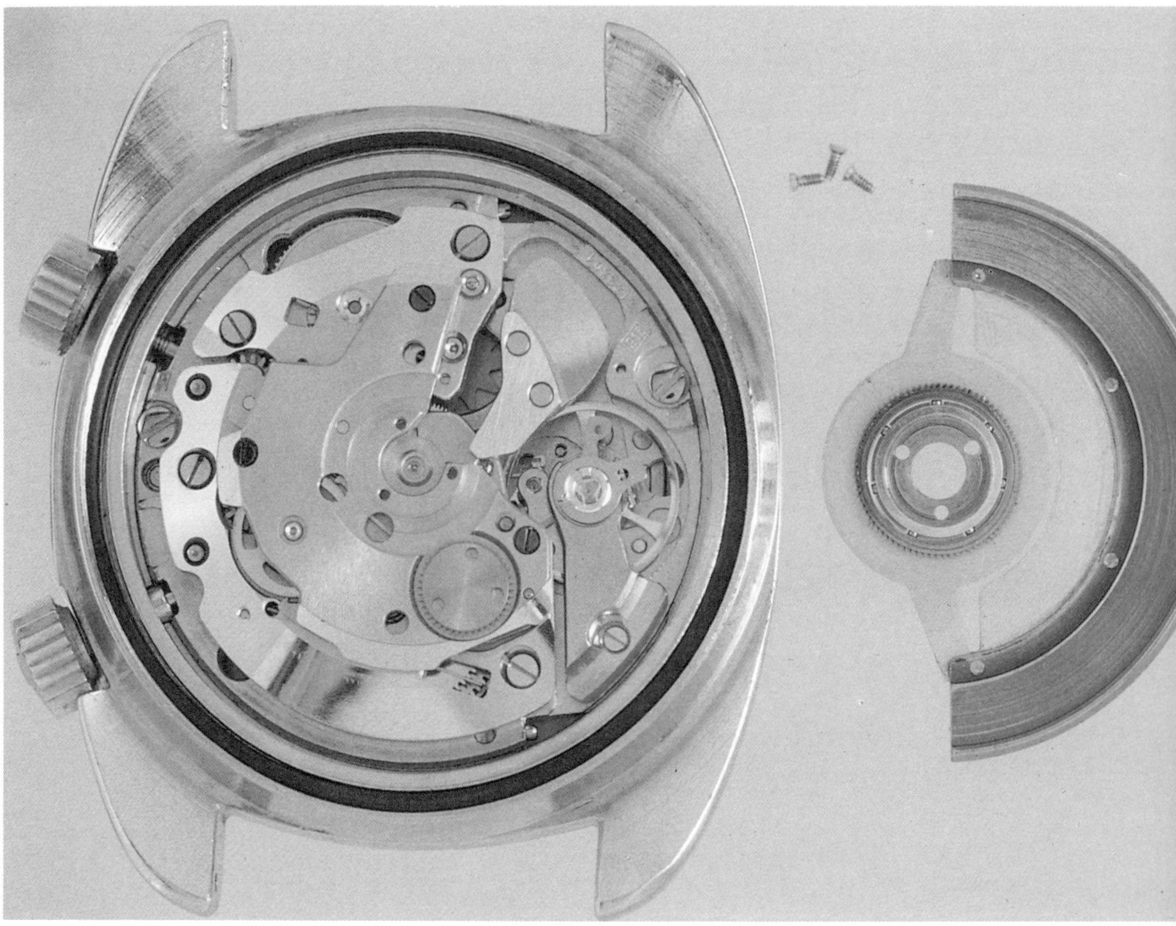

Caliber List: Jaeger Le Coultre

Caliber	Diameter Lignes	mm	Height mm	Other	BPH	Jewels	Running Reserve	Notes	Year	Photo
476	13	29,6	6,15	SCI	18.000	17	46		1947	
481	"	"	"	SCI, RMG	"	"	40	Powermatic	1948	xx
493	9 1/2	21,3	4,85	SCI	18.000	17			1951	xx
497	13 3/4	31,3	6,3	s, ST, RMA	18.000	17	34	Futurematic	1953	xx
812	12 1/2	28,0	4,55	s	19.800	17			ca. 55	xx
813	"	"	5,25	s, QG	"	19				
815	14	31,6	7,05	SCI, REV	18.000	17	55	Memovox	1956	
817	14	31,6		s, ST, RMG	18.000			Futurematic	1955	xx
825	14	31,6	7,5	SCI, QG, REV	18.000	17	55	Memovox	1958	xx
827	13 3/4	31,3	6,3	s, ST, RMA	18.000			See Caliber 497	1955	
832	7 1/4	16,2	4,25		19.800	17/25				
833	"	"	4,75	SCI	"	"				
837	13 3/4	31,3	6,3	s, ST, RMG	18.000					
880 G	11 1/2	26,0	5,77	SCD	19.800	17/23	45		1959	
880 S	"	"		SCD, ST	"		"		1959	
881	"	"	6,14	SCD, QG	"	17	"		1959	xx
881 G/S	"	"		SCD, QG, ST	"	23	"		1959	
882/S	"	"	5,77	SCD/SCD, ST	"	17	"			
883/S	"	"	"	SCD, QG/SCD, ST	"	"	"			
888	12 1/2	28,4	3,25	SCI, QG, RDR, ST	21.600	33			1967	xx
889	"	"	"		"	"			1967	
916	13	30,0	7,45	SCD, QG, REV	28.800	17		"Perpetual" Grand	1969	xx
919	"	"	8,30	SCD, REV, „ew." Kal.	"	31	44	Reveil	1989	

Caliber Notes

Caliber 481

D = 29.6 mm (13 lignes), H = 6.15 mm, 17 jewels

AUTOMATIC MODULE set atop the small basic movement, hammer with ruby bearings, winding in one direction via click-carrier rocker with combined winding and reduction wheel on crown wheel.

BASIC MOVEMENT: two-spoke Glucydur screw balance, Nivarox hairspring, 18,000 BPH, JLC shock resistance, Sirius mainspring. Differential drive to indicate the running reserve integrated into the basic movement.

Caliber 493

D = 21.3 mm (9.5 lignes), H = 4.85 mm, 17 jewels

AUTOMATIC MOVEMENT screwed to the basic movement as a complete unit (three blued screws), central rotor with heavy metal ring, supported by four steel balls that run in a groove under cover of the central section of the rotor, winding in both directions via switching rocker and two reduction wheels, the first with sawteeth and locking click, on the double crown wheel, crown-wheel coupling to uncouple during hand-winding.

BASIC MOVEMENT: two-spoke Glucydur screw balance, Nivarox hairspring, 18,000 BPH, Kif-Protechoc shock resistance.

Caliber 497

D = 31.3 mm (13.75 lignes), H = 6.3 mm, 17 jewels

AUTOMATIC MODULE integrated into the basic movement, two-section hammer with heavy-metal inlay swings about 90 degrees on both sides of the bottom plate and is limited by two leaf springs, winding in both directions via switching rocker, two reduction wheels (the first with sawteeth and spring click) on the locking wheel, hammer blocking by winding-dependent axially moveable spring-bolt nut that swings a lever out far enough to engage the catching hook of the swinging mass.

Running reserve indication was by parallel transmission of the limiting lever mechanism by a toothed-segment lever and indicator pinion.

BASIC MOVEMENT: two-spoke Glucydur screw balance, Nivarox hairspring, 18,000 BPH, Kif-Protechoc shock resistance, eccentric seconds hand, mainspring with eyelet minus sliding stud pre-wound to about 1.5 turns, no hand-winding, second stop.

Caliber 812
D = 28.0 mm (12 lignes), H = 5.55 mm, 17 jewels
AUTOMATIC MODULE screwed to the basic movement, hammer with ruby bearings and buffer springs, winding in one direction via click-carrier rocker with combined winding and reduction wheel on crown wheel.
BASIC MOVEMENT: two-spoke Glucydur screw balance, Nivarox hairspring, 19,800 BPH, Kif-Flector shock resistance, eccentric seconds hand.

Caliber 817
D = 31.3 mm (13.75 lignes), H = 6.3 mm, 17 jewels
See Caliber 497, but with running reserve and second indication by marked discs in round dial apertures.

Caliber 825
D = 31.6 mm (13 lignes) H = 7.5 mm, 17 jewels
AUTOMATIC MODULE screwed to basic movement, hammer with ruby bearings and heavy metal ring, buffer springs, winding in one direction via click-carrier rocker with combined winding and reduction wheel on crown wheel.
BASIC MOVEMENT: two-spoke Glucydur screw balance, Nivarox hairspring, 18,000 BPH, Kif-Flector shock resistance. Alarm mechanism integrated into basic movement and wound only by hand, with a second barrel, alarm hammer strikes against a pin in the center of the case bottom.

Caliber 881
D = 26.0 mm (11.5 lignes), H = 6.14 mm, 17 jewels
AUTOMATIC MODULE screwed to the basic movement, central rotor with heavy metal ring, winding in both directions via switching rocker, two reduction wheels (the first with spring click) on the upper coupling locking wheel to uncouple during hand-winding, lower coupling wheel (under the dial) to uncouple hand-winding mechanism during automatic winding.
BASIC MOVEMENT: two-spoke screwless Glucydur balance, Nivarox hairspring, 19,800 BPH, Kif-Flector shock resistance, moveable hairspring block carrier, direct center seconds, date indication with quick change.

Caliber 888
D = 28.4 mm (12.5 lignes), H = 3.25 mm, 33 jewels
AUTOMATIC MODULE integrated into basic movement, central rotor with ball bearings and heavy metal ring, winding in both directions via intermediate wheel with ball bearings, switching rocker, three reduction wheels (the first with triangular teeth and locking click, the third with sprung-disc coupling to uncouple during hand-winding) on the locking wheel.
BASIC MOVEMENT: two-spoke screwless Glucydur balance, Nivarox hairspring, 21,600 BPH, Kif-Ultraflex shock resistance, Spirofin fine adjustment with moveable hairspring block carrier, escape wheel mounted in combined jewel housings, Eloxal barrel, second stop, date indication with quick change and correction.

Caliber 916
D = 30.0 mm (13 lignes), H = 7.45 mm, 17 jewels
AUTOMATIC MODULE set atop basic movement and alarm mechanism, central rotor with ball bearings and heavy metal ring, winding in both directions by one eccentric wheel and two independently sprung winding clicks on fine-toothed click wheel (with uncoupling device for hand-winding) on locking wheel.
BASIC MOVEMENT: three-spoke screwless Glucydur balance, Nivarox hairspring, 28,800 BPH, Kif-Elastor shock resistance, Spirofin fine adjustment, direct center seconds.
Alarm mechanism integrated into the basic movement, separate barrel wound only by hand, alarm hammer strikes against the automatic plate. Date indication with quick change.

Caliber Development

The Kelek S.A. was founded at La Chaux-de-Fonds in 1896. The firm devoted itself chiefly to the completion of complicated pocket and wristwatches. The automatic chronograph Caliber TDB 1376, which came on the market in 1974, was only completed by Kelek. It was a joint development of the Buren and Brac firms (Breitenbach, Canton Solothurn), which produced the basic movement and the automatic module, and the firm of Dubois-Déprax of Le Lieu (Vallée de Joux), which built the chronograph module, and it is one of the smallest of its kind.

The automatic mechanism was integrated into the basic movement. The rotor wound in both directions via a switching device with click wheels resembling those of the first Eta-Eterna patent.

Caliber Notes

Caliber TDB 1376
D = 26.0 mm (11.5 lignes), H = 7.6 mm, 17 jewels
AUTOMATIC MODULE integrated into the basic movement, central rotor with heavy metal ring and gliding bearings, winding in both directions via twin click wheels, two reduction wheels on the crown wheel.
Chronograph module (with coulisse lever switching) screwed onto the basic movement.
BASIC MOVEMENT: three-spoke screwless nickel balance, Nivarox hairspring, 19,800 BPH, Incabloc shock resistance, moveable hairspring block carrier, center seconds wheel, date indication with quick change.

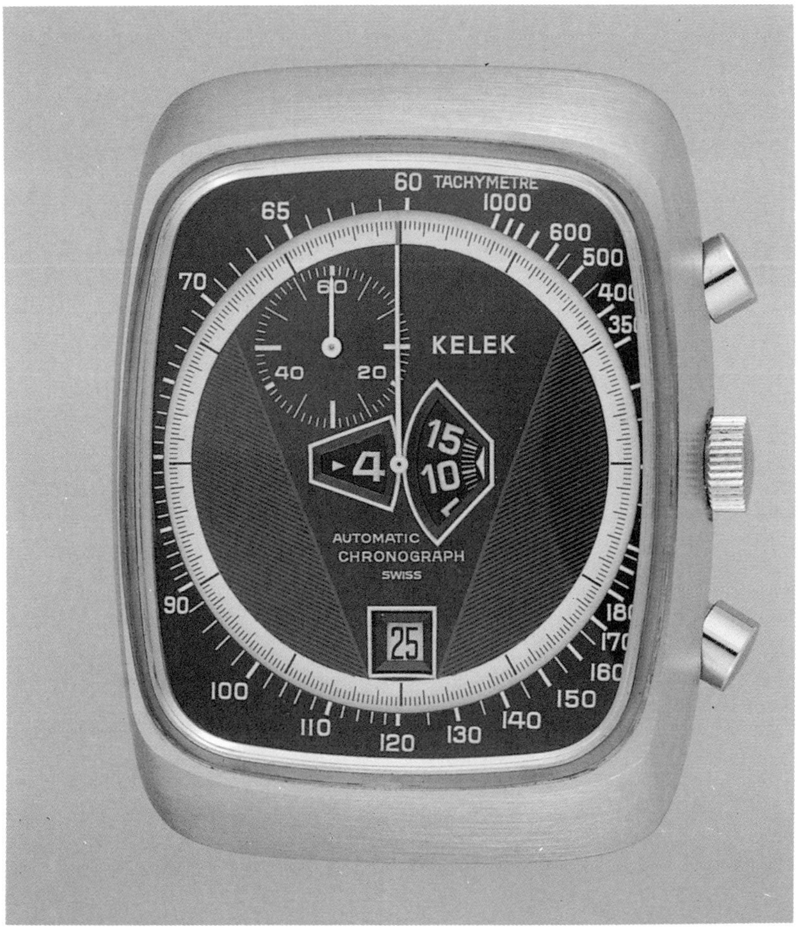

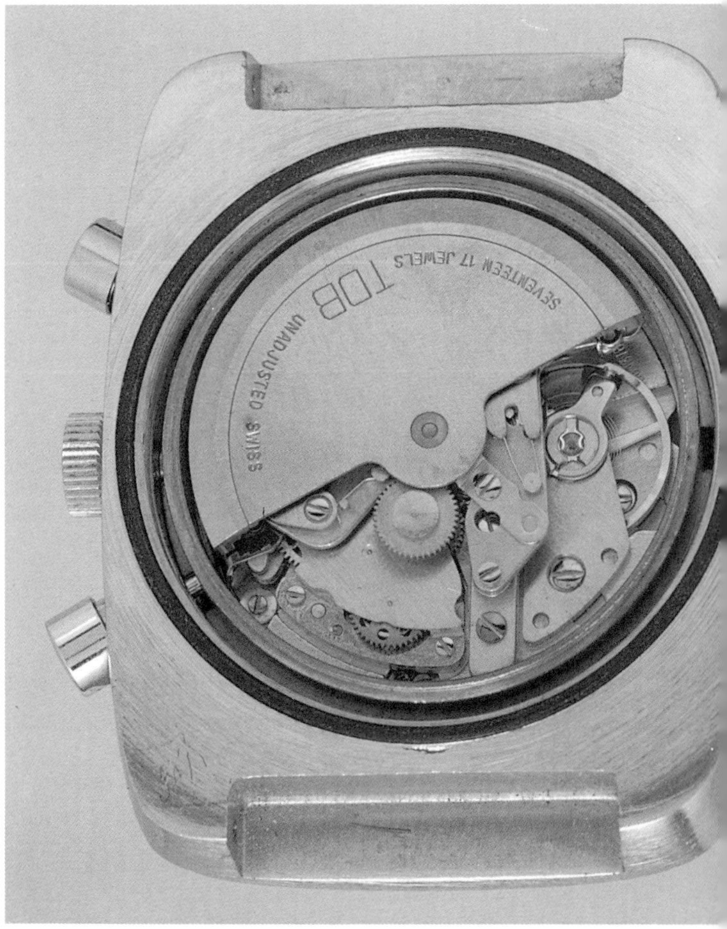

Kelek Caliber TDB 1376

Automatic chronograph made by Kelek in 1974, with TDB Caliber 1376.
Rounded, tapered square stainless steel case with screw bottom and mineral glass.
Digital indication of hours and minutes, center seconds hand, 60-minute counter on the auxiliary dial by the 11.
Rotor with heavy metal ring, winding in both directions via click-wheel switcher (at 6:30). Automatic drive integrated into the basic movement.
Chronograph module with coulisse-lever switching.
Screwless nickel balance, Nivarox hairspring, Incabloc shock resistance.

Caliber Development

Johann Kottmann founded the firm as a raw-movement factory at Langendorf in 1873. Under Karl Kottmann, complete watches were made as of 1893; the factory was now called "Langendorf Watch Co" (Lanco). In postwar years it developed into one of the largest Swiss watch producers, employing up to 1500 workers, and in 1965 it merged with the firms of Cortebert, Favre, Lemania, Omega and Tissot to found the SSIH (Sociéte Suisse pour l'Industrie Horlogère).

At the end of the 1940s the firm's first automatic watch, the "Lanco-Matic" (Caliber 1232), was introduced. It worked with a click-wheel mechanism constructed similarly to that of the later IWC rotor automatic, and ranks with the Revue Caliber 80, the BFG "Bovimatic" and the "Futurematic" among the few watches with hammer winding in two directions. The "Lanco-Rotor" of 1951 was the company's first rotor automatic. While the first series used a centrally mounted switching rocker (Caliber 1233), this was later replaced by two reciprocally acting gliding pinions (Caliber 1233 N).

In the ensuing years, one or two similarly constructed automatic movements followed until in 1967 the Caliber 1146, a modern, relatively flat design, was introduced. Here the rotor was equipped with miniature ball bearings and springless twin click wheels functioning as mechanical equalizers.

Caliber Notes

Caliber 1232
D = 28.0 mm (12.5 lignes), H = 6.1 mm, 17 jewels
AUTOMATIC MODULE screwed to the basic movement as a complete unit, hammer winding in both directions via a rocker lever with two reciprocally acting clicks located at either side of the lever pivot point, combined winding and reduction wheel on the crown wheel.
BASIC MOVEMENT: two-spoke nickel screw balance, Nivarox hairspring, 18,000 BPH, Incabloc shock resistance, direct center seconds.

Caliber List: Lanco

Caliber	Diameter Lignes	Diameter mm	Height mm	Other	BPH	Jewels	Running Reserve	Notes	Year	Photo
1033	10 1/2			SCD	18.000	17				
1035	"			SCD, QG	"	"				
1105	11 1/2			SCD	19.800					
1106	"			SCD, QG	"					
1133	11			SCD	18.000					
1145	11 1/2			SCD	19.800		45		1967	
1146	"			SCD, QG	"		"		1967	xx
1231	12				18.000					
1232	12 1/2	28,0	6,1	SCD	"	17		Lanco-Matic	ca. 49	xx
1233	12	27,2	5,4	SCD	18.000	17		Lanco-Rotor	1951	
1233 N	"	"	"	SCD, QG	"	"		"		xx
1235	"	"	"	SCD, QG	"					
1235 N	"	"	"	SCD, QG	"					
1259	12			SCD, QG	18.000					
1261	"			SCD	"					
1261 N	"			SCD	"					
1262	"			SCD, QG	"					
1262 N	"			SCD, QG	"					
1278	12			SCD	18.000					
6146	11 1/2			SCD, QG	19.800					

Lanco Caliber 1232
Men's automatic, "Lanco-Matic", made by Lanco circa 1949, with Caliber 1232. Stainless steel case with screw bottom. Hammer winding in both directions via rocker lever. Hand-winding drive uncoupled only when the winding crown is pushed in all the way (three positions). Nickel screw balance, Nivarox hairspring, Incabloc shock resistance (dial and case with signs of use).

Caliber 1233 N
D = 27.2 mm (12 lignes), H = 5.4 mm, 17 jewels
AUTOMATIC MODULE set atop the basic movement, central rotor with heavy metal ring, winding in both directions via two gliding pinions under separate potences, one of them contacting the first (with spring click) of two reduction wheels on the locking wheel directly, the other via an intermediate pinion. Hand-winding drive uncoupled by a sprung gliding pinion between the crown wheel and the pinion of the second reduction wheel.
BASIC MOVEMENT: two-spoke nickel screw balance, Nivarox hairspring, 18,000 BPH, Incabloc shock resistance, moveable hairspring block carrier, direct center seconds, date indication.

Caliber 1146
D = 11.5 lignes, 21 jewels
AUTOMATIC MODULE integrated into the basic movement under a separate potence, central rotor with ball bearings, winding in both directions via springless twin click wheels and two reduction wheels on the crown wheel.
BASIC MOVEMENT: three-spoke screwless Glucydur balance, Nivarox hairspring, 19,800 BPH, Incabloc shock resistance, moveable hairspring block carrier, direct center seconds, date indication.

Lanco Caliber 1233 N

Men's automatic made by Lanco circa 1952, with "Lanco-Rotor" Caliber 1233 N.

Stainless steel case with screw bottom. Rotor with heavy metal ring, winding in both directions via gliding-pinion switcher, uncoupling of hand-winding. Nickel screw balance, Nivarox hairspring, Incabloc shock resistance.

Lanco Caliber 1146
Men's automatic made by Lanco circa 1970, with Caliber 1146.
Gilded case with stainless steel screw bottom, winding crown with Lanco symbol.
Monometallic rotor with ball bearings, winding in both directions via click-wheel switcher.
Screwless Glucydur balance, Nivarox hairspring, Incabloc shock resistance.

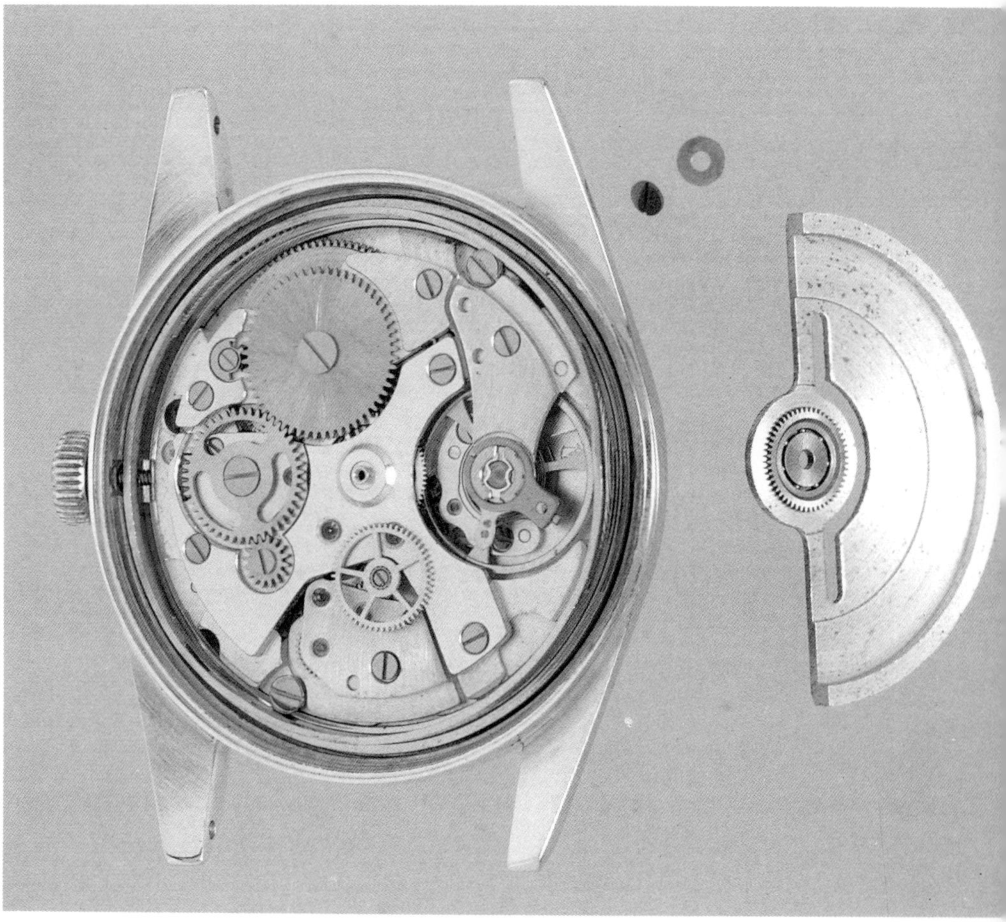

Magazine advertisement (from JSH, 1950)

Caliber Development

In 1978 Jean Lassale, director of the firm of Bouchet-Lassale S.A. in Geneva, presented what was then the flattest automatic caliber (No. 2000) in the world, with a movement height of 2.08 mm. This low height was achieved only by leaving out the formerly customary two-sided wheel bearings and using one-sided "flying" wheels with the help of miniature ball bearings. Even the barrel was kept "swimming" by three ball bearings along its rim. The resulting elimination of the two-sided mainspring boss bearings allowed the barrel's height to expand over the entire thickness of the basic movement. An additional bearing in the inside improved the friction coefficient of the mainspring boss. Only the anchor and the fast beat balance were mounted in ruby bearings under separate bridges. Because of the watch's low weight and the short distance to the plate or the balance bridge, shock resistance could be dispensed with. Another design specialty was the arrangement of almost all the wheels on the dial side of the monoblock plate.

The basic movement was also installed as an independent unit (Caliber 1200) with a height of 1.20 mm, and included nine ruby jewels and 14 ball bearings. When the complete automatic module was attached, the result was Caliber 2000 with a total of 18 ball bearings, with a very flat central rotor of platinum winding in one direction.

This first-class micromechanical product could be realized only by thoroughgoing utilization of the most modern work processes, new-type materials and the experience of the Swiss watch industry. The concentration of the Swiss watch manufacturers, under the pressure of a crisis at that time, had a beneficial effect. The firm of Tavescap S.A. of Tavannes is especially worthy of note as the producer of the miniature ball bearings, which had to be made to previously inconceivable minimal dimensions for this model.

Unfortunately, the firm of Bouchet-Lassale also fell victim to the quartz-watch offensive launched by the East. The name was bought by Seiko for its flat quartz watches, while the remaining raw movements were bought up by Piaget in 1983.

Caliber Notes

Caliber 2000
D = 20.4 mm (9 lignes), H = 2.08 mm. 9 jewels, 18 ball bearings
AUTOMATIC MODULE screwed to the basic movement as a complete unit. Platinum central rotor with ball bearings, winding in one direction via a sprung gliding pinion, three reduction wheels (the first with a locking spring), one intermediate wheel on the locking wheel (drives reduced by 1:150).
BASIC MOVEMENT (Caliber 1200, H = 1.20 mm): Screwless Glucydur balance with three spokes, bent at right angles, under a balance bridge, self-compensating flat hairspring, 28,800 BPH, no shock resistance, large wheel mounted out of the center, barrel with "swimming" mount (by three peripheral ball bearings) with 44-hour running reserve.

Caliber Development

Paul Brack founded the "Lavina S.A." factory at Villeret in 1852. The firm's only automatic development was introduced in 1957 as the "Vimatic" (Caliber 150 A). The basic movement used was the hand-wound Caliber 150 (10.5 lignes), on which a complete automatic unit with a central rotor, winding in both directions via springless click wheels, was mounted. During hand-winding, the automatic drive was uncoupled by a crown-wheel coupling to protect it.

Caliber List: Lavina

Caliber	Diameter Lignes	mm	Height mm	Other	BPH	Jewels	Running Reserve	Notes	Year	Photo
150 A	11 1/2	26,3	6,05	SCD	18.000	21	40	Vimatic	1957	

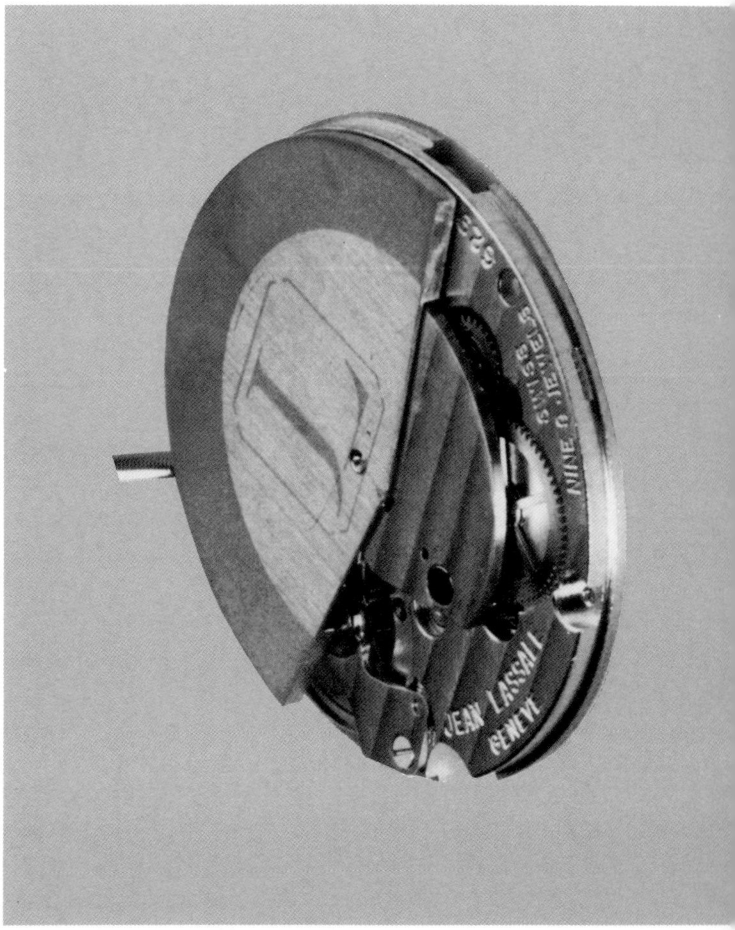

Lassale Caliber 2000

Men's model, extremely flat, made by the firm of Jean Lassale in 1978, with Caliber 2000, the flattest automatic caliber to date.

Screwed case with integrated stainless steel or 18-karat gold band.

Platinum rotor with ball bearings, winding in one direction via gliding pinion.

Lower right: the basic movement (Caliber 1200) without automatic module: screwless Glucydur balance, self-compensating flat hairspring, no shock resistance.

All wheels are mounted on only one side, or "flying", and run on a total of 18 miniature ball bearings.

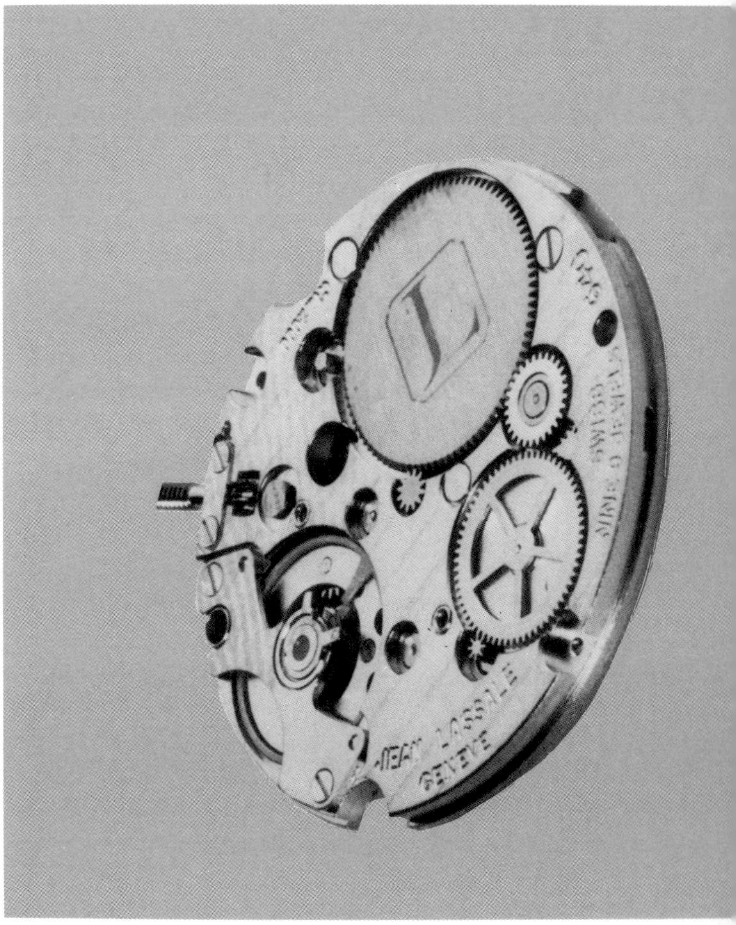

Caliber Development

Founded by Alfred Lugrin at L'Orient in 1884, Lemania merged with Omega and Tissot in 1932 to found the SSIH e(Socité Suisse pour l'Industrie Horlorère S.A.).

Marketed in 1953, the Caliber 3600 was Lemania's first automatic watch. It was a simply built but sturdy design with a rotor winding in one direction and eccentric seconds hand.

The firm's next automatic development, the Caliber 4650 of 1955, was also built according to the same functioning principles. It was a relatively small movement, though, especially designed for use in the square watch models that were in style then. Although it included a center seconds, it was flatter than its predecessor. The functioning of its central pinion, mounted under the rotor, is noteworthy since it functioned as both a winding and a locking wheel.

After the Zenith firm had introduced the first automatic chronograph in 1969, it was natural for Lemania as a renowned manufacturer of chronographs to develop one of their own—the Caliber 1341 of 1971. Although a strong mainspring was needed to power the chronograph mechanism, it was thought that a rotor winding in one direction would be sufficient. For reasons of space, the Lemania engineers had to make a special design for the reduction drive: Formed as a Breguet coupling and under the winding click wheel, the first reduction gear (made of plastic), is mounted in the form of a planetary gear.

On the basis of this design the firm also built the movement for their sailing automatic, the "Regatta", using Caliber 1345 also known as "Aquastar".

Today Lemania, using its last Caliber 5100 developed around 1980, ranks among the four Swiss constructors of automatic chronographs. The automatic module of this chronograph is constructed very much like that of the first Lemania automatic, Caliber 3600. But the chronograph mechanism is hard to get at here, located as it is between the dial and the bottom plate.

Caliber Notes

Caliber 3600
D = 27.0 mm (12 lignes), H = 6.5 mm, 21 lignes
AUTOMATIC MODULE screwed onto the basic movement under a potence, monometallic central rotor, winding in one direction via a leaf spring on the rotor and the central pinion, two reduction wheels (the first with locking spring) on the double locking wheel. Complete reciprocal uncoupling of hand- and automatic winding by the locking-wheel coupling.
BASIC MOVEMENT: two-spoke Glucydur screw balance, Nivarox hairspring, 18,000 BPH hour, Incabloc shock resistance, eccentric seconds hand.

Caliber 3610
See Caliber 3600, but with a locking click in the reduction drive.

Caliber 4650
D = 23.3 mm (10.5 lignes), H = 6.0 mm, 21 jewels
AUTOMATIC MODULE screwed to the basic movement as a complete unit, monometallic central rotor, winding in one direction with the help of a leaf spring on the rotor and the centrally mounted winding pinion, which is also contacted at the same time by the locking spring, via two reduction wheels of beryllium bronze on the locking wheel, uncoupling of the hand-winding drive by a toothed rocker between the crown and locking wheels.
BASIC MOVEMENT: two-spoke Glucydur screw balance, Nivarox hairspring, 18,000 BPH, Incabloc shock resistance.

Caliber 1341
D = 31.0 mm (13.75 lignes), H = 8.0 mm, 17 jewels
AUTOMATIC MODULE set on the basic movement beside the chronograph mechanism (with coulisse switching), central rotor with ball bearings and heavy metal ring, winding in one direction via one intermediate wheel, one click wheel with Breguet coupling and locking spring, two reduction wheels (the first as planetary wheel drive) on the crown wheel.
BASIC MOVEMENT: three-spoke screwless nickel balance, Nivarox hairspring, 28,000 BPH, Kif-Ultraflex shock resistance, moveable hairspring block carrier.

Caliber 1345
See Caliber 1341, but with simplified chronograph mechanism.

Caliber List: Lemania

Caliber	Diameter Lignes	mm	Height mm	Other	BPH	Jewels	Running Reserve	Notes	Year	Photo
1341	13 3/4	31,0	8,0	s, QG, RDR, CHR	28.800	17			1971	xx
1345	"	"	"	SCI, CHR	"	"		Aquastar	1971	xx
2980	13 3/4	30,8	6,8	SCD, QG, RDR, REV	21.600	19		s. Omega „Memomatic“	ca. 72	(xx)
3600	12	27,0	6,5	s	18.000	21			1953	xx
3610	"	"	"	s	"	"			1954	xx
4650	10 1/2	23,3	6,0	SCI	18.000	21			1955	xx
4651	"	"	"	SCI	"					
5082		31,0	8,2	s, QG, JG, RDR, CHR	21.600		45			
5100		"	"	s, QG, JG, RDR, CHR	28.800	17	50		ca. 80	

***Lemania* Caliber 3600**

Men's automatic made by Lemania in 1953, with Caliber 3600.
18-karat gold case with press-in bottom. Brass rotor winding in one direction via Harwood-type drive. Winding click in form of a leaf spring on the rotor, small drive potence visible (at 12:30).
Glucydur screw balance, Nivarox hairspring, Incabloc shock resistance (winding crown not original).

Lemania Caliber 3610

Men's automatic made by Lemania in 1954, with Caliber 3610.
Gilded case with stainless steel screw bottom, winding crown with Lemania emblem.
Movement identical to Caliber 3600 (see page 182) except that the locking spring has been replaced by a locking click (at 1:00).

Lemania Caliber 4650

Men's automatic made by Lemania in 1955, with Caliber 4650.

Square 18-karat gold case, gold hands and dial inlays.

Brass rotor winding in one direction via Harwood-type drive, winding click in form of a leaf spring on the rotor, locking spring (at 3:00) working parallel to it on the moveable central pinion.

Glucydur screw balance, Nivarox hairspring, Incabloc shock resistance (dial flecked, case angles rounded by polishing, second hand and winding crown not original).

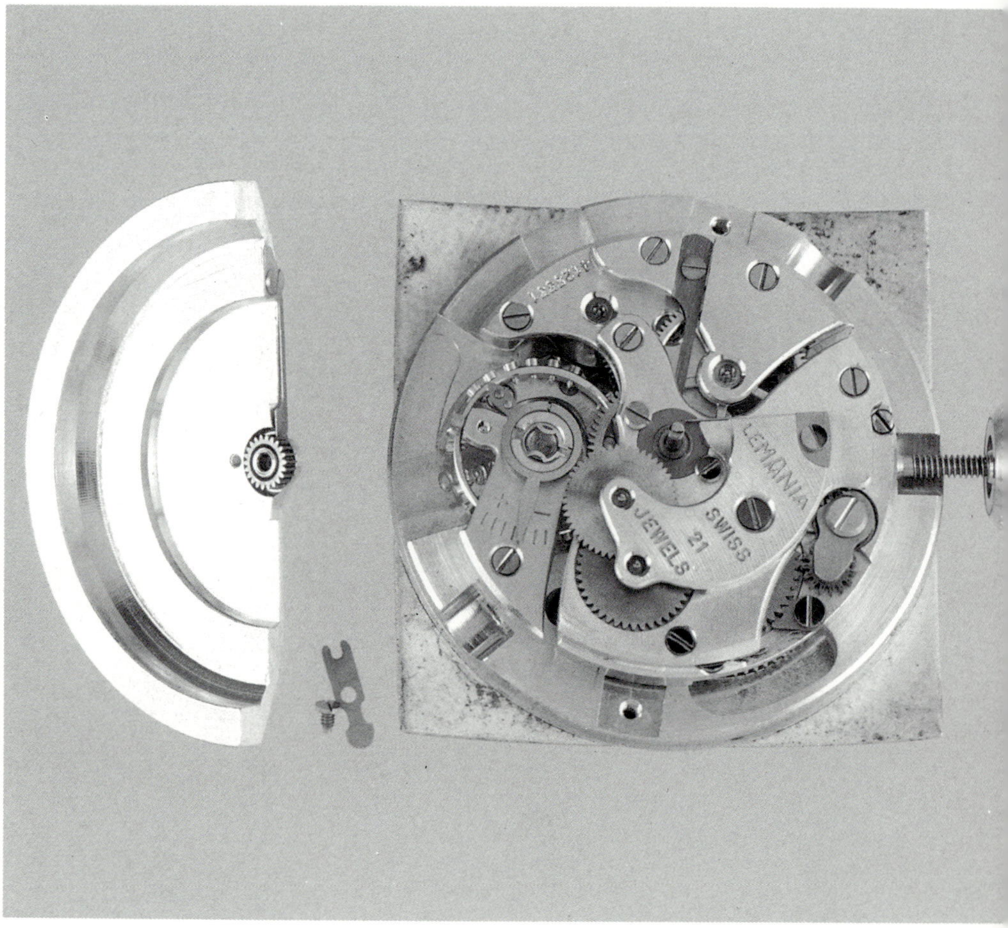

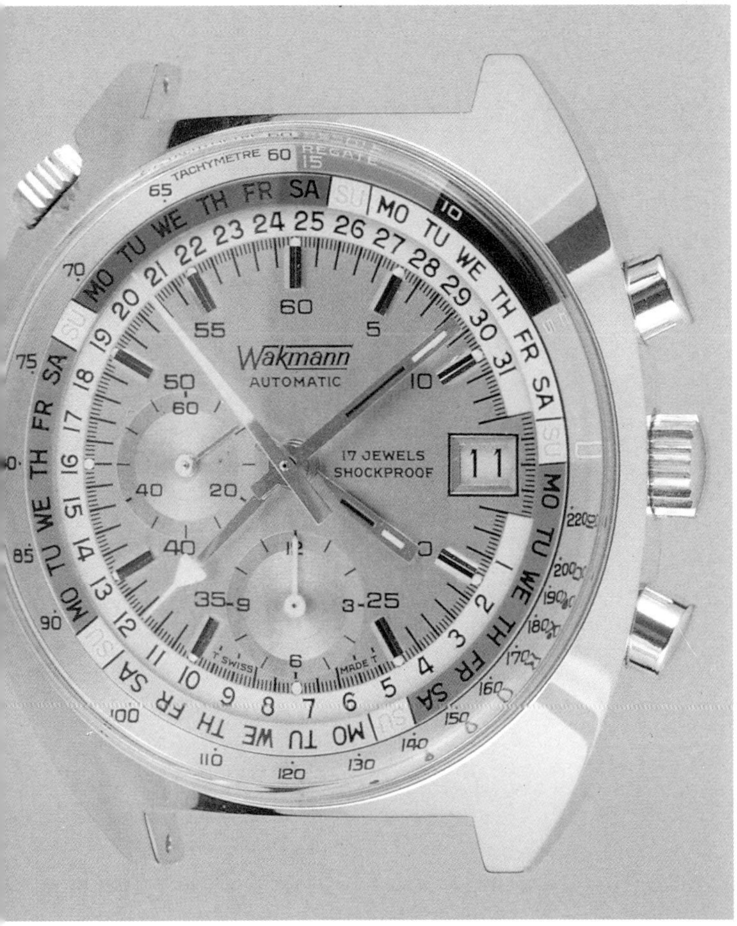

Lemania Caliber 1341
Automatic chronograph of the "Wakmann" brand, with Lemania Caliber 1341, made in 1971.
Gilded case with press-in bottom.
Crown by the 11 to set date and weekday. Rotor with ball bearings and heavy metal ring, winding in one direction via Breguet coupling. Automatic drive partly integrated into basic movement (at 11:00).
Chronograph movement with coulisse-lever switching.
Screwless nickel balance, Nivarox hairspring, Kif-Ultraflex shock resistance.

Lemania Caliber 1345

"Regatta" sailing automatic made by Lemania in 1971, with Caliber 1345 ("Aquastar"). Barrel-shaped stainless steel case with screw bottom.

15 minutes before a regatta starts, the chronograph mechanism, which counts backward from 15 to 0, is released. After every five minutes, the color in all five dial apertures changes from blue to red or red to gray, giving warning about the time when the sailboats are to cross the start line.

Automatic drive and movement identical to Caliber 1341 (see page 185), but with simplified chronograph mechanism.

Magazine advertisement (from JSH, 1955)

Caliber Development

The Longines company was founded by Ernest Francillon in 1867 at St. Imier in the Swiss Jura. He built his first factory on a long piece of land called "Les Longines".

For its first automatic wristwatch, Longines developed an eccentric system that utilized both of the rotor's directions for winding. In this Caliber 22 A, introduced in 1945, the customary spring clicks on the winding wheel are replaced by a springless toothed locking wheel. To protect the thin rotor axle against shocks, the peripheral heavy-metal ring is suspended elastically in the rotor disc.

In its successor, Caliber 19 A of 1952, this eccentric winding is found again. Its winding capability had been improved so that the smaller and lighter rotor could work without shock resistance.

Six years later, Longines brought out a new development, Caliber 290, the construction of which showed similarities to AS movements. Correspondingly, a completely different switching principle was found in this caliber. It had twin click wheels with springless locking anchors. The basic movement was built with a direct second and an indirectly driven minute hand to save space. In addition, the entire surface under the flat central rotor was available for the basic movement, so that a disadvantage of the earlier Longines automatic watches was eliminated, namely the relatively small basic movement. To comply with buyers' interests at that time, this caliber was also produced in small numbers with up and down indicators, numbered either 292 or 294.

Around 1960 a raw movement was purchased from the Cyma firm. With its very interesting eccentric winding, it could be regarded as a further development of the first Longines automatic. It was installed in the "Flagship" model and given caliber number 380.

In 1962 Longines was able to introduce its own flat automatic movement, Caliber 340, with a large rotor mounted slightly off-center and coupled to the drive by an outer ring geared on the inside. This was integrated into the basic movement and functioned with a new, patented satellite-wheel lock as a switcher, its function as a kind of double click wheel based on the firm's older patent on the toothed locking wheel.

Presumably the production costs of this very elaborately designed caliber were no longer competitive and the firm was forced to bring out a considerably simpler movement (Caliber 501) in 1965. Its type of design is also reminiscent of the typical AS designs, i.e., the switching wheels equipped with double clicks, and it is quite possible that this was a raw movement made by AS and reserved for Longines.

With the Caliber 430 "Ultrachron", the firm introduced a so-called "fast beat" escapement at a rate of 36,000 BPH in 1967, following the "HF Chronometer" of Girard Perregaux. Both types were fitted with the "Clinergic 21" anchor escapement developed by the Fabriques d'Assortiments Réunies, which provided a mean daily error of three seconds without any major adjustment.

Since the long-term reliability of these fast beat escapements caused even greater problems, they soon gained a reputation as poor sellers. Until Longines could produce its own improved second-generation high-frequency movement in 1972, Caliber 6641 with only 28,800 BPH, in 1972, the firm found itself compelled to utilize modified Eta movements (Caliber 633.1).

Because of their greater need for energy, basically all fast beat escapements require a stronger and thus thicker mainspring. This means a smaller running reserve if the barrel dimensions remain the same. The Longines engineers solved this problem by installing a second barrel in Caliber 890, which was ready to go into production in 1975. The total number of turns to full winding of the two barrels, set up to work one after the other, now rose from 6 to 17 which increased the running reserve from 36 hours (as in Caliber 6641) to 48 hours. This driving principle, tested by Alfred Hellwig in building chronometers, also offers additional advantages in an automatic watch. For along with its improved isochronism (evenness) of running thanks to the more evenly released driving energy, the stronger turning moment caused by the increased barrel speed, which led to premature wear of the drive could be reduced. The result was regained driving power with a lighter rotor and a smaller reduction ratio of the automatic drive.

Though this movement design, with its barrels set concentrically one atop the other and its rotor winding functioning in both directions, is still relatively high, the height of the subsequent Caliber 980, intróduced two years later, had been reduced to 2.95 mm by placing the two barrels side by side and simplifying the automatic drive to wind in one direction. This movement, with center seconds and date indication, ranked for almost a decade as not only the flattest automatic caliber of this type but also can be regarded as one of the most modern in terms of design.

Caliber Notes

Caliber 22 AS
D = 28.8 mm (13 lignes), H = 6.5 mm
AUTOMATIC MODULE set atop the basic module, central rotor with peripheral spring-mounted heavy metal ring, winding in both directions via a rotor eccentric which moves the sprung click carriers over a roller, combined winding and reduction wheel on the locking wheel. Winding and locking "clicks" in the form of small moveable toothed discs against fixed toothed segments.
BASIC MOVEMENT: two-spoke Glucydur spring balance, Nivarox hairspring, 18,000 BPH, Shock Resist system.

Caliber 19 ASD
D = 25.3 mm (11.25 lignes), H = 6.95 mm, 19 jewels
AUTOMATIC MODULE screwed to the basic movement, central rotor with heavy metal ring, winding in both directions via a rotor eccentric that moves the sprung click-carrier rocker over a roller, combined winding and reduction wheel on the locking wheel.
BASIC MOVEMENT: two-spoke Glucydur screw balance, Nivarox hairspring, 18,000 BPH, Incabloc shock resistance, date indication.

Caliber 14.17
D = 14.2 x 17.3 mm (6.25 x 7.5 lignes), H = 5.25 mm, 17 jewels
AUTOMATIC MODULE screwed to the basic movement as an independent unit, central rotor with heavy metal ring, winding in both directions via springless twin click wheels with satellite-wheel lock, one reduction wheel, one intermediate wheel on the locking wheel.
BASIC MOVEMENT: two-spoke Glucydur screw balance, Nivarox hairspring, 19,800 BPH, Incabloc shock resistance, no second hand, hand-setting and winding by rocker mechanism.

Caliber 291
D = 25.6 mm (11.5 lignes), H = 6.0 mm, 24 jewels
AUTOMATIC MODULE screwed to the basic movement, heavy metal central rotor with ruby bearings, winding in both directions via springless twin click wheels, one reduction wheel on the locking wheel.
BASIC MOVEMENT: two-spoke Glucydur screw balance, Nivarox hairspring, 19,800 BPH, Incabloc shock resistance, direct center seconds, date indication with instant change.

Caliber 294 See Caliber 291, but with additional running reserve indication by relative movement of a central disc (unwinding) and a concentric dial ring (winding) without differential drive.

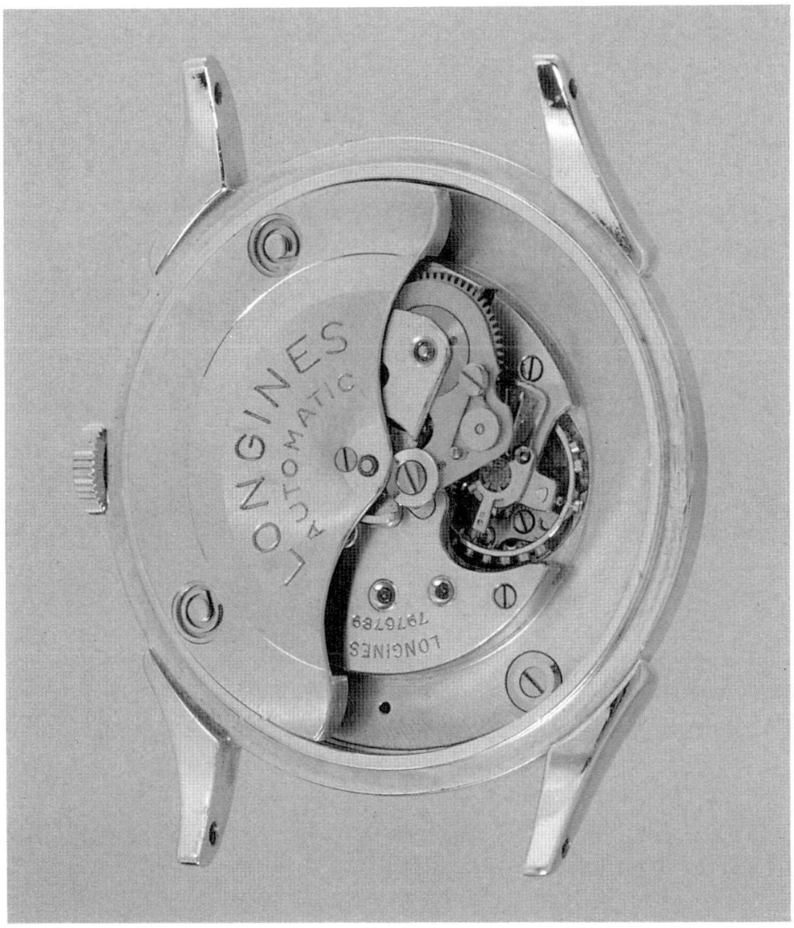

Longines Caliber 22 AS

Men's model made by Longines circa 1948, with Caliber 22 AS, the firm's first automatic caliber, made since 1945. Gilded (80 u) case with press-in bottom. Rotor with spring-suspended heavy metal ring, winding in both directions via patented eccentric switcher (at 2:00). Glucydur screw balance, Nivarox hair-spring, Shock-Resist system (winding crown not original).

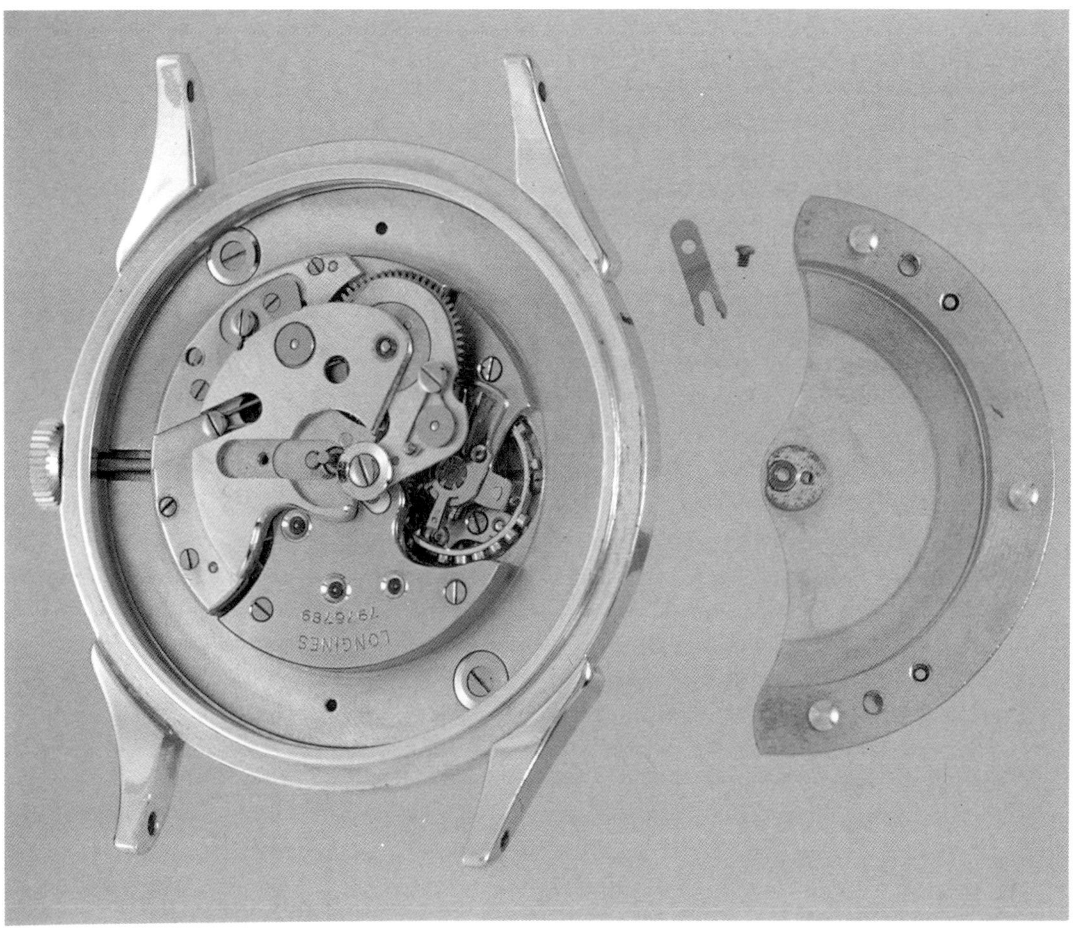

Longines Caliber 19 ASD

Men's automatic made by Longines in 1956 as "Conquest Calendar" model, with Caliber 19 ASD.

Eighteen karat gold case with screw bottom (with worked-in blue enamel plate, see page 187), gold hands, dial inlays and winding crown with Longines monogram.

Rotor with heavy metal ring, winding in both directions via eccentric switcher (at 11:00).

Glucydur screw balance, Nivarox hairspring, Incabloc shock resistance (dial with fine flecks marring color pigment).

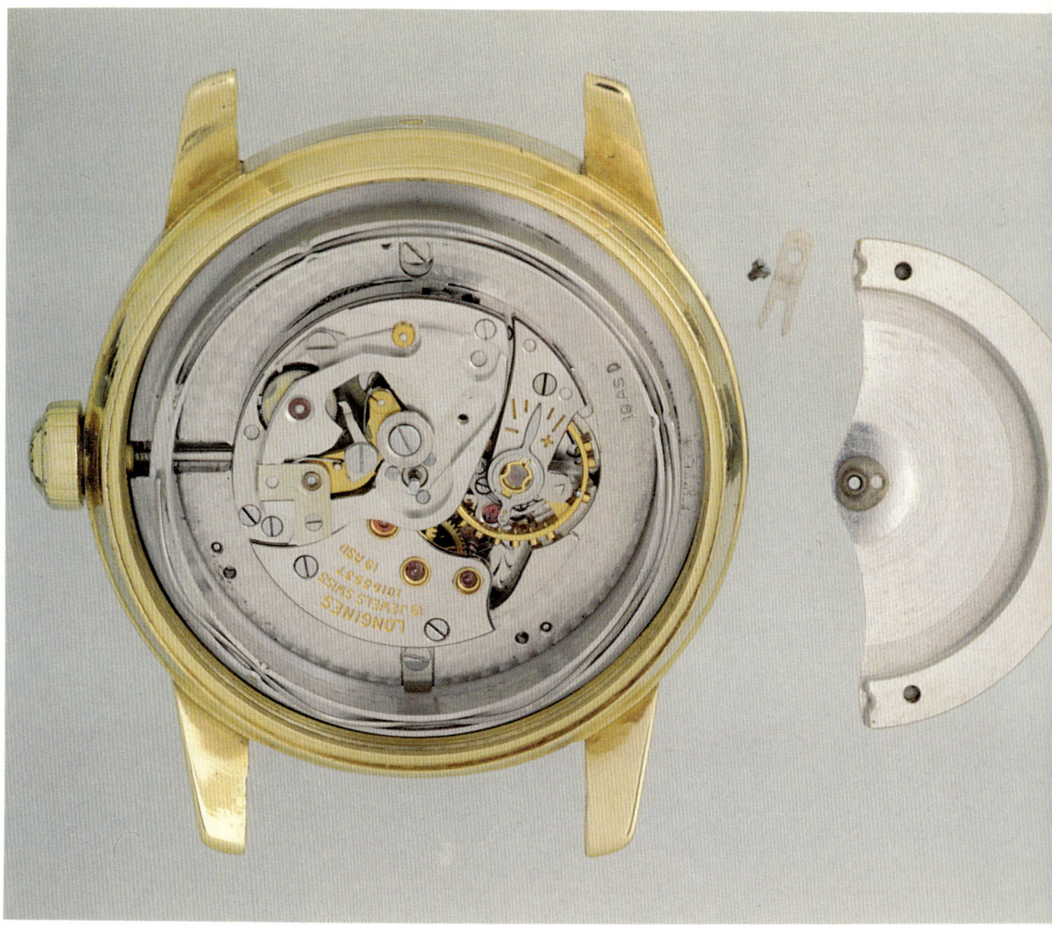

Longines Caliber 14.17
Ladies' automatic made by Longines in 1956, with formed Caliber 14.17.
Stainless steel case with press-in bottom, winding crown with Longines monogram.
Rotor with heavy metal ring, winding in both directions via click-wheel switcher (at 9 and 12:00).
Glucydur screw balance, Nivarox hairspring, Incabloc shock resistance (dial rim with traces of pressure).

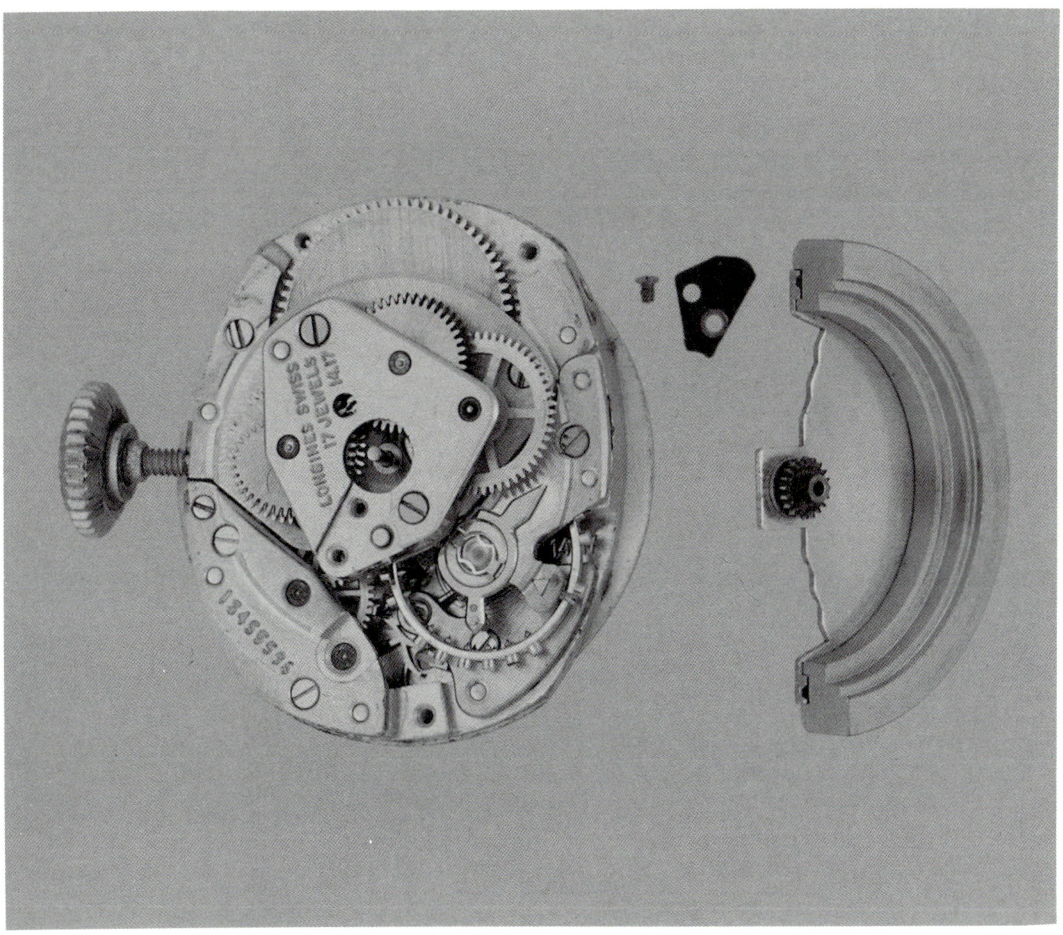

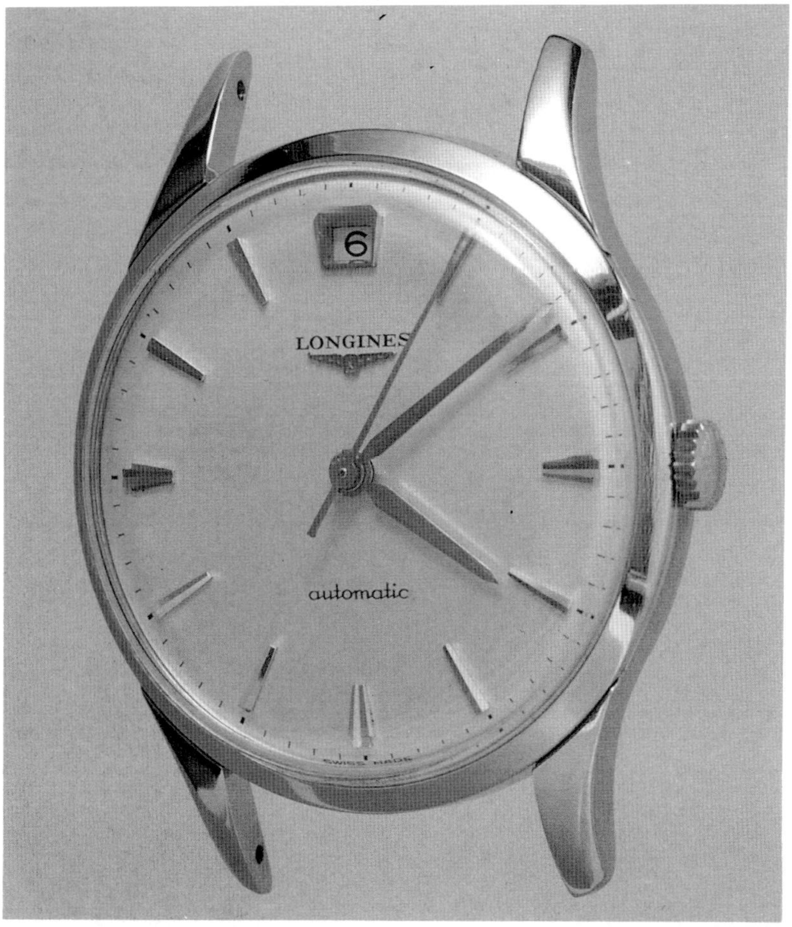

Longines Caliber 291

Men's automatic made by Longines in 1958, with Caliber 291. Stainless steel case with press-in bottom, winding crown with Longines monogram. Rotor with heavy metal ring and fast unbolting, winding in both directions via click-wheel switcher (at 5:00). Glucydur screw balance, Nivarox hairspring, Incabloc shock resistance (replaced glass without date loupe).

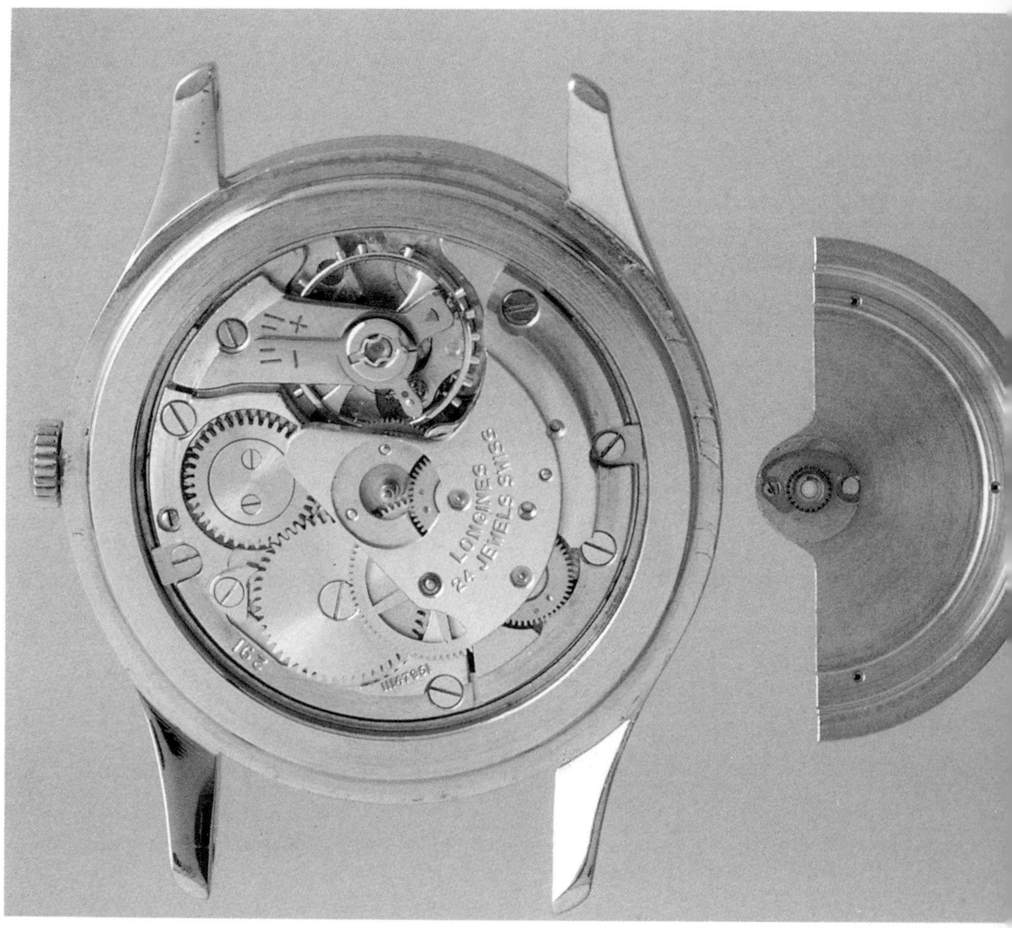

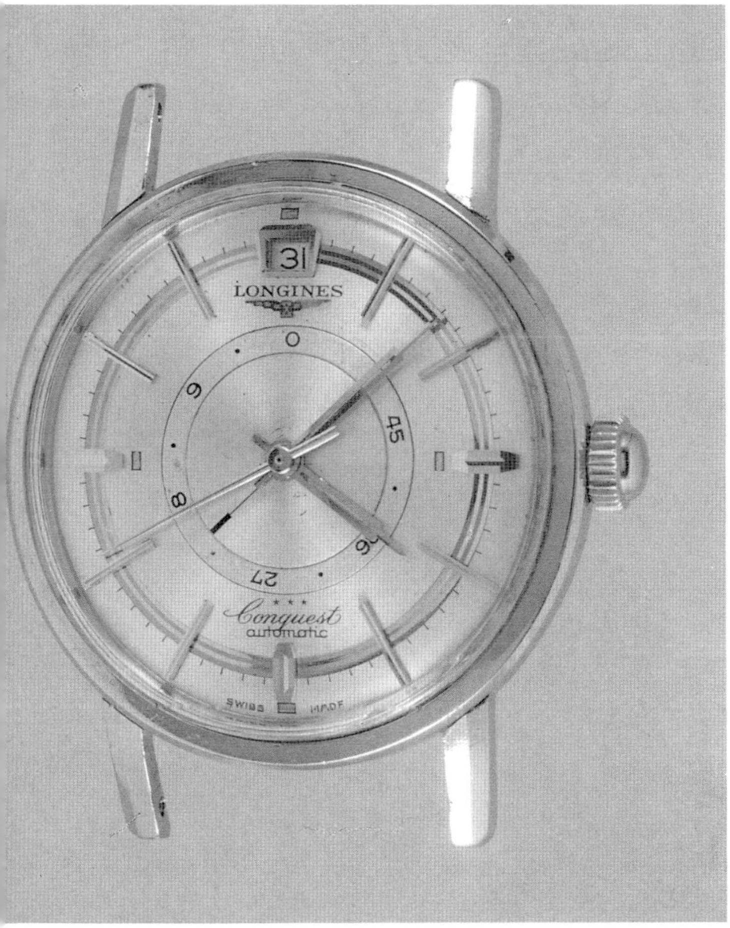

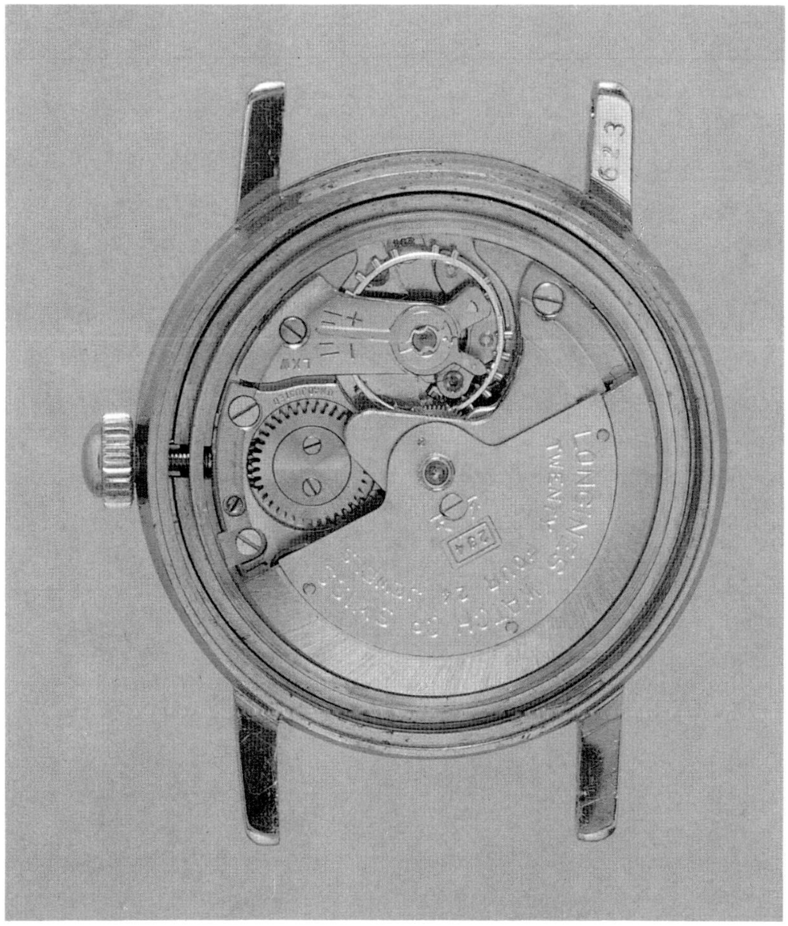

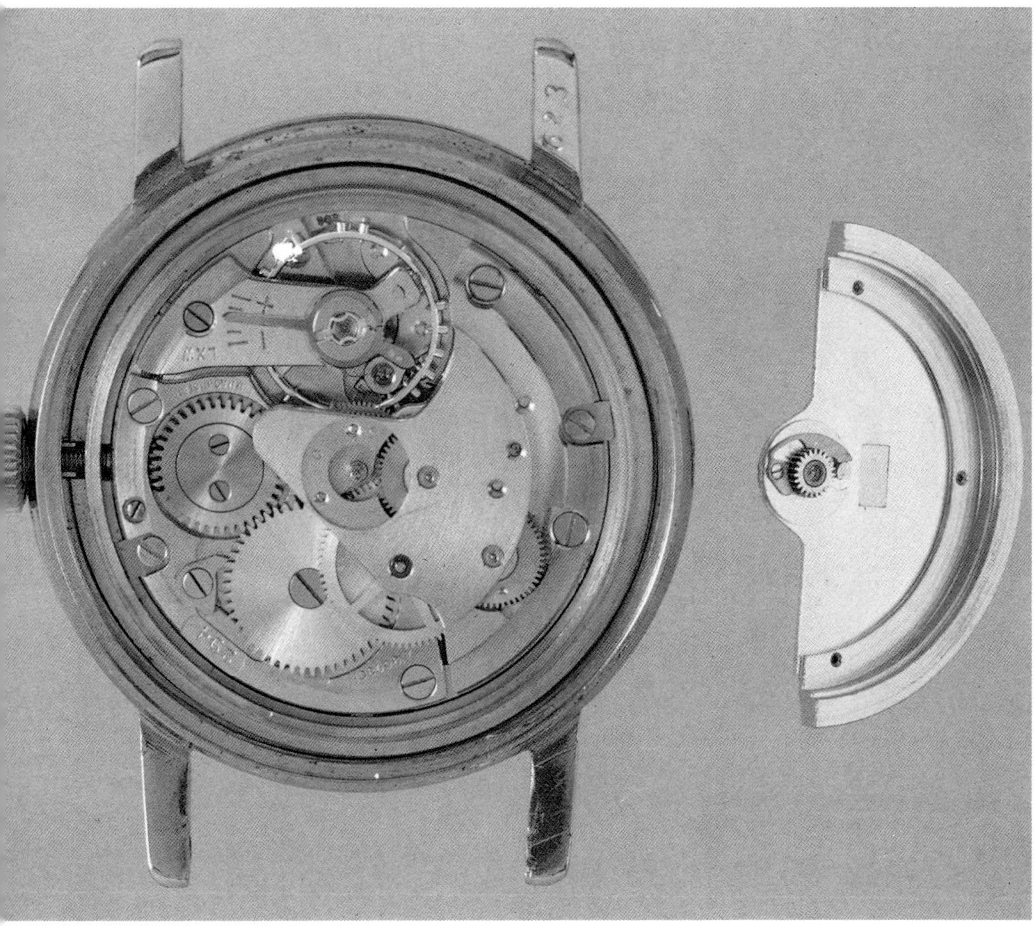

Longines Caliber 294
Men's automatic, "Conquest" model, made by Longines in 1958, with Caliber 294.
Stainless steel case with screw bottom (with worked-in green enamel plate, see page 187).
Up and down indication by central disc and rotating scale ring.
Movement identical to Caliber 291 (see page 192)); (winding crown not original, case with light signs of wear, replaced glass without date loupe).

Longines Caliber 380

Men's automatic, "Flagship" model, made by Longines circa 1960 (Caliber 380) with Cyma raw movement (compare with page 88).

Stainless steel case with gold bezel and hood, stainless steel screw bottom with engraved sailing ship, winding crown with Longines monogram, gold hands and dial inlays.

Rotor with heavy metal ring and shock resistance (by sprung arbor and safety ring on movement), winding in both directions via eccentric switcher (visible in center), uncoupling of the hand-winding drive.

Glucydur screw balance, Nivarox hairspring, Kif-Flector shock resistance.

Longines Caliber 342

Men's automatic made by Longines circa 1965, with Caliber 342.

Gilded case with stainless steel press-in bottom, winding crown with Longines monogram.

Heavy metal rotor, mounted on ball bearings somewhat out of the center, with toothed outer ring, winding in both directions via patented satellite-wheel switcher.

Screwless Glucydur balance, Nivarox hairspring, Kif-Ultraflex shock resistance, fine regulator adjustment by eccentric screw.

Longines Caliber 505
Men's automatic, "Admiral" model, made by Longines circa 1965, with Caliber 505.
Eighteen karat gold case with screw bottom and emblem, winding crown with Longines monogram, gold hands and dial inlays.
Rotor with heavy metal ring, winding in both directions via click-wheel switcher (at 5 and 7:00).
Screwless Glucydur balance, Nivarox hairspring, Kif-Ultraflex shock resistance (case with slight signs of wear, worn winding crown).

Longines Caliber 431

Men's automatic, "Admiral" model, made by Longines in 1967, with fast-swinging Caliber 431.
Stainless steel case with screw bottom, winding crown with Longines emblem.
Rotor with ball bearings and heavy metal ring, winding in both directions via patented satellite-wheel switcher (at 2:00), uncoupling of the hand-winding drive.
Automatic drive integrated into the basic movement under a separate bridge.
Screwless Glucydur balance, self-compensating flat hairspring, Kif-Ultraflex shock resistance, fine regulator adjustment by eccentric screw.

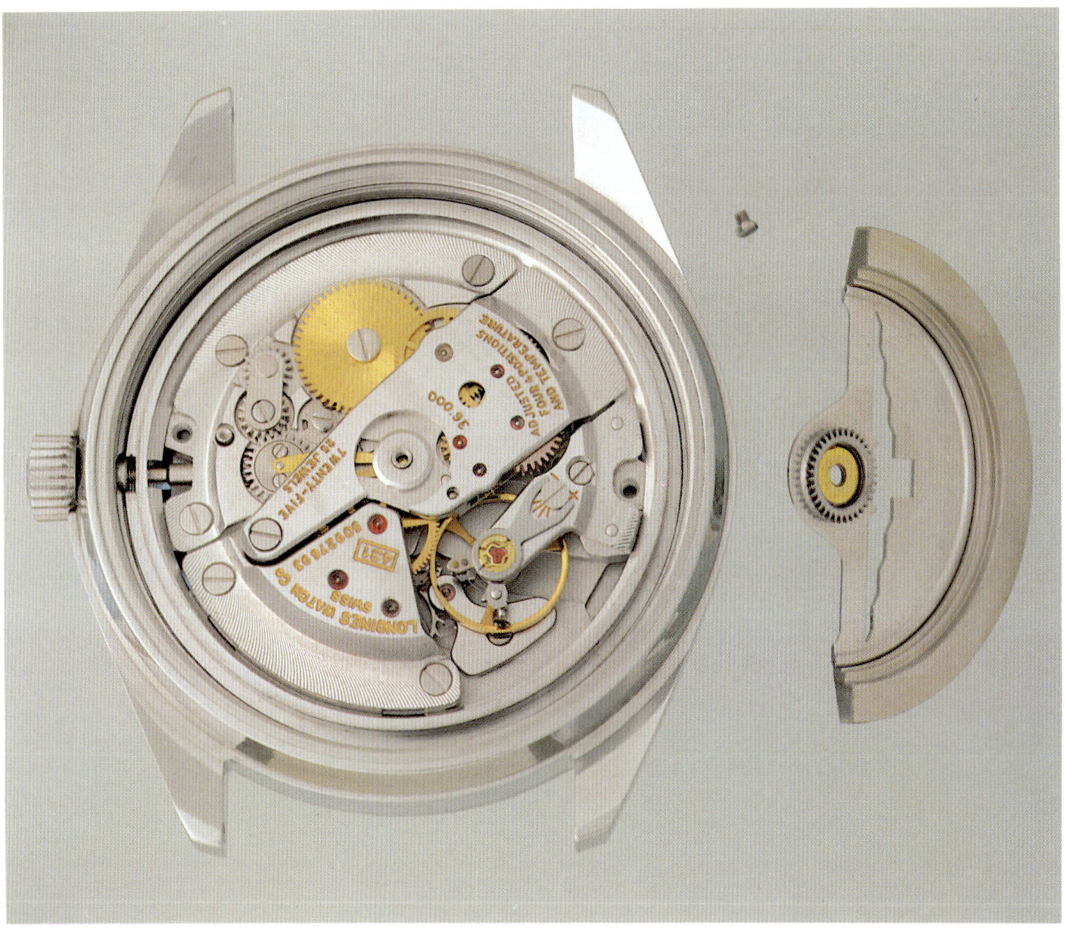

Longines Caliber 633.1

Men's automatic made by Longines circa 1972 (Caliber 633.1), with fast-swinging Eta raw movement.

Gilded case with stainless steel screw bottom, winding crown with Longines emblem.

Rotor with ball bearings and heavy metal ring, winding in both directions via click-wheel switcher (partly concealed at 2:00), uncoupling of the hand-winding drive. Screwless Glucydur balance, Incabloc shock resistance, Triovia fine regulation.

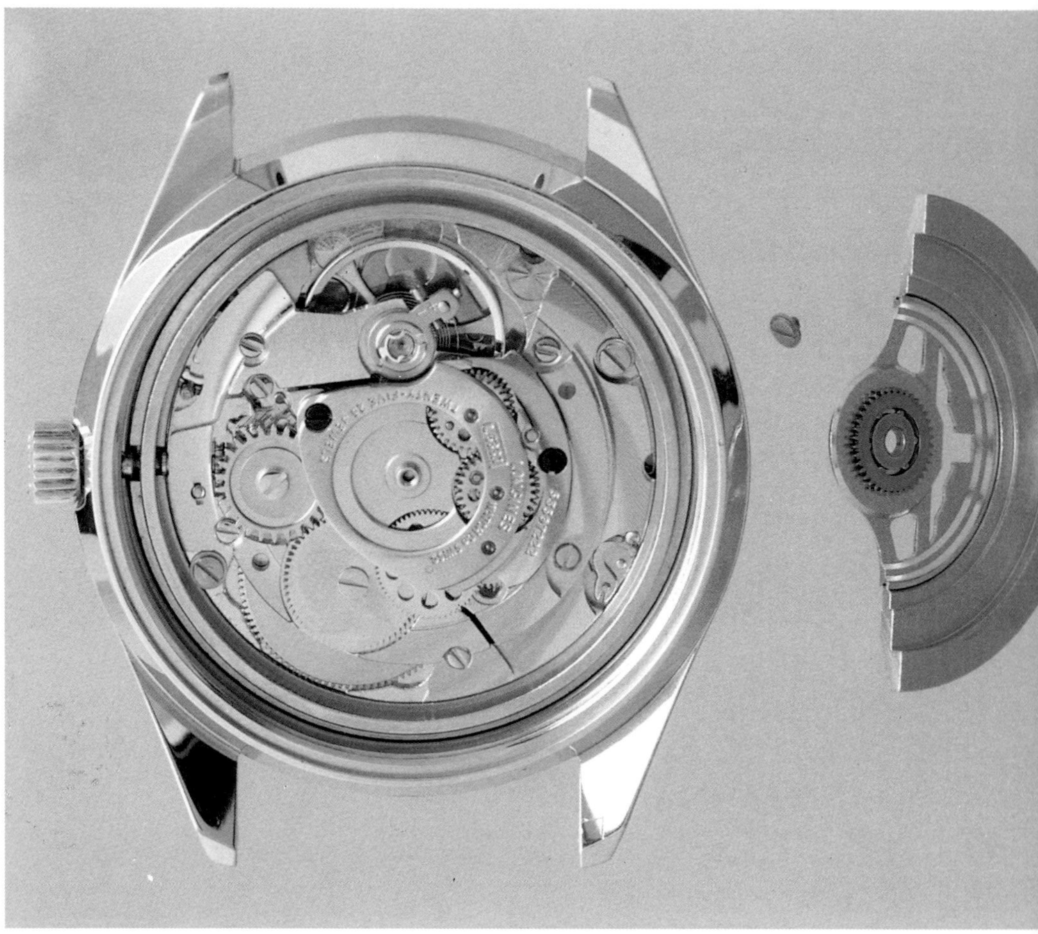

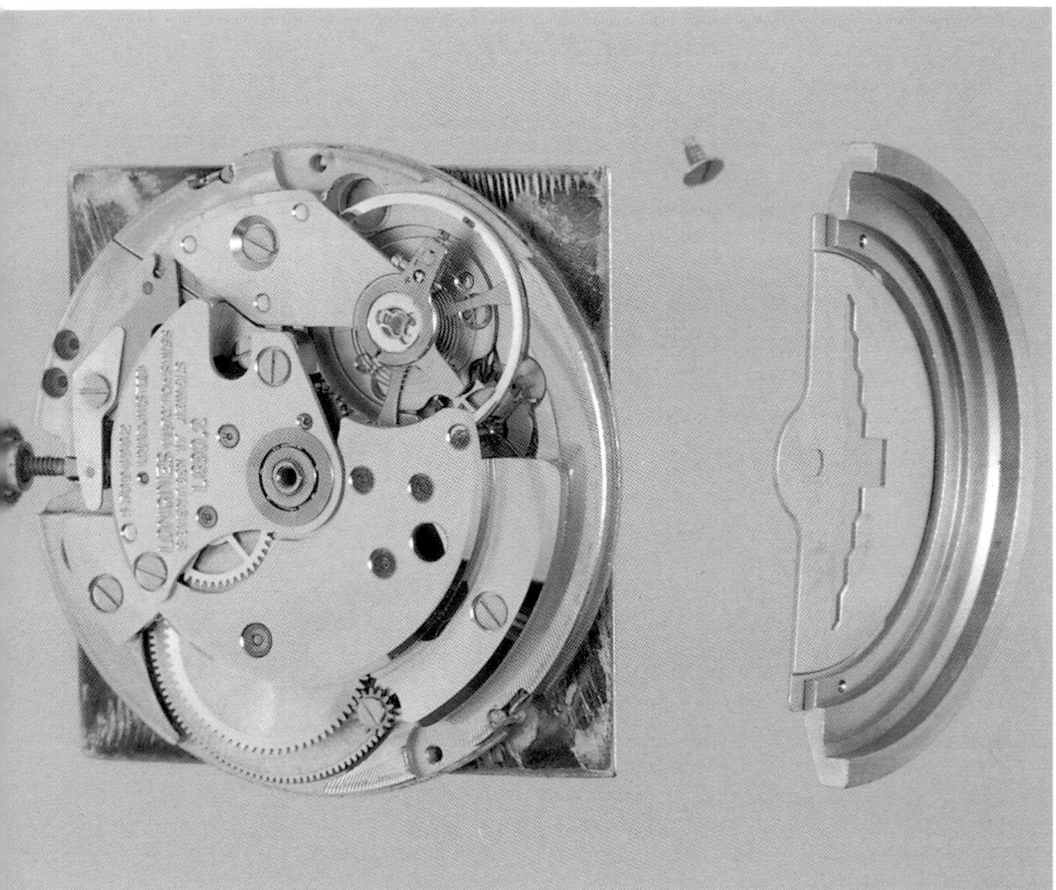

Longines Caliber 890.2

Men's automatic, "Admiral" model, made by Longines in 1975, with two-barrel Caliber 890.2.

Square 14-karat gold case, gold hands and dial inlays.

Rotor with ball bearings and heavy metal ring, winding in both directions via click-wheel switcher, uncoupling of the hand-winding drive.

Screwless Glucydur balance, Nivarox hairspring, Kif-Ultraflex shock resistance, Spirofin fine adjustment, two barrels working one after the other but located one over the other (at 7:00); (dial rim shows signs of pressure, case angles have been rounded by polishing).

Longines Caliber 980.1

Men's flat automatic made by Longines in 1978, with fast beat escapement two-barrel Caliber 980.1.

Gilded case with stainless steel screw bottom, winding crown with Longines monogram and emblem.

Heavy metal rotor with ball bearings, winding in one direction via click wheel, with barrels switched one after the other (at 5 and 7:00).

Screwless Glucydur balance, self-compensating flat hairspring, Kif-Ultraflex shock resistance, Spirofin fine adjustment.

Caliber List: Longines

Caliber	Diameter Lignes	mm	Height mm	Other	BPH	Jewels	Running Reserve	Notes	Year	Photo
14.17	6 1/4 x	14,2 x	5,25		19.800	17			1956	xx
	7 1/2	17,3								
19 A	11 1/4	25,3	5,05	s	18.000		36		1952	
19 AD	"	"		QG	"		"			
19 AS	"	"	6,0	SCI	"		"	Conquest	1954	
19 ASD	"	"	6,95	SCI, QG	"		"	Conquest Calendar	1956	xx
22 A	13	28,8	5,4	s	18.000		36		1945	
22 AS (AN)	"	"	6,5	SCI	"		"			xx
290	11 1/2	25,6	6,0	SCD	19.800	24	45		1958	
291	"	"	"	SCD, QG	"	"	"			xx
292	"	"	"	SCD, QG, RM	"	"	"		ca. 58	
294	"	"	"	SCD, QG, RMA	"	"	"			xx
340	12	27,0	4,45	SCD	19.800	17			1962	
341	"	"	4,85	SCD, QG	"	"				
342	"	"	4,6	SCD	"	"				xx
343	"	"	5,1	SCD, QG	"	"				
345	"	"	"	SCD, QG	"	"				
350	"	"	4,45	o./m.s	"	"				
351	"	"	4,85	o./m.s, QG	"	"				
352	"	"	4,6	o./m.s	"	"				
353	"	"	5,1	o./m.s, QG	"	"				
355	"	"	"	QG	"	"				
380	12 1/2	28,0	5,2	SCD	18.000	27	40	Cyma-Ebauche 480	ca. 60	xx
								Flagship		
430	11 1/2	25,6	4,3	SCI	36.000	17	36		1967	
431	"	"	4,8	SCI, QG	"	"	"	Admiral		xx
432	"	"			"	"	"			
433	"	"		QG	"	"	"			
470	6 3/4				19.800			Certina-Ebauche		
472	"				21.600			Certina-Ebauche		
501	11 1/2			SCD, QG	19.800	17				
503	"			SCD, QG, JG	"	"				
505	"			SCD, QG, RDR	"	"			ca. 65	xx
507	"			SCD, QG, RDR, ST	"	"				
L 561	7 3/4	17,2	4,8	SCD, QG, RDR, ST	28.800	17/25				
L 633.1	11 1/2		4,6	SCD, QG, RDR, ST	28.800	25		Eta-Ebauche	ca. 72	xx
6641	11 1/2	25,6	4,3	SCI	28.800	25		Basic Caliber 430		
6642	"	"	"		"	17				
6651	"	"	4,8	SCI, QG	"	25				
6652	"	"	"	SCI, QG	"	17				

Caliber List: Longines

Caliber	Diameter Lignes	mm	Height mm	Other	BPH	Jewels	Running Reserve	Notes	Year	Photo
L 890.2	12	27,0	5,2	SCI, QG, RDR, ST	28.800	17	46	2 barrels	1975	xx
L 892.1	"	"		SCI, ST	"	25	"	"	1975	
L 892.2	"	"		SCI, ST	"	17	"	"		
L 893.1	"	"	5,2		"	25	"	"		
L 893.2	"	"	"		"	17	"	"		
L 990.1	11 1/2	25,6	2,95	SCI, QG, RDR, ST	28.800	25	44	2 barrels	1977	xx
L 990.2	"	"	"	SCI, QG, RDR, ST	"	17	"	"		
L 992.1	"	"	"	SCI, ST	"	25	"	"		
L 992.2	"	"	"	SCI, ST	"	17	"	"		
L 993.1	"	"	"		"	25	"	"		
L 993.2	"	"	"		"	17	"	"		
L 994.1	"	"	"	QG, RDR	"	25	"	"		
L 994.2	"	"	"	QG, RDR	"	17	"	"		

Caliber 380
Cyma Ebauche, D = 28.0 mm (12.5 lignes), H = 5.2 mm, 27 jewels
AUTOMATIC MODULE screwed to basic movement as a complete unit, central rotor with ruby bearings, heavy metal ring and special shock resistance (arbor springs and tipping limited by movement ring), winding in both directions by rotor eccentric via two reciprocally operating clock wheels with ruby-bearing rollers on a centrally mounted winding wheel, one reduction wheel on the locking wheel, uncoupling of hand-winding drive by a toothed-wheel rocker between the crown and reduction wheels.
BASIC MOVEMENT: two-spoke Glucydur screw balance, Nivarox hairspring, 18,000 BPH, Kif-Flector shock resistance, direct center seconds.

Caliber 342
D = 27 mm (12 lignes), H = 4.45 mm, 17 jewels
AUTOMATIC MODULE integrated into the basic movement, large rotor with ball bearings and heavy metal ring mounted out of the center, shock resistance on the arbor, winding in both directions via ring attached to periphery of rotor and inside gearing, double satellite-wheel lock, two reduction wheels on the locking wheel.
BASIC MOVEMENT: three-spoke screwless Glucydur balance, Nivarox hairspring, 19,800 BPH, Kif-Ultraflex shock resistance, moveable hairspring block carrier, fine adjustment by eccentric screw, direct center seconds via double intermediate wheel, hand-winding and setting by rocker mechanism.

Caliber 505
D = 25.6 mm (11.5 lignes), 17 jewels
AUTOMATIC MODULE partly integrated into the basic movement, central rotor with heavy metal ring, winding in both directions via springless twin click wheels, one reduction wheel on the crown wheel.
BASIC MOVEMENT: three-spoke screwless Glucydur balance, Nivarox hairspring, 19,800 BPH, Kif-Ultraflex shock resistance, moveable hairspring block carrier, direct center seconds, eloxied GL barrel, date indication with quick change.

Caliber 431
D = 25.6 mm (11.5 lignes), H = 4.8 mm, 25 jewels
AUTOMATIC MODULE integrated into the basic movement as a complete unit, central rotor with ball bearings and heavy metal ring, winding in both directions via double satellite-wheel lock, two reduction wheels on the locking wheel, uncoupling of hand-winding mechanism by toothed-wheel rocker between the crown and locking wheels.
BASIC MOVEMENT: three-spoke screwless Glucydur balance, Nivarox hairspring, 36,000 BPH, Kif-Ultraflex shock resistance, moveable hairspring block carrier, fine adjustment by eccentric screw, Clinergic 21 anchor escapement, date indication with quick change.

Caliber 633.1
Eta Ebauche, D = 25.6 mm (11.5 lignes), H = 4.6 mm, 25 jewels
AUTOMATIC MODULE screwed to the basic movement as a complete unit, central rotor with ball bearings and heavy metal ring, winding in both directions via springless twin click wheels, two reduction wheels on the locking wheel, uncoupling of hand-winding mechanism by moveable crown wheel.
BASIC MOVEMENT: three-spoke screwless Glucydur balance, Nivarox hairspring, 28,800 BPH, Incabloc shock resistance, Triovis fine regulation, direct center seconds, second stop, date indication with quick change and correction.

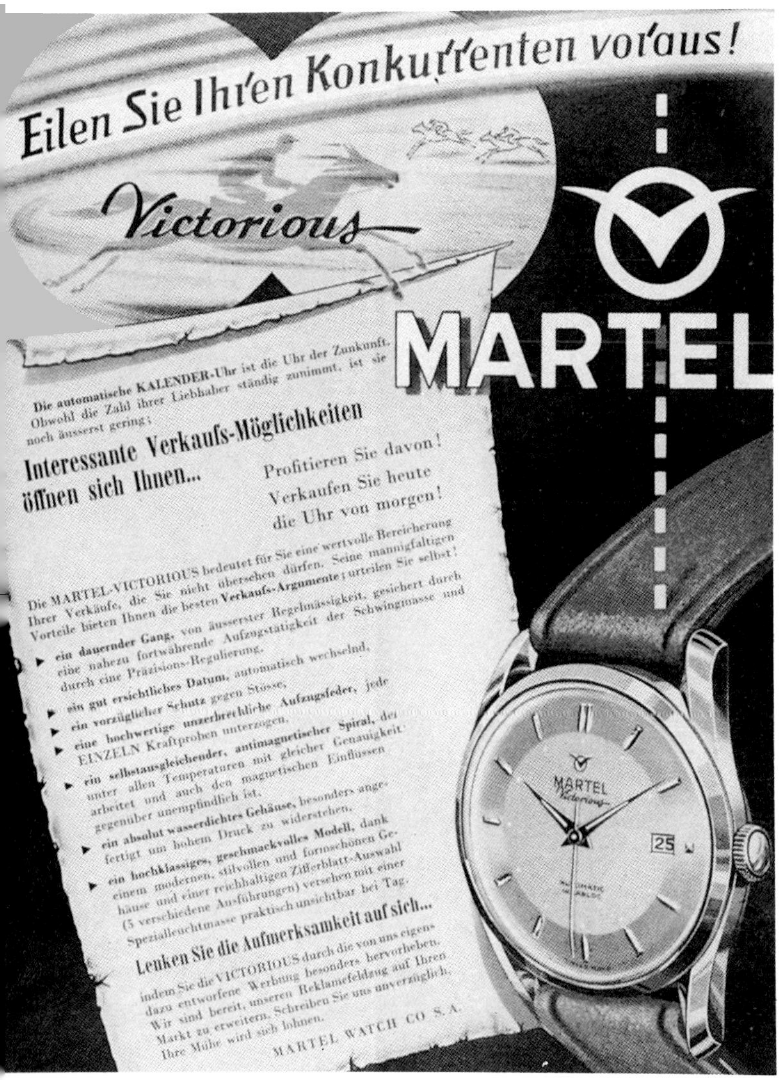

Magazine advertisement (from JSH, 1956)

Caliber 890.2
D = 27.0 mm (12 lignes), H = 5.2 mm, 17 jewels
AUTOMATIC MODULE integrated into the basic movement, central rotor with ball bearings and heavy metal ring, winding in both directions via double springless click-wheel lock, one reduction wheel on upper barrel, uncoupling of hand-winding mechanism by a rocker wheel.
BASIC MOVEMENT: three-spoke screwless Glucydur balance, Nivarox hairspring, 28,800 BPH, Kif-Ultraflex shock resistance, Spirofin fine regulation, two coaxial barrels, one above the other, working in series, second stop, date indication with quick change and correction.

Caliber 990.1
D = 25.6 mm (.11.5 lignes), H = 2.95 mm, 25 jewels
AUTOMATIC MODULE integrated into the basic movement, central rotor with ball bearings and heavy metal ring, winding in one direction via a combined click and locking wheel, one reduction wheel on the locking wheel of the smaller barrel.
BASIC MOVEMENT: three-spoke Glucydur screw balance, Nivarox hairspring, 28,800 BPH, Kif shock resistance, Spirofin fine regulation, two barrels side by side, working in series (17 turns in all), second stop, date indication with quick change and correction.

Caliber Development

The Martel Watch & Co. S.A. firm was founded by Georg Pellaton-Steudler at Les Ponts-de-Martel in 1911. In the latter half of the 1950s, the firm developed one automatic caliber each for ladies' (Caliber 613) and men's watches (Caliber 1115 or 1118). Both are equipped with a central rotor and based on a basic movement with jewel-anchor escapement and direct center seconds.

Caliber List: Martel

Caliber	Diameter Lignes	Diameter mm	Height mm	Other	BPH	Jewels	Running Reserve	Notes	Year	Photo
613	6 3/4	15,3	4,6						ca. 58	
1115 1118	11 1/2 "	25,6 "	5,3	SCD SCD, QG	18.000 "			„Victorious" "	ca. 56	

Magazine advertisement (from UJ, 1966)

Caliber Development

The Marvin Watch Co. S.A. was founded in 1850 at St. Imier by the brothers M. and E. Ditesheim. In 1894 the firm's headquarters was moved to La Chaux-de-Fonds.

In addition to raw movements and complete watches with hand-winding, Marvin also developed a single automatic movement, Caliber 580, introduced in 1955. This very solidly built design has a rotor winding in both directions and a high-quality basic movement, and was also offered with a chronometer certificate as Caliber 580 P. A noteworthy feature is the reciprocal uncoupling device between the hand-winding and automatic mechanisms.

The firm's other watch models have used movements from outside; for example, AS Caliber 1250 was used around 1946.

Caliber Notes

Caliber 580 P
D = 26 mm (11.5 lignes), H = 5.8 mm, 25 jewels
AUTOMATIC MODULE screwed to the basic movement, central rotor with ruby bearings and heavy metal ring, winding in both directions via switching rocker, one tension wheel with locking click, two reduction wheels on the crown wheel, reciprocal uncoupling between automatic and hand-winding drives by a double crown wheel or cam coupling of the last reduction wheel.
BASIC MOVEMENT: two-spoke Glucydur screw balance, Nivarox hairspring, 18,000 BPH hour, Incabloc shock resistance, direct center seconds.

Caliber List: Marvin

Caliber	Diameter Lignes	mm	Height mm	Other	BPH	Jewels	Running Reserve	Notes	Year	Photo
580	11 1/2	26	5,8	SCD	18.000	25				
580 C	"	"	6,6	SCD, QG	"	"		Autodate	1955	
580 P	"	"	5,8	SCD	"	"		Flying Dutchman (Chronometer)	1956	xx

Marvin Caliber 580 P

Men's automatic, "Flying Dutchman" model, made by Marvin in 1956, with Caliber 580 P in chronometer version. Barrel-shaped 18-karat gold case with screw bottom.

Rotor with ruby bearings and heavy metal ring, winding in both directions via switching rocker (at 5:00), complete reciprocal uncoupling of the automatic and hand-winding drives.

Glucydur screw balance, Nivarox hairspring, Incabloc shock resistance (dial surface marred, winding crown not original).

Caliber Development

This firm was founded at Biel in 1917 by G. Schaeren. In 1937 it introduced the first waterproof wrist chronometer, known as the "Multi-Centerchrono".

In the early 1940s, Mido became one of the first Swiss watchmaking firms to stress the production of sporting automatic wristwatches, and thus was especially successful overseas. By using movements from outside, which came without exception from the raw-movement producer AS, high-quality but reasonably priced waterproof and robust automatic watches could be offered.

The firm's first watch model with one-direction hammer winding, known as the "Multifort-Superautomatic", appeared in 1935. The automatic module was rather complex, consisting of sixteen individual components (without screws), but on the other hand, it offered an easily serviced arrangement of winding and locking clicks under separate plates.

Under the same model name, Mido offered its first rotor automatic for ladies (Caliber 707) in 1951, followed by a men's model with rotor (Caliber 917 P), known as the "Powerwind", in 1954. On the basis of this movement design, there arose a whole array of calibers in varying sizes, from 18 mm (Caliber 707 P) to 28 mm in diameter (Caliber 917 PC), in which the number of components comprising the automatic module had been reduced to seven.

The new "Ocean Star" model assortment aroused much interest as the 1960 Basel Watch Fair, for their one-piece cases with pressed-in Permafit glass guaranteed better protection against condensation water buildup and water.

The exterior of this watch, with its recessed crown and short band attachments, was dominated by its striking dial and still prevails today as the unchanging design of Mido watches. In this type of case, the movement can be reached with only tension-ring pliers that allow the special glass with its inner pressure ring to be lifted out, whereupon a certain position of the crown allows the movement to fall out. The smallest of all automatics was the "Minimido", a ladies' watch introduced in 1968 and equipped with the six-ligne AS Caliber 1775.

Caliber Notes

Caliber 817
AS Ebauche
D = 24.5 mm (10.5 lignes)
AUTOMATIC MODULE mounted on the basic movement, hammer winding in one direction via winding click on the hammer, centrally mounted winding wheel, two reduction wheels (the first with locking click, the second as a Breguet coupling for uncoupling during hand-winding) on the locking wheel.
BASIC MOVEMENT: two-spoke Glucydur screw balance, Nivarox hairspring, 18,000 BPH hour, Incabloc shock resistance.

Caliber 0917
AS Ebauche
D = 25.8 mm (11 lignes), H = 5.5 mm, 17 jewels
AUTOMATIC MODULE screwed to basic movement under two potences, hammer winding in one direction via winding click on the hammer, centrally mounted winding wheel, two reduction wheels, the first with locking click, the second with Breguet coupling for uncoupling during hand-winding, on the crown wheel. All clicks reachable under separate plates.
BASIC MOVEMENT: two-spoke Glucydur screw balance, Nivarox hairspring, 18,000 BPH, Incabloc shock resistance, Incastar fine adjustment with moveable hairspring block carrier.

Caliber 917 P
AS Ebauche
D = 28 mm (12 lignes), H = 5.5, 17 jewels
AUTOMATIC MODULE screwed to the basic movement, monometallic central rotor, winding in both directions via springless twin click wheels with locking anchor, one reduction wheel on the locking wheel.
BASIC MOVEMENT: two-spoke Glucydur screw balance, Nivarox hairspring, 18,000 BPH, Incabloc shock resistance, Incastar fine adjustment with moveable hairspring block carrier.

Caliber 1117 PC
AS Ebauche
D = 13 lignes, H = 4.5 mm, 17 jewels
AUTOMATIC MODULE integrated into the basic movement, large rotor with heavy metal ring mounted outside the center of the movement, winding in both directions via an intermediate wheel, springless twin click wheels with locking anchor, two reduction wheels on a spring-boss wheel under the dial corresponding to the locking wheel.
BASIC MOVEMENT: two-spoke Glucydur screw balance, Nivarox hairspring, 18,000 BPH, Incabloc shock resistance, moveable hairspring block carrier, fine adjustment by toothed disc, direct center seconds, date indication with quick change.

Cal 0617 P
AS Ebauche
D = 15.3 mm (6.75 lignes), 17 jewels
AUTOMATIC MODULE screwed onto the basic movement as a complete unit, central rotor with heavy metal ring, winding in both directions via springless twin click wheels with locking anchor, one reduction wheel on the locking wheel.
BASIC MOVEMENT: two-spoke screwless Glucydur balance, Nivarox hairspring, 18,000 BPH, Incabloc shock resistance, Incastar fine adjustment with moveable hairspring block carrier.

Mido Caliber 817

Medium automatic made by Mido as of 1941, "Multifort" model, with AS raw movement.
Stainless steel case with screw bottom.
Hammer winding in one direction via Harwood-type drive, uncoupling during hand-winding.
Glucydur balance, Nivarox hairspring, Incabloc shock resistance.

Caliber 1127 OC
AS Ebauche
D = 13 lignes, 17 jewels
AUTOMATIC MODULE integrated into the basic movement, heavy-metal central rotor with ball bearings, winding in one direction via the first reduction wheel, one click wheel with spring click, second reduction wheel on crown wheel.
BASIC MOVEMENT: three-spoke screwless Glucydur balance, Nivarox hairspring, 18,000 BPH, Incabloc shock resistance, moveable hairspring block carrier, fine adjustment with screw and swan's-neck spring, direct center seconds Eloxal GL barrel, date indication with quick change.

Caliber 1147 OCD
See Caliber 1127 OC, but with 19,800 BPH and additional weekday indication.

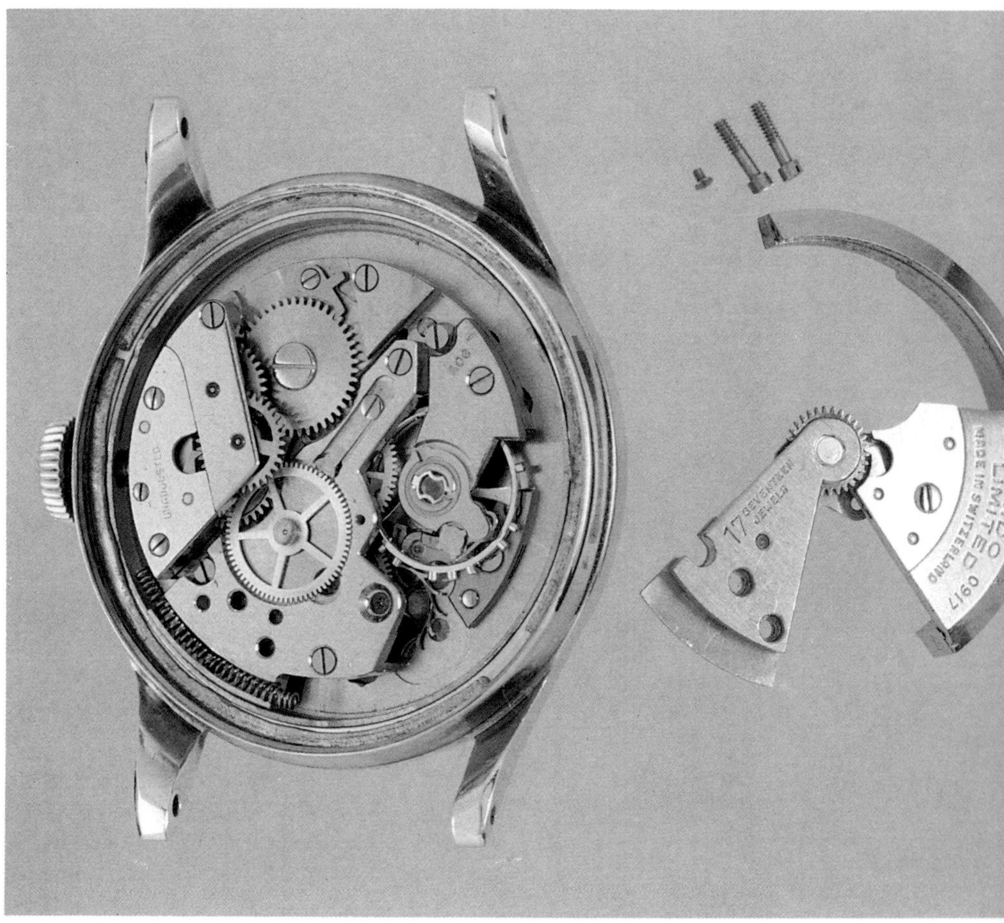

Mido Caliber 0917

Men's automatic, "Multifort" model, made by Mido circa 1943, with AS raw movement.
Fourteen karat gold case with screw bottom.
Hammer winding in one direction via Harwood-type drive, locking click under separate plate (at 9:00), uncoupling during hand-winding.
Glucydur balance, Nivarox hairspring, Incabloc shock resistance, Incastar fine regulation (dial and hands corroded).

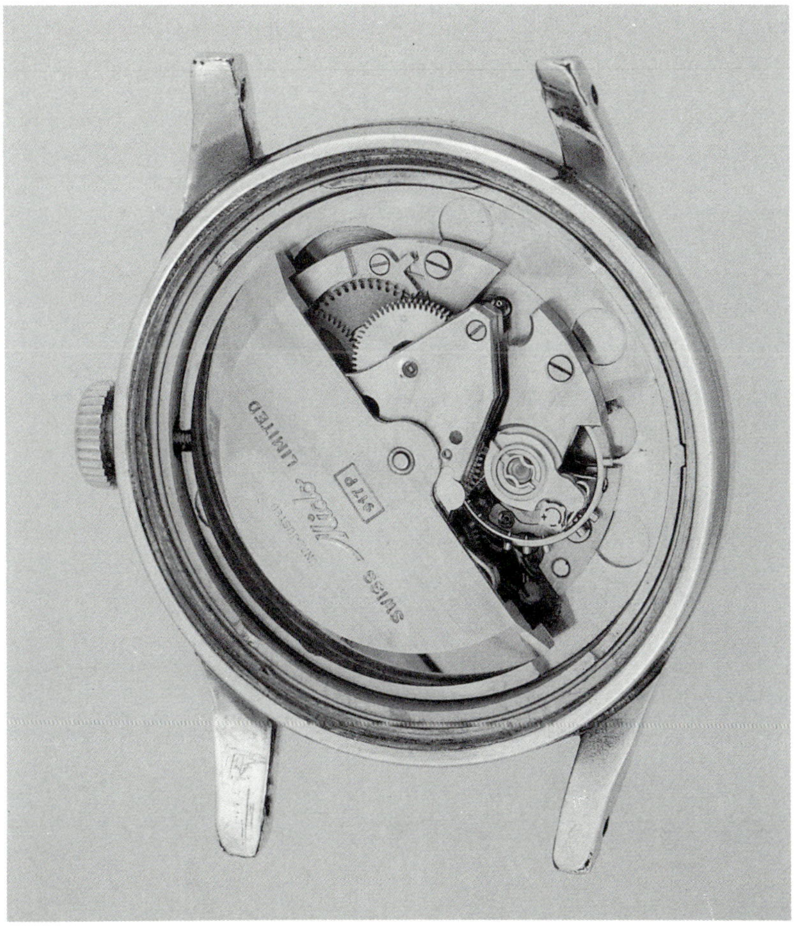

Mido Caliber 917 P

Men's automatic, "Multifort" model, made by Mido in 1954, with the first "Powerwind" Caliber 917 P (exclusive AS raw movement).

Gilded case with stainless steel screw bottom, winding crown with Mido monogram.

Monometallic rotor, winding in both directions via click-wheel switcher (at 9 and 12:00).

Glucydur balance, Nivarox hairspring, Incabloc shock resistance, Incastar fine regulation (case with signs of wear, band attachments bent).

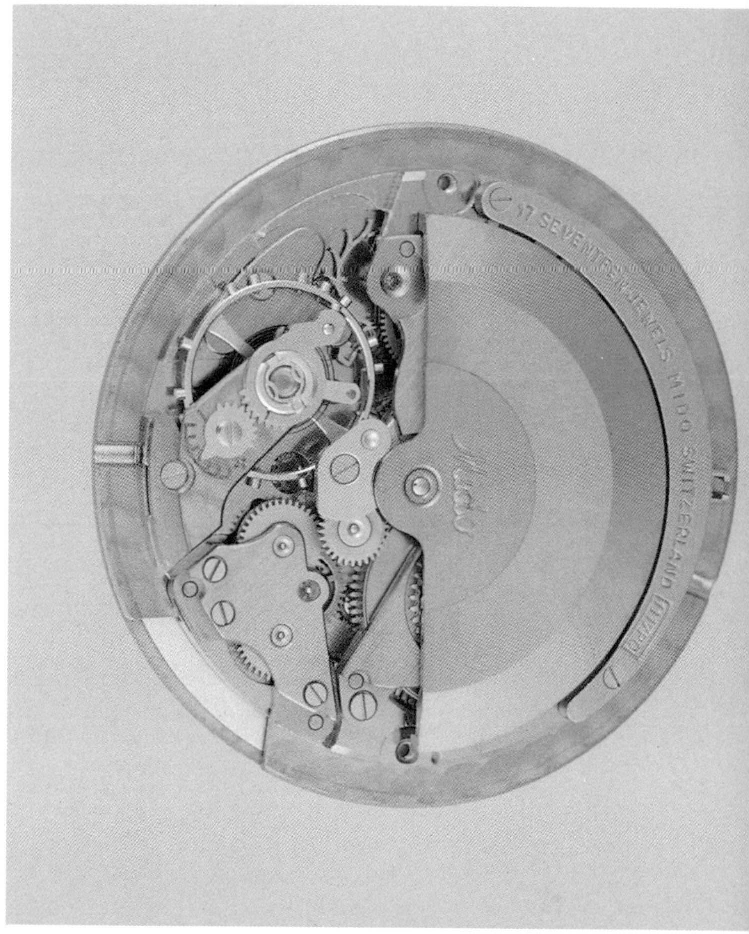

Mido Caliber 1117 PC

Men's automatic, "Ocean Star" model, made by Mido in 1960, with Caliber 1117 PC (AS raw movement).

One-piece stainless steel case with Permafit pressed-in glass and divided winding stem, winding crown with Mido monogram (access to movement only with help of special glass-removing pliers).

Movement identical to AS Caliber 1656 (see page 36), but with fine regulation by toothed disc.

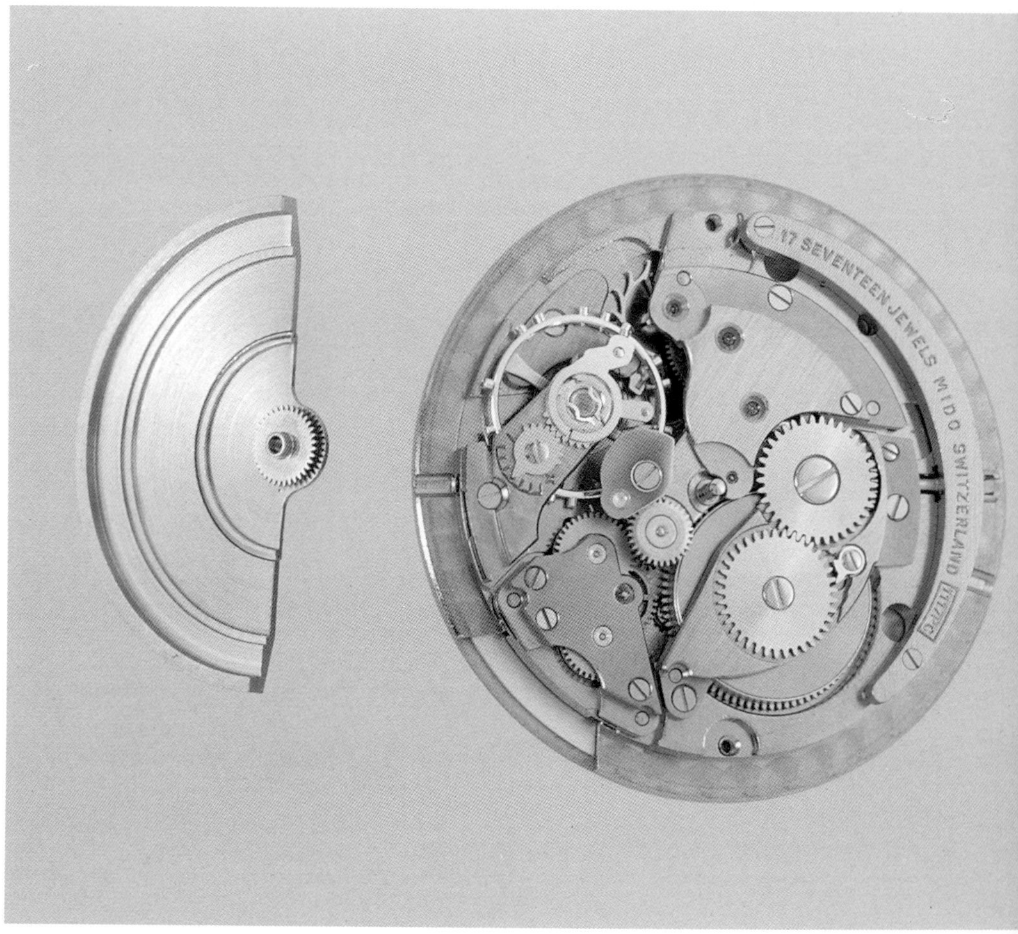

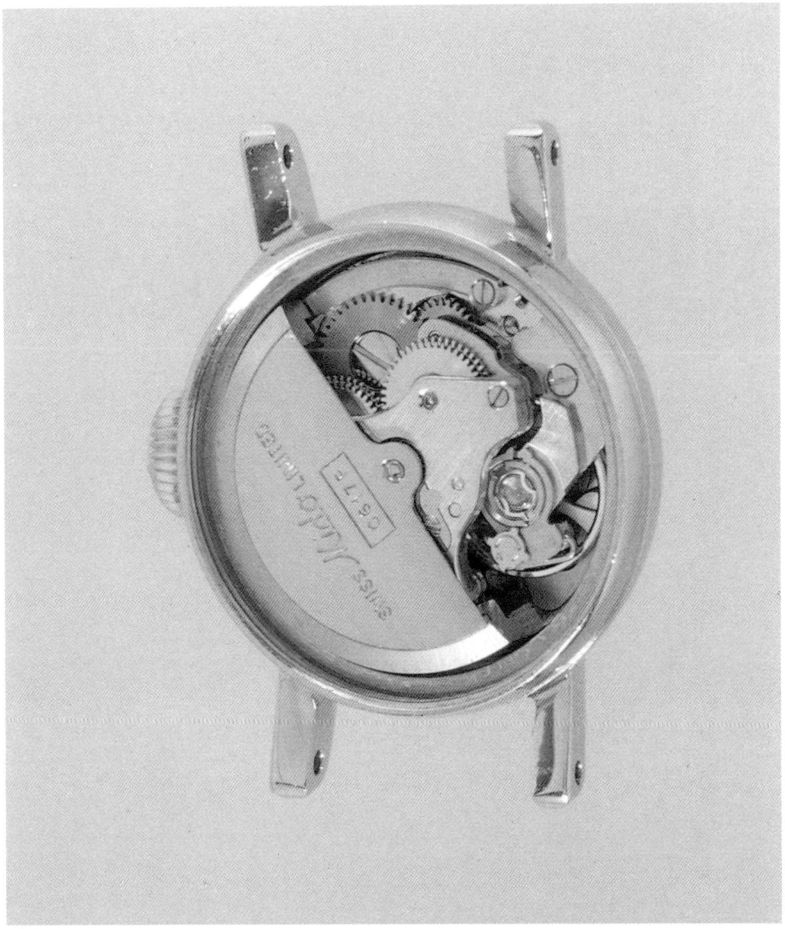

Mido Caliber 0617 P
Ladies' automatic, "Multifort" model, made by Mido circa 1960, with Caliber 0617 P (AS raw movement).
Stainless steel case with screw bottom, winding crown with Mido monogram.
Rotor with heavy metal ring, winding in both directions via click-wheel switcher (at 10 and 12:00).
Screwless Glucydur balance, Nivarox hairspring, Incabloc shock resistance, Incastar fine regulation.

Mido Caliber 1127 OC

Men's automatic, "Ocean Star" model, made by Mido circa 1963, with Caliber 1127 OC (AS raw movement).

One-piece 14-karat gold case with Permafit pressed-in glass, removable stem (after glass is removed from front), winding crown with Mido monogram, gold hands and dial inlays.

Heavy metal rotor with ball bearings, winding in one direction via click wheel, automatic drive recessed in an oval plate cutout (at 6:00).

Screwless Glucydur balance, Nivarox hairspring, Incabloc shock resistance, swan's-neck fine regulation.

Mido Caliber 1147 OCD
Men's automatic, "Ocean Star" model, made by Mido circa 1965, with "datoday" Caliber 1147 OCD.
One-piece stainless steel case with Permafit pressed-in glass and divided winding stem, winding crown with Mido monogram.
Movement identical to Caliber 1127 OC (see page 212), but with BPH increased to 19,800 BPH.

Caliber List: Mido

Caliber	Diameter Lignes	Diameter mm	Height mm	Other	BPH	Jewels	Running Reserve	Notes	Year	Photo
01147 O	12 1/2			SCD	19.800					
01147 OC	"			SCD, QG	"					
01147 OCD	"			SCD, QG, JG	"					
001147	11 1/2			SCD	19.800					
001147 C /OC	"			SCD, QG	"					
001147 OCD	"			SCD, QG, JG	"					

Caliber List: Mido

Caliber	Diameter Lignes	mm	Height mm	Other	BPH	Jewels	Running Reserve	Notes	Year	Photo
707	7 1/2	17,0	5,2	SCI	18.000	17			1951	
707 P 707 PC	7 3/4 "	18 "		SCI SCI, QG	18.000 "	17 "				
717	9		5,05	SCI	18.000	17			1938	
737 737 C /C-1 /C-2 737 OC-1	7 3/4 " "			SCD SCD, QG SCD, QG	21.600 " "					
816 817	10 1/2 "	24,5 "		 SCI	18.000 "			AS 1081 "	1939 1941	 xx
916 916 P	12 "	 28,0	4,85 5,5	s s	18.000 "	17 "		As 1081		
0916 P	11	25,8	5,5	s	18.000	17				
00916 P	10 1/2	24,5	5,5	s	18.000	17			1938	
917	12	28,0	5,3	SCI	18.000	17			1942	
917 P /PO 917 PC /PCO	12 "	28,0 "	5,5	SCI SCI, QG	18.000 "	17 "		Powerwind	1954	xx
0917 /P 0917 PC	11 "	25,8 "	5,5	SCI SCI, QG	18.000 "	17 "		AS 1191		xx
00917 P	10 1/2	24,5	5,5	SCI	18.000	17				
D 917	12			SCI, QA	18.000	17				
1016 1017	12 "			 SCI	18.000 "					
1117 P /P-1 1117 PC /PC-1	13 "			SCD SCD, QG	18.000 "	17 "		 AS 1678	1959 1960	 xx
01117 P /P-1 01117 PC /PC-1	12 "			SCD SCD, QG	18.000 "	17 "				
1127 O 1127 OC	13 "			SCD SCD, QG	18.000 "	17 "		 AS 1717 (11 1/2 ")	 ca. 63	 xx
1137 O 1137 OC	13 "			SCD SCD, QG	19.800 "					
01137 O 01137 OC	12 1/2 "			SCD SCD, QG	19.800					
001137	11 1/2			SCD	19.800					
1147 /O 1147 C /OC 1147 OCD	13 " "			SCD SCD, QG SCD, QG, JG	19.800 " "				 ca. 67	 xx
01147 O 01147 OC 01147 OCD	12 1/2 " "			SCD SCD, QG SCD, QG, JG	19.800 " "					
001147 001147 C /OC 001147 OCD	11 1/2 " "			SCD SCD, QG SCD, QG, JG	19.800 " "					

Caliber Development

The firm was founded at La Chaux-de-Fonds in 1881 by Achilles Ditesheim under his own name, and renamed "Movado Watch Co." in 1905.

The wristwatch (Caliber 220) introduced in 1946 under the name "Tempomatic" was the first Movado-made automatic.

The swinging mass was a hammer, winding in one direction. The automatic module was screwed to the relatively large basic movement as a complete unit, so that a correspondingly dimensioned barrel with what was then an unusually high running reserve of 56 hours could be included.

In the ensuing years, several models based on this caliber were equipped with additional features. The "Calendoplan" has date indication, the "Calendomatic" has a simple calendar, with indications of the date, weekday and month.

A new development, Caliber 115, introduced in 1950 and also known as the "Futuramic", was the flattest automatic of the time. This was accomplished by a particularly small peripheral hammer weight of heavy metal and the joint mounting of the hammer and the minute wheel in one bearing jewel. This in turn required effective shock resistance for this bearing to avoid breakage. The spoke of the swinging mass is therefore in S form, so as to be able to have a cushioning effect. For reasons of height reduction, this movement is also equipped with only a small seconds hand, in whose dial area of the "Celandolux" (as of 1952) the date and weekday are indicated in a separate aperture. The first Movado automatics with rotors appeared in 1956 as the "Queenmatic" as a ladies' caliber (165) and in men's size as the "Kingmatic" (Caliber 431). Here the rotor's turning circle is not beside, but right over the basic movement, so that the whole area of the case interior is available for it. The balance and barrel are appropriately large, and thus chronometer quality is achieved (Caliber 436). In these movements too, the heavy metal ring of the rotor is fitted with the typical Movado S-shaped spoke for shock resistance. Winding is done in both directions, via a carefully made switching rocker with ruby bearings. The hand-winding mechanism is uncoupled during automatic winding by a swinging toothed-wheel rocker between the crown and locking wheels, in order to reduce friction losses.

In 1960 this somewhat obsolescent caliber for men was equipped with a faster oscillating balance (Caliber 531), which allowed its running regularity to improve further. With the development of the new Caliber 380 ff, a further decrease in the movement thickness was attained. For the first time in a Movado watch, the rotor had ball bearings, and the entire hand-winding mechanism was mounted under the dial. The barrel consisted of an eloxied aluminum-alloy Antikorodal 15B component, guaranteeing regular functioning of the sliding stud over the years; the mainspring was unbreakable and covered with a self-lubricating coating.

For economic reasons, the firm was compelled to cooperate with the Zenith firm in 1968. The two firms worked together to develop the first automatic chronograph, which was introduced by Movado in 1969 as the "Datachron", using the jointly produced Caliber 3019 PHC.

Caliber Notes

Caliber 220
D = 28.8 mm (12.75 lignes), H = 5.7 mm, 17 jewels
AUTOMATIC MODULE screwed to the basic movement as a complete unit, hammer with ruby bearings, winding in one direction via a sprung winding click on the rotor, centrally mounted winding wheel, three reduction wheels (the first with a sprung locking click) on the double locking wheel for uncoupling during hand-winding.
BASIC MOVEMENT (10.5 lignes): two-spoke Glucydur screw balance, self-compensating Breguet hairspring, 18,000 BPH, Incabloc shock resistance.

Caliber 118
D = 26.5 mm (11.5 lignes), 17 jewels
AUTOMATIC MODULE set atop the basic movement, hammer with heavy metal ring, lower hammer mount shared with the minute wheel, slim S-shaped swinging-mass spoke for shock resistance, winding in one direction via click-carrier rocker, combined winding and reduction wheel on the crown wheel.
BASIC MOVEMENT: two-spoke Glucydur screw balance, Nivarox hairspring, 18,000 BPH, Incabloc shock resistance, eccentric seconds hand, date and weekday indication in cutouts near the small second hand.

Caliber 223 A
D = 28.8 mm (12.75 lignes), 17 jewels
See Caliber 220 for movement construction, but with simple calendar (date, day and month indication).

Caliber 438
D = 25.5 mm (11.25 lignes), 28 jewels
AUTOMATIC MODULE screwed to the basic movement as a complete unit, central rotor with ruby bearings and heavy metal ring, wide but flat S-shaped swinging-mass spoke for shock resistance, winding in both direction via switching rocker with ruby bearings, three reduction wheels (the first with spring click mounted in ruby bearings) on the double locking wheel, locking-wheel coupling for uncoupling from the hand-winding drive, toothed-wheel rocker between the crown and lower locking wheels for uncoupling during hand-winding.
BASIC MOVEMENT: two-spoke Glucydur screw balance, Nivarox hairspring, 18,000 BPH, Incabloc shock resistance, direct center seconds date indication. with slow change.

Caliber 531
D = 26.5 mm (11.75 lignes), 28 jewels
Movement almost identical to Caliber 438, but with 21,600 BPH.

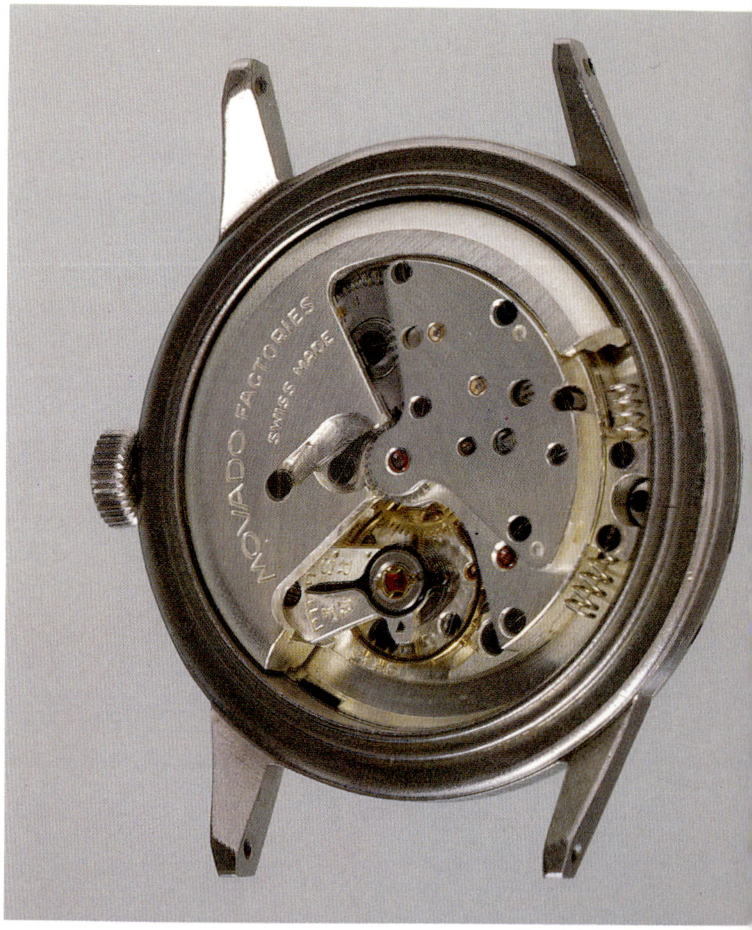

Movado Caliber 220
Men's automatic, "Solidomatic" model, made by Movado in 1946, with "Tempomatic" Caliber 220.
Stainless steel case with press-in bottom and dust cover.
Hammer with ruby bearings, winding in one direction via Harwood-type drive, uncoupling during hand-winding.
Glucydur screw balance, self-compensating Breguet hairspring, Incabloc shock resistance.

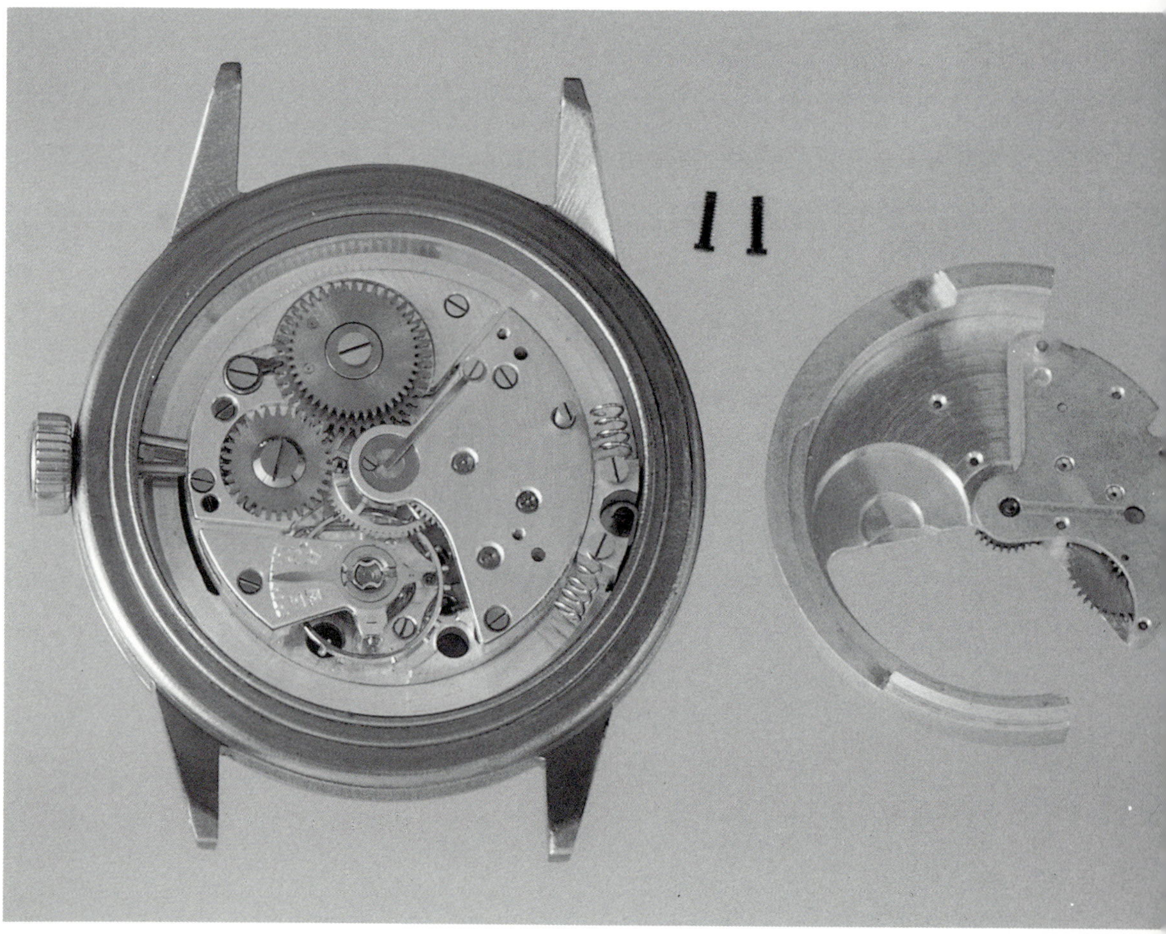

Movado Caliber 118

Men's automatic, "Calendolux" model, made by Movado in 1953, with Caliber 118, flattest of its time. Gilded case (40 u) with press-in bottom.
Date and weekday indication in area of the small seconds hand. Quick date correction button by the "4". Hammer with ruby bearings, heavy metal ring and sprung S-shaped spoke, winding in one direction via click-carrier rocker (at 8:00).
Glucydur screw balance, Nivarox hairspring, Incabloc shock resistance (second hand not original).

Movado Caliber 223 A

Men's automatic, "Calendomatic" model, made by Movado circa 1952, with Caliber 223 A.

14-karat gold case with screw bottom, winding crown with Movado emblem.

Date indication with peripheral date circle and quick correction button by the 8, weekday and month in apertures (quick month correction button by the 9).

Hammer with ruby bearings and heavy metal ring, winding in one direction via Harwood-type drive, uncoupling during hand-winding. Automatic

Movado Caliber 438

Men's automatic, "Kingmatic Calendar" model, made by Movado in 1959, with Caliber 438, the firm's first rotor caliber.

Stainless steel case with screw bottom, winding crown with Movado "Hand" emblem.

Rotor with ruby bearings, heavy metal ring and sprung S-shaped spoke, winding in both directions via switching rocker (at 11:00), complete reciprocal uncoupling of automatic and hand-winding drives.

Glucydur screw balance, Nivarox hairspring, Incabloc shock resistance.

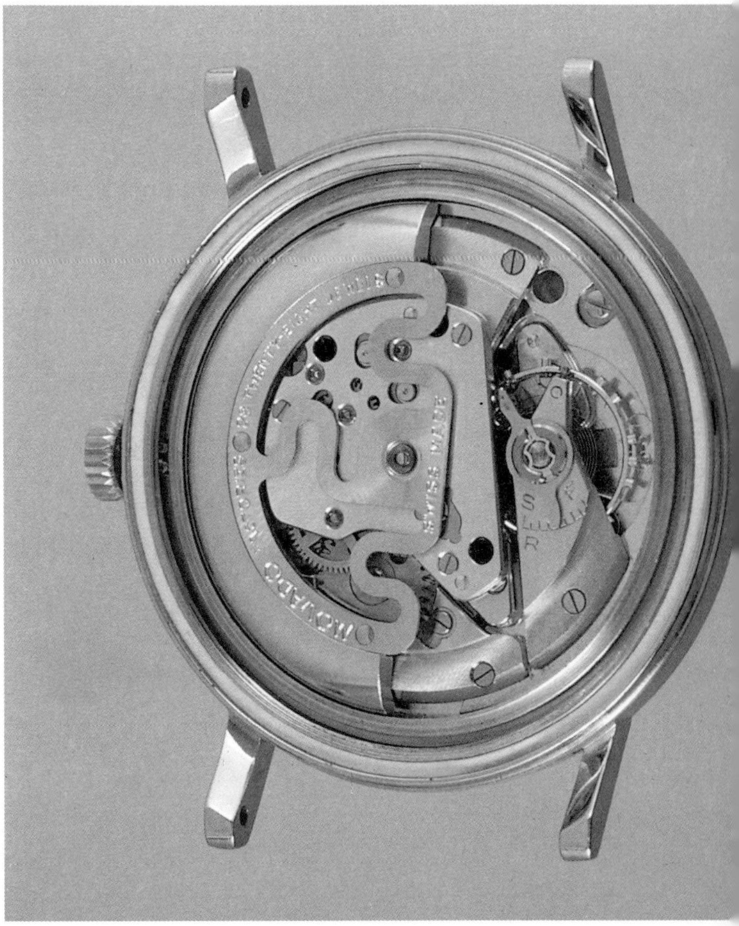

Movado Caliber 531

Men's automatic, "Kingmatic" model, made by Movado circa 1959, with Caliber 531.

Stainless steel case with screw bottom, winding crown with modern Movado emblem.

Movement identical to Caliber 438 (see page 219), but with faster beat and different rotor shape.

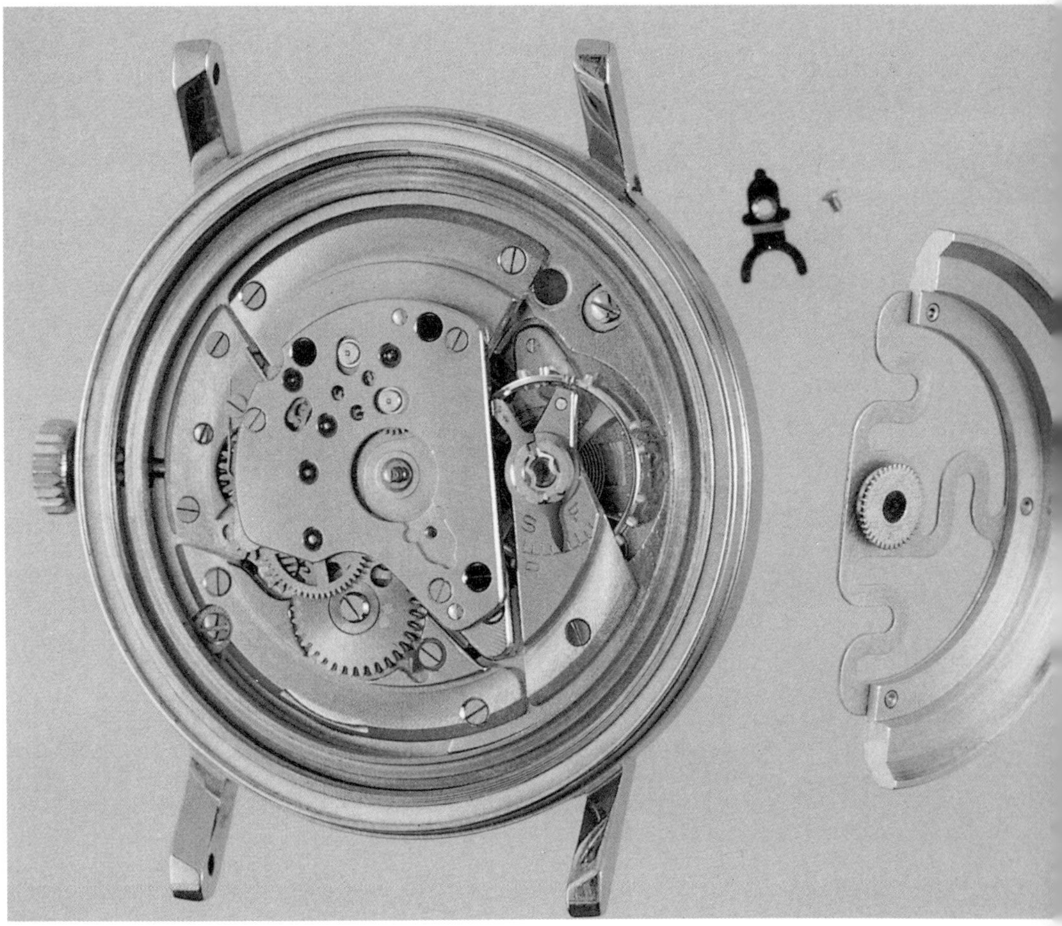

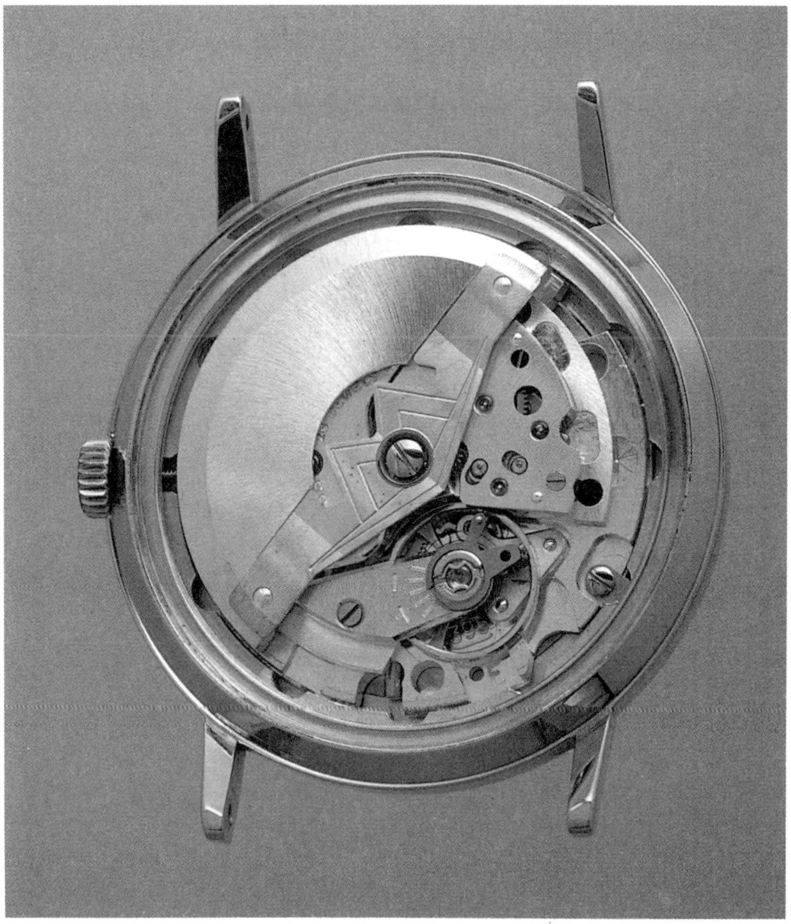

Movado Caliber 395 E

Men's automatic, "Kingmatic S" model, made by Movado in 1965, with Caliber 395 E.

Gilded case with stainless steel screw bottom, winding crown with modern Movado emblem.

Heavy metal rotor with ball bearings, winding in both directions via switching rocker (at 3:00).

Screwless Glucydur balance, Nivarox hairspring, Incabloc shock resistance (case with signs of wear).

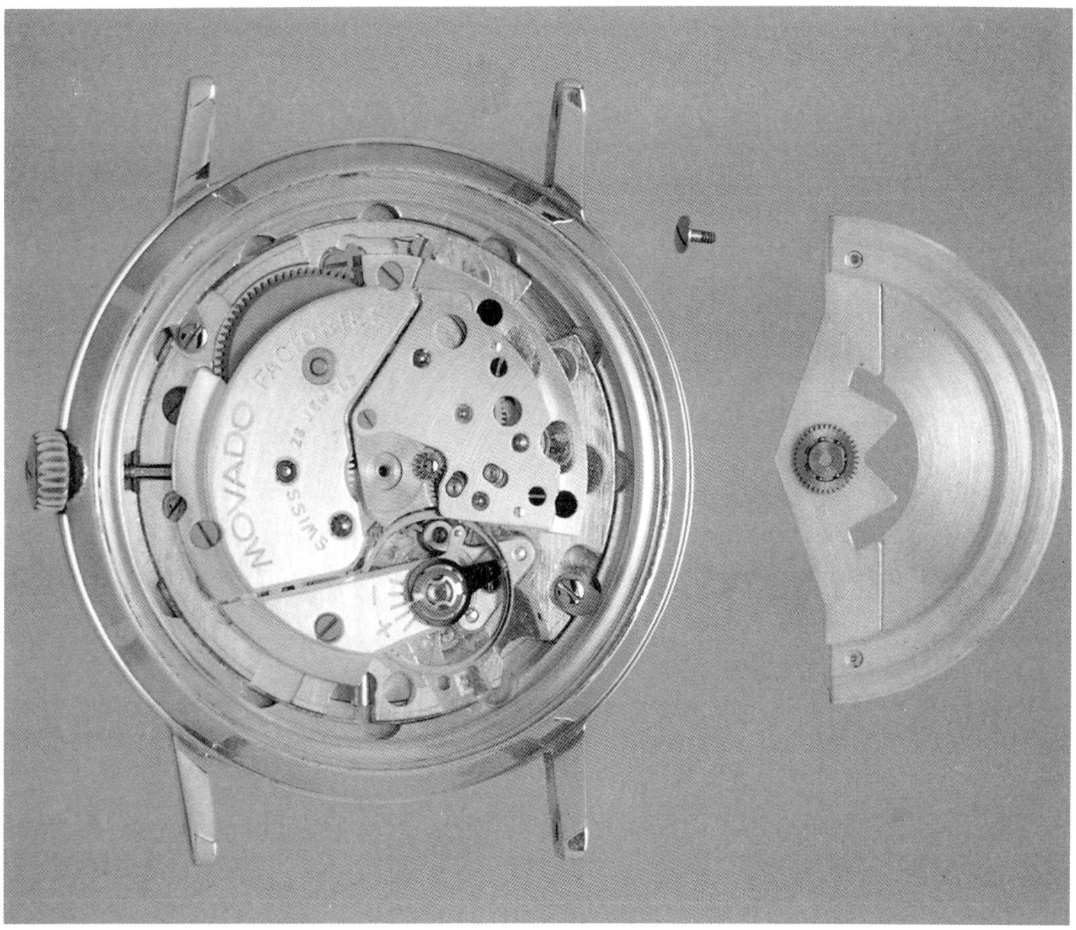

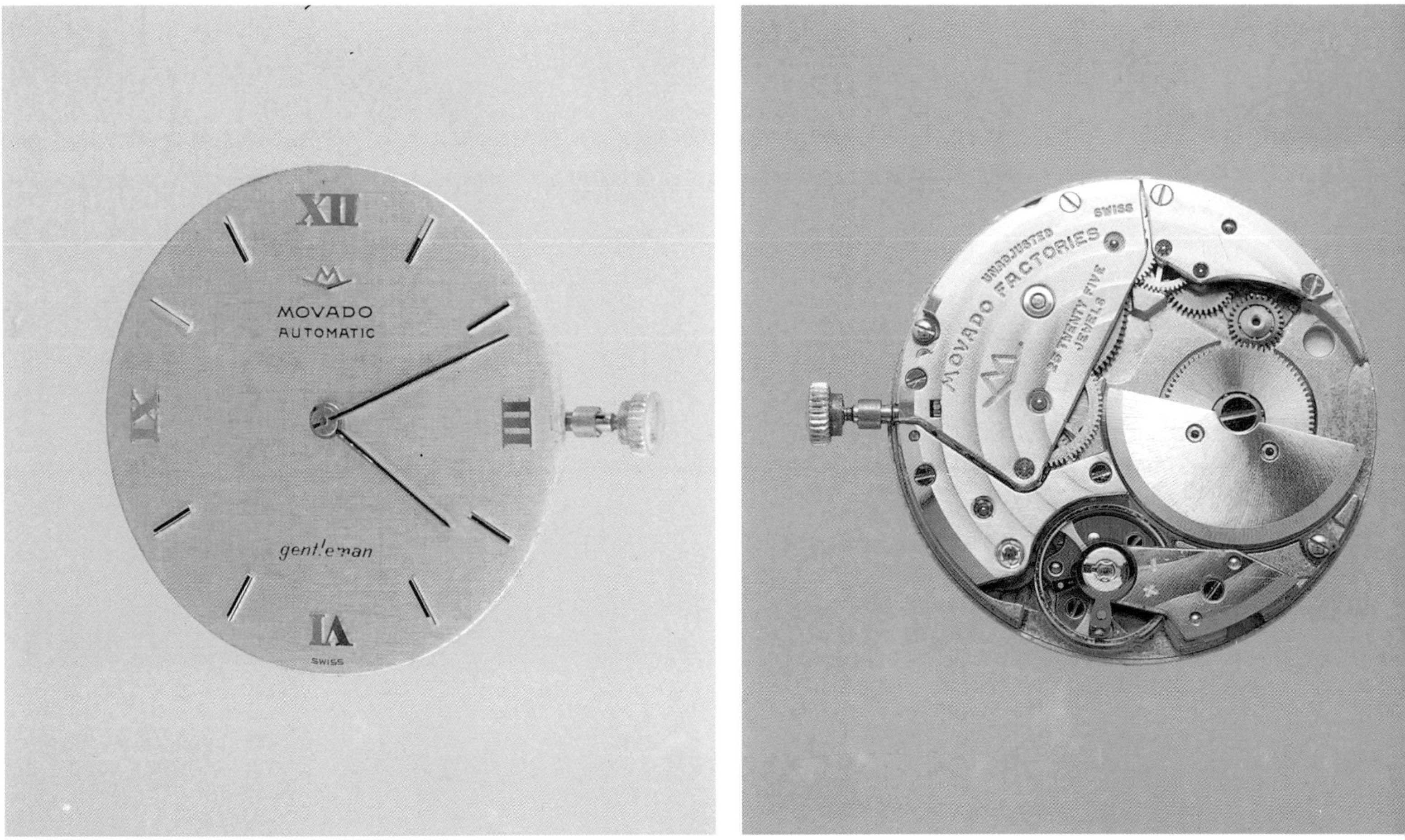

Movado Caliber 230
Automatic movement for the flat "Gentleman" men's model made by Movado circa 1966 (Caliber 230), with microrotor caliber made by Universal (see page 308).

Caliber 395 E
D = 25.3 mm (11.5 lignes), H = 3.95 mm, 28 jewels
AUTOMATIC MODULE integrated into the basic movement as a complete unit, central rotor with ball bearings and heavy metal ring, winding in both directions via shifting rocker with ruby bearings, two intermediate wheels, two reduction wheels (the first with spring click in ruby bearings) on the locking wheel (under the dial).
BASIC MOVEMENT: three-spoke screwless Glucydur balance, Nivarox hairspring, 21,600 BPH, Incabloc shock resistance, direct center seconds, eloxied Antikorodial 15B barrel, rocker mechanism for hand-winding and play-free hand-setting via double wheels.

Caliber 230
Universal ebauche (see there, Caliber 2-66)
D = 27.6 mm (12.25 lignes), H = 2.5 mm, 25 jewels

Caliber List: Movado

Caliber	Diameter Lignes	mm	Height mm	Other	BPH	Jewels	Running Reserve	Notes	Year	Photo
115	11 1/2	26,5	4,3	s	18.000	17		Futuramic	1950	
116	"	"	"	s	"	21		Chronometer		
117	"	"		s, OG, JG	"	17		Calendolux	1953	xx
165/166	7 1/4	16,5			18.000	30		Queenmatic	1956	
167/169	"	"		SCI	"	"		"		
220	12	27,0		SCI	18.000	17	56	Tempomativ	1946	xx
220 M	12 3/4	28,8		SCI	"	"	"	"		
221	12	27,0		SCI	"	"	"	Heavy metal ring		

Caliber List: Movado

Caliber	Diameter Lignes	mm	Height mm	Other	BPH	Jewels	Running Reserve	Notes	Year	Photo
221 A	12 3/4	28,8	5,7	SCI	18.000	17	56	(Heavy metal ring)		
222	12	27,0		SCI, QG	"	"	"	Calendoplan	ca. 52	
223	12	"		SCI, QA, JG, MG	"	"	"	Calendomatic	1948	
223 A	12 3/4	28,8		SCI, QA, JG, MG	"	"	"	"	ca. 52	xx
224 /A	12	27,0		SCI, QG	"	"	"	Calendoplan	ca. 53	
225	12	"		SCI, QA, JG, MG	"	"	"	Calendomatic		
225 M	12 3/4	28,8		SCI, QA, JG, MG	"	"	"	"		
226 /A	12 3/4	"		SCI	"	"	"	Tempomatic		
228 /A	12 3/4	"		SCI, QA, JG, MG	"	"	"	Calendomatic		
230	12 1/4	27,6	2,5		19.800	25	48	Gentalman (Universal Eubache)	ca. 66	xx
331	11 1/2	26,5	4,3	s	18.000	17		s. Cal 115	1951	
331	"	"		s, QG, JG	"	"		Calendolux, s. Cal 118	1953	
380	11 1/4	25,3	3,95	SCD	21.600	28	50	Kingmatic S	1963	
388	"	"		SCD, QG	"	"	"	"		
389	"	"		SCD, QG	"	"	"	"		
395 /A....F	"	"	3,95	SCD	"	"	"	"	1965	xx
400	11 1/4			SCD	36.000					
405	"			SCD, QG, JG	"					
408	"			SCD, QG	"					
421	7 1/4				19.800					
423	"			SCI	"					
425	"				"					
427	"				"					
431 /A	11 1/4	25,5		SCD	18.000	28		Kingmatic	1956	
436	"	"		SCD	"	"		Chronometer		
438	"	"		SCD	"	"		Kingmatic Calendar	1959	xx
531	11 3/4	26,5		SCD	21.600	28		Kingmatic	ca. 59	xx
532	"	"		SCD	"	"		"		
536	"	"		SCD	"	"		Kingmatic Chronometer		
538	"	"		SCD, QG	"	"		Kingmatic Calendoplan		
608	11 1/2			SCD, QG	18.000	25		(Eta-EubacheCal 2472)		
1425	7 1/4				19.800					
1427	"			SCI	"					
2552 PC	11 1/2	25,6	5,63	SCD, QG	21.600	23		(Zenith Eubache Cal 2552 PC)		
2562 PC	"	"	"	SCD, QG	"	"		(Zenith Eubache Cal 2562 PC)		
3019 PHC		30	6,5	S, QG, RDR, CHR	36.000	31	50	Datachron, s. Zenith	1969	(xx)
3019 PHF		"		S, QG, JG, MG, PH RDR, CHR	"	"	"	Datron, s. Zenith		

Caliber Development

The firm was founded at La Chaux-de-Fonds in 1848 by Louis Brandt and moved to Biel in 1880. In 1903 the former name "Louis Brandt & Fils" was changed to "Omega Watch Co.", named after a watch caliber developed in 1894. Intensive collaboration with the firm of "Charles Tissot & Fils" began at Le Locle in 1925. In 1930 the world economic crisis compelled the firm to merge with Cortebert, Hamilton, Lanco, Lemania, Ravville and Tissot to form the SSIH (Holding Société Suisse pour l'Industrie Horlogère). As of 1971, the struggle against the Far Eastern export offensive required further collaboration, with the lower-priced brands of Agon, Buler, Continental and Ferex joining the ESTH (Economic Swiss Time Holding), and somewhat later, the founding of the SMH (Société Microelectronique Horlogère), the holding company of the Swiss watch industry, including the brands of Longines, Omega, Rado, Swatch and Tissot.

In 1943 Omega introduced Caliber 28.10 RA PC, its first automatic, equipped with one-direction hammer winding, as was then the rule.

With a few design improvements, such as the covered, loosely inserted buffer springs, replacing the screwed-in visible coil springs that often broke, and additional features such as the center seconds and date indication, calibers of this basic model were produced in great numbers until 1955. They were characterized by dependable functioning and great running reliability, and beginning in 1950 they won first prizes in the precision competitions at the Geneva and Neuenburg observatories.

Thus it was only in 1955 that Omega felt it necessary to equip its watch calibers with rotors, as its competitors had been doing for years with growing success.

The improved winding performance of a rotor working in both directions allowed Omega to offer its first ladies' automatic, the "Ladymatic" (Caliber 450/455). Using a specially designed click-carrier rocker, a running reserve of 36 hours could be guaranteed despite the remarkably small movement dimensions of 7.75 lignes.

The men's calibers, 470 and 490, were also equipped with rotor winding in two directions, but here a switching rocker was used. Because of strong signs of wear, especially on the first reduction wheel that was equipped with a sprung locking click, it was replaced in 1959 by a more modern design, the remarkably successful 550 ff caliber family. By integrating the automatic drive into the basic movement, which could be accomplished only with the help of a patented direction changer (a springless double satellite-wheel lock), the thickness of the movement could be reduced considerably. At the same time, the developmental engineers replaced the screw balance with a smooth four-spoke ring balance. By utilizing precise manufacturing and control processes, the manufacturers were able to put increased production emphasis on making automatic watches on the basis of this design. With the "Constellation", Omega could then turn out greater numbers of automatic wrist chronometers than the Rolex firm and in 1963, using this caliber, the firm was the only brand to break all precision records in the wristwatch category.

With a new design, Caliber 590 ff, the firm tried in 1960 to offer a reasonably-priced automatic movement that was designed especially for large-series production and turned out to be considerably thicker. The satellite-wheel lock was replaced here by simpler twin click wheels with leaf springs.

In 1963 Omega offered another automatic caliber for ladies' watches (Caliber 660), in which improved winding performance was attained by the double satellite-wheel lock. Its somewhat larger version, Caliber 683 ff, could even be offered with a chronometer certificate.

Following the trend of the times toward ever-flatter watches, Omega developed a very low automatic, Caliber 710, in 1966. Its movement height of only 3 mm was attained by a new design in the basic movement, including an indirectly driven minute hand, and by especially close production tolerances, which allowed further decreases in the distances between the individual wheels and the parts of the escapement.

In 1969 Omega built the "Memomatic", the first fully automatic alarm wristwatch in which, unlike the alarm watches built by Jaeger LeCoultre, the alarm mechanism was also wound automatically by using a shared barrel. A clever mechanism allowed the alarm to function with only one turn of the barrel, leaving a sufficient running reserve for the movement.

The alarm time could be set to the minute by two index discs in the center of the dial. Here, as in the Caliber 1040 automatic chronograph introduced two years later, Omega used raw movements made by Lemania.

In 1969 too, Omega introduced Caliber 1000 ff, its first automatic with a fast beat balance, but its reliability was not good enough to be a convincing market factor. In the firm's last automatic development, Caliber 1010 ff, the movement height had to be increased from 4.0 mm to 4.8 mm.

In retrospect it can be seen that the Omega watches were less noteworthy for spectacular innovations in automatic development than for the reliability and precision that won this brand of Swiss watches their fame.

Only the two best-known models of the "Constellation" and the "Seamaster", which were marked by a symbol stamped into their case bottoms, will be noted. While the "Constellation" is sold only as a precision watch with a chronometer certificate (from an observatory), the "Seamaster" is a robust, waterproof sporting model. References to the calibers installed in them, though, cannot be made here.

Omega Caliber 28.10 RA PC
First men's automatic made by Omega, in 1943, with Caliber 28.10 RA PC. Nine karat gold case with press-in bottom.
Hammer with ruby bearings, winding in one direction via click-carrier rocker. Glucydur screw balance, Nivarox hairspring, Incabloc shock resistance (winding crown renewed).

Omega Caliber 28.10 RA SC PC
Eighteen karat gold case with press-in bottom, gold hands and dial inlays. Movement identical to Caliber 28.10 RA PC except for center seconds, see page 225. (Dial and movement somewhat corroded, case with signs of wear).

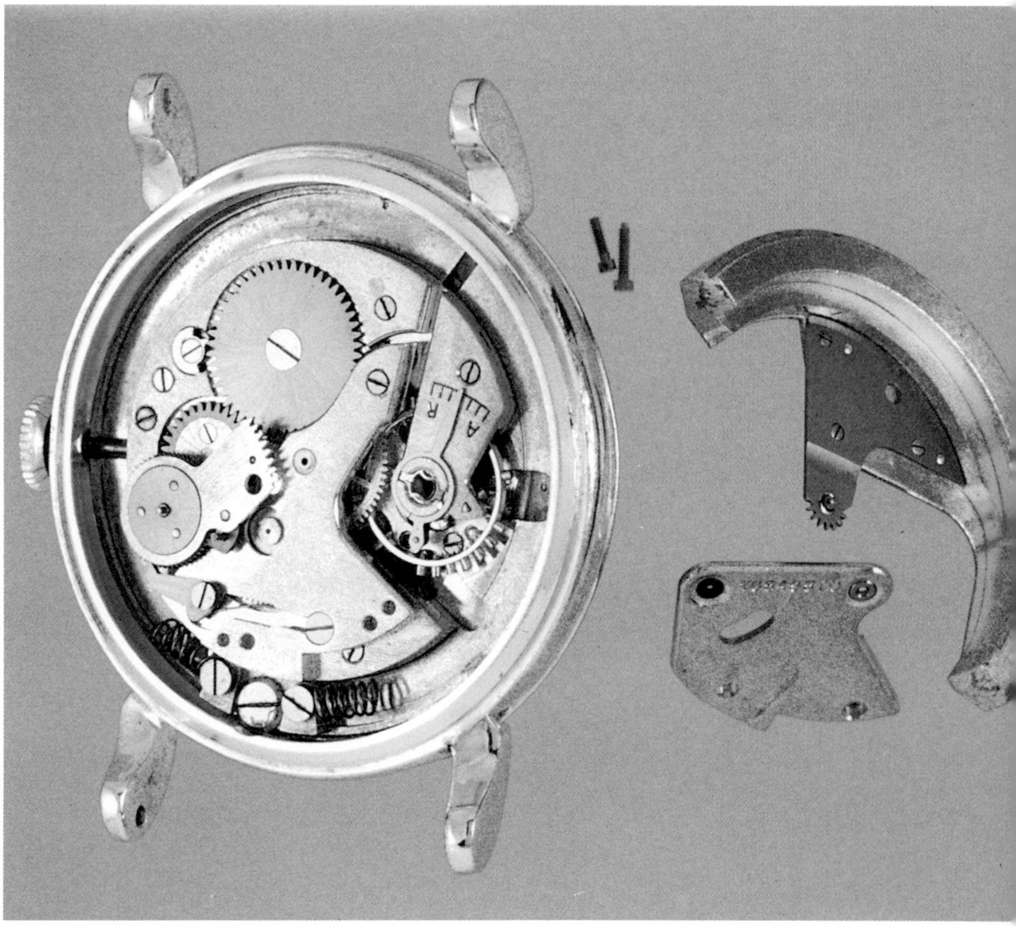

Omega Caliber 342

Men's automatic, "Seamaster" model, made by Omega in 1949, with Caliber 342.

Eighteen karat gold case with screw bottom, winding crown with Omega symbol, gold hands and dial inlays. Movement identical to Caliber 28.10 RA PC (see page 225) but with concealed buffer springs (at 6:00).

Omega Caliber 354

Men's automatic made by Omega in 1953, with Caliber 354.
Eighteen karat gold case with screw bottom, gold winding crown with Omega symbol, gold hands and dial inlays.
Movement identical to Caliber 28.10 RA SC PC (see page 226), but with concealed buffer springs and swan's-neck fine regulation (movement with slight traces of corrosion).

Omega Caliber 355

Men's automatic, "Seamaster Calendar" model, made by Omega in 1953, with Caliber 355.

Stainless steel case with gold hood and screw bottom, winding crown with Omega symbol, gold hands and dial inlays.

Movement identical to Caliber 28.10 RA SC PC (see page 226), but with date and concealed buffer springs as well as swan's-neck fine adjustment (movement and case with slight signs of wear).

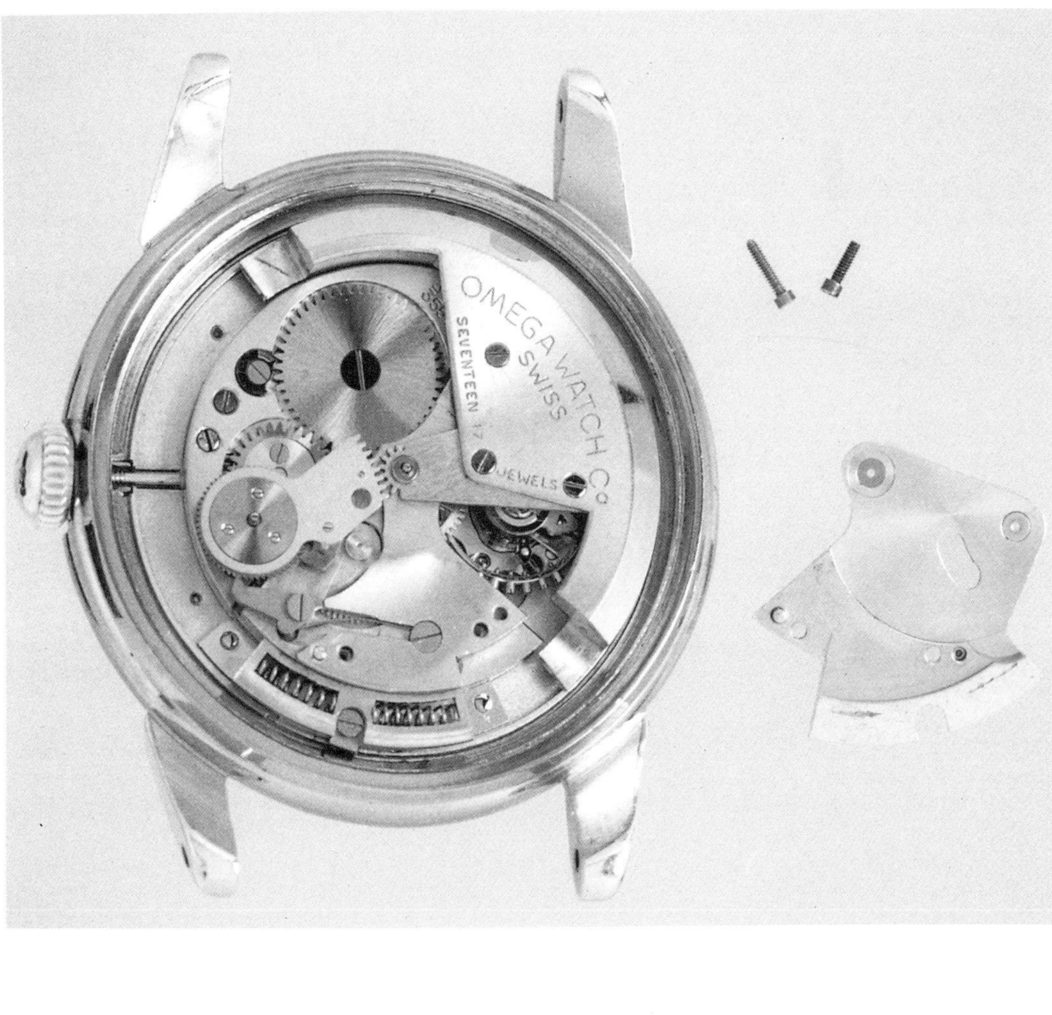

Omega Caliber 455

"Ladymatic", first ladies' model made by Omega, in 1955, with Caliber 455.
Square 18-karat gold case with press-in bottom, winding crown with Omega symbol, gold dial inlays.
Monometallic rotor, winding in both directions via patented click-carrier rocker (see lower right).
Glucydur screw balance, Nivarox hairspring, Kif-Flector shock resistance (dial and hands strongly corroded, case with clear signs of wear).

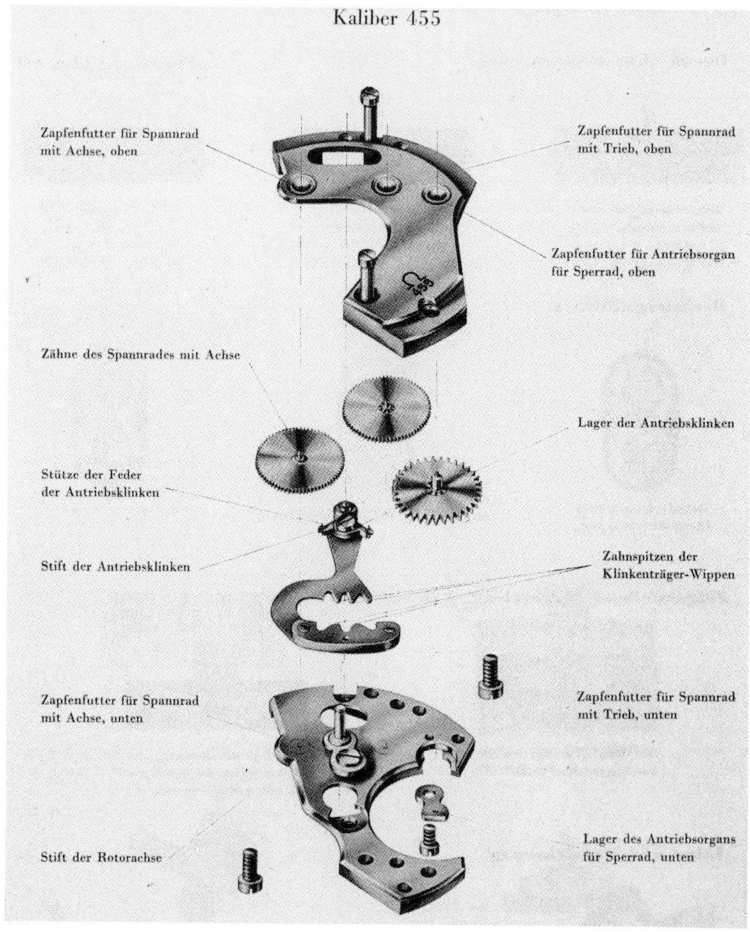

Omega Caliber 471

Men's automatic, "Seamaster" model, from 1955 with Omega's first rotor Caliber 471.

Stainless steel case with thickened press-in bottom, winding crown with Omega symbol.

Rotor with heavy metal ring, winding in both directions via switching rocker (at 6:00), uncoupling of the hand-winding drive.

Glucydur screw balance, Nivarox hairspring, Incabloc shock resistance, fine regulator adjustment by side screw.

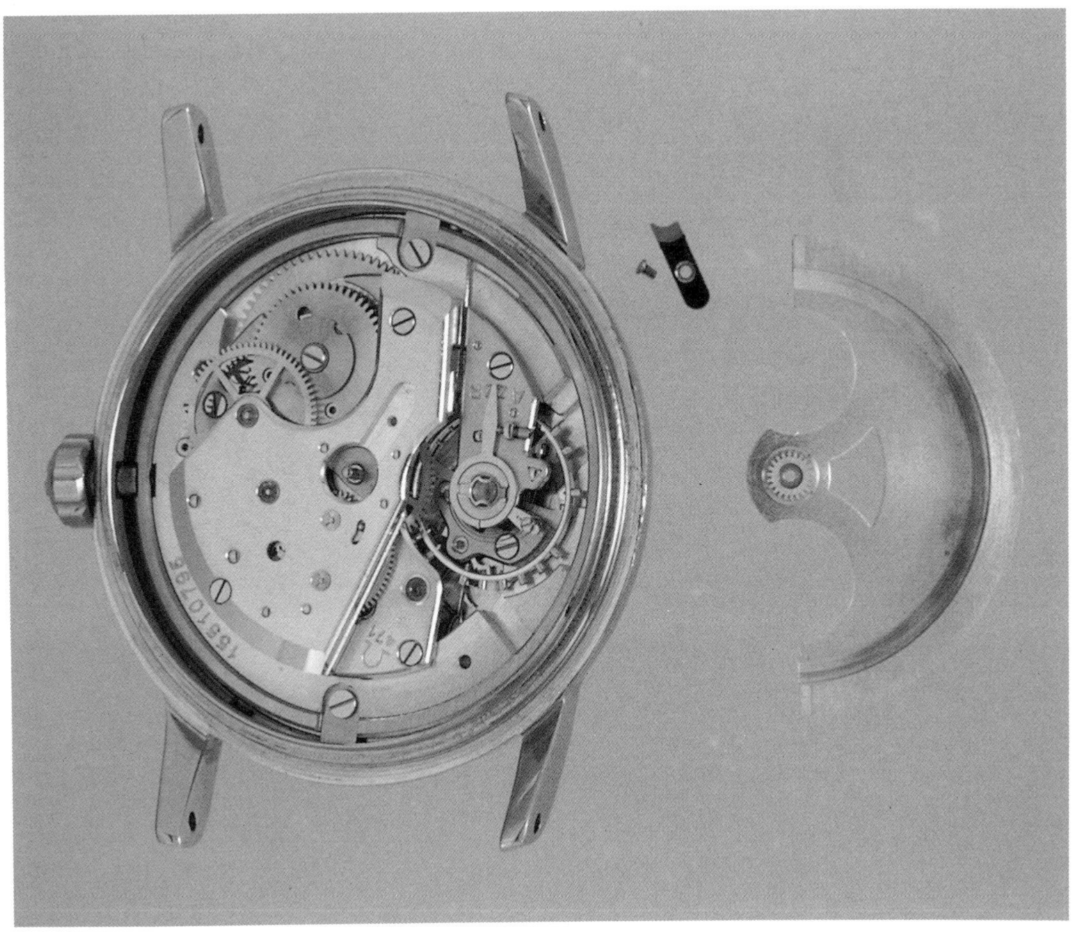

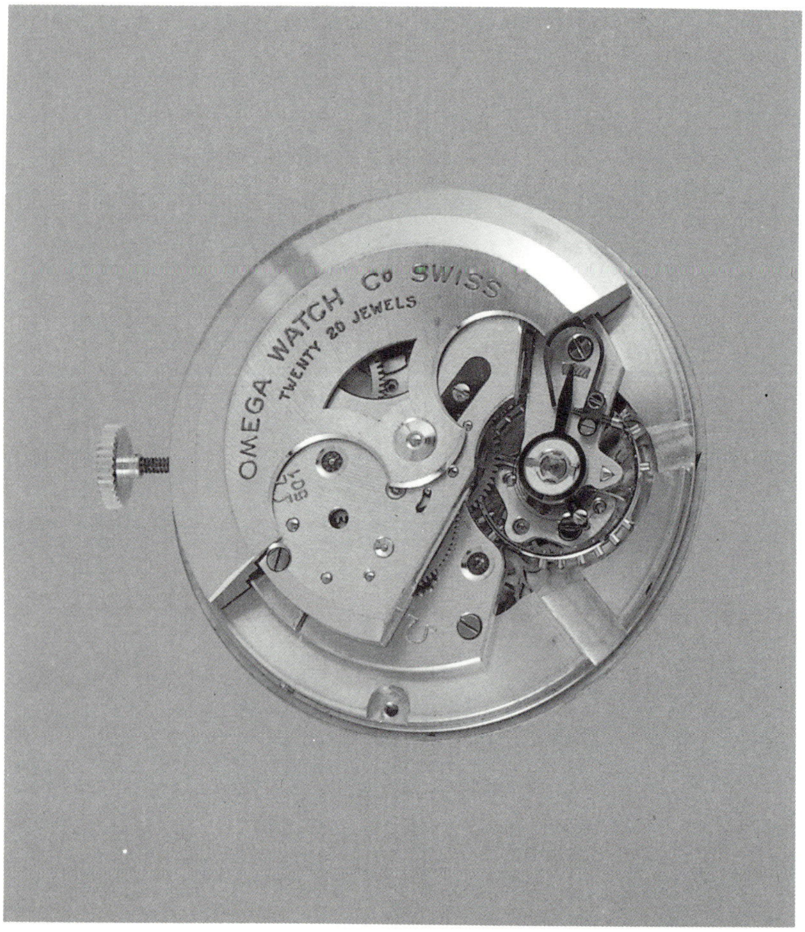

Omega Caliber 501
Automatic Movement made by Omega in 1956, with Caliber 501 for men's watches.
Movement mainly identical to Caliber 471 (see page 231), but with fine regulator adjustment by swan's-neck screw.

Omega Caliber 561

Men's automatic as "Constellation" chronometer model, made by Omega as of 1959, with Caliber 561.

Eighteen karat gold case with screw bottom (stamped observatory symbol), gold hands and dial.

Rotor with heavy metal ring, winding in both directions via patented satellite-wheel switcher, uncoupling of the hand-winding drive.

Screwless large Glucydur balance, Nivarox hairspring, Incabloc shock resistance, fine regulator adjustment by swan's-neck screw (unsuitable winding crown, rotor surface corroded).

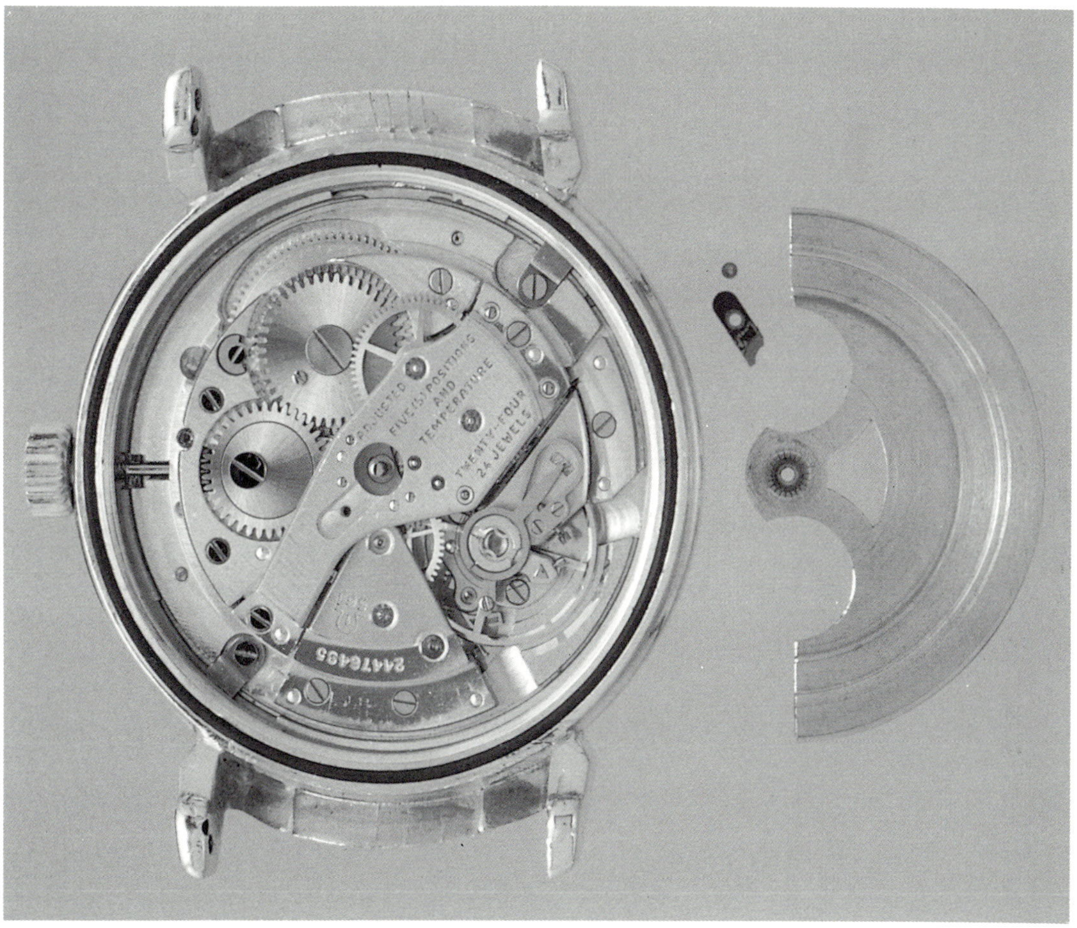

Omega Caliber 564

Men's automatic, "Seamaster" chronometer model made by Omega in 1966, with Caliber 564.

Stainless steel case with screw bottom (stamped sea-horse symbol), winding crown with Omega symbol.

Movement identical to Caliber 561 (see page 233), but with quick date correction.

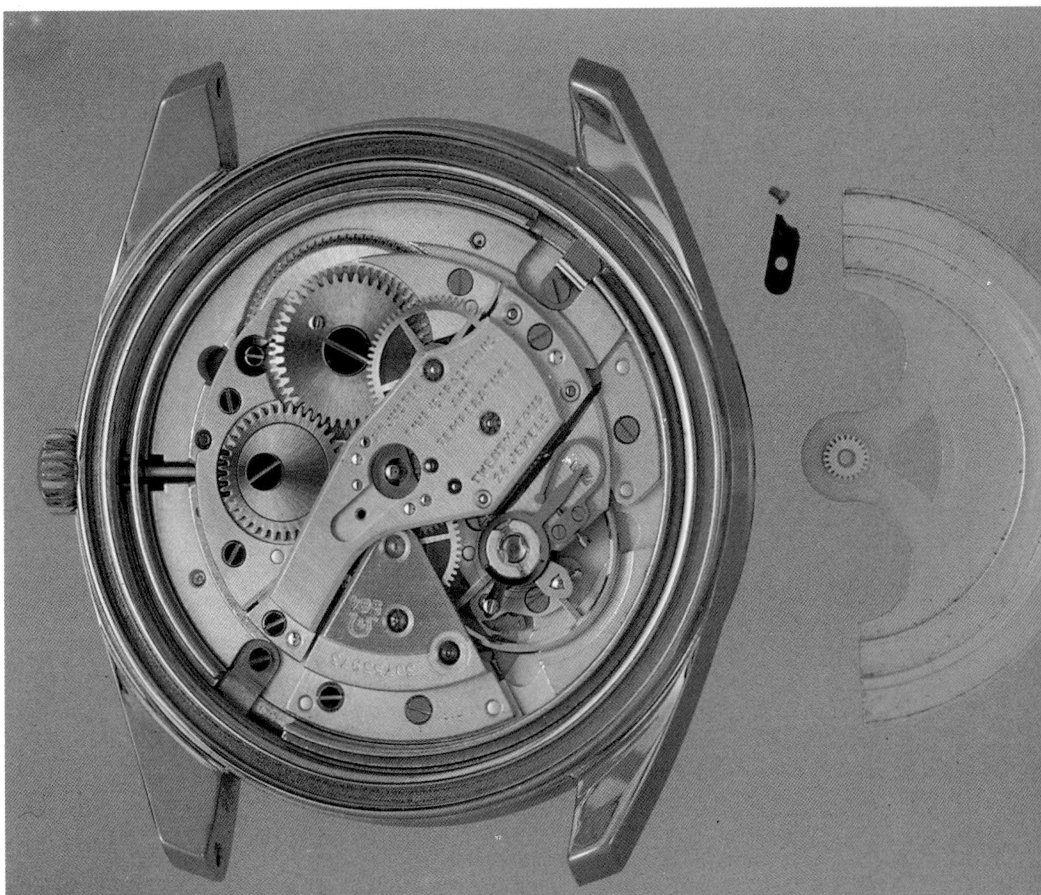

Omega Caliber 685

Ladies' automatic made by Omega in 1971, with Caliber 685.
Stainless steel case with screw bottom, winding crown with Omega symbol.
Rotor with heavy metal ring, winding in both directions via satellite-wheel switcher, uncoupling of the hand-winding drive.
Screwless Glucydur balance, Nivarox hairspring, Incabloc shock resistance, fine regulator adjustment by side screw.

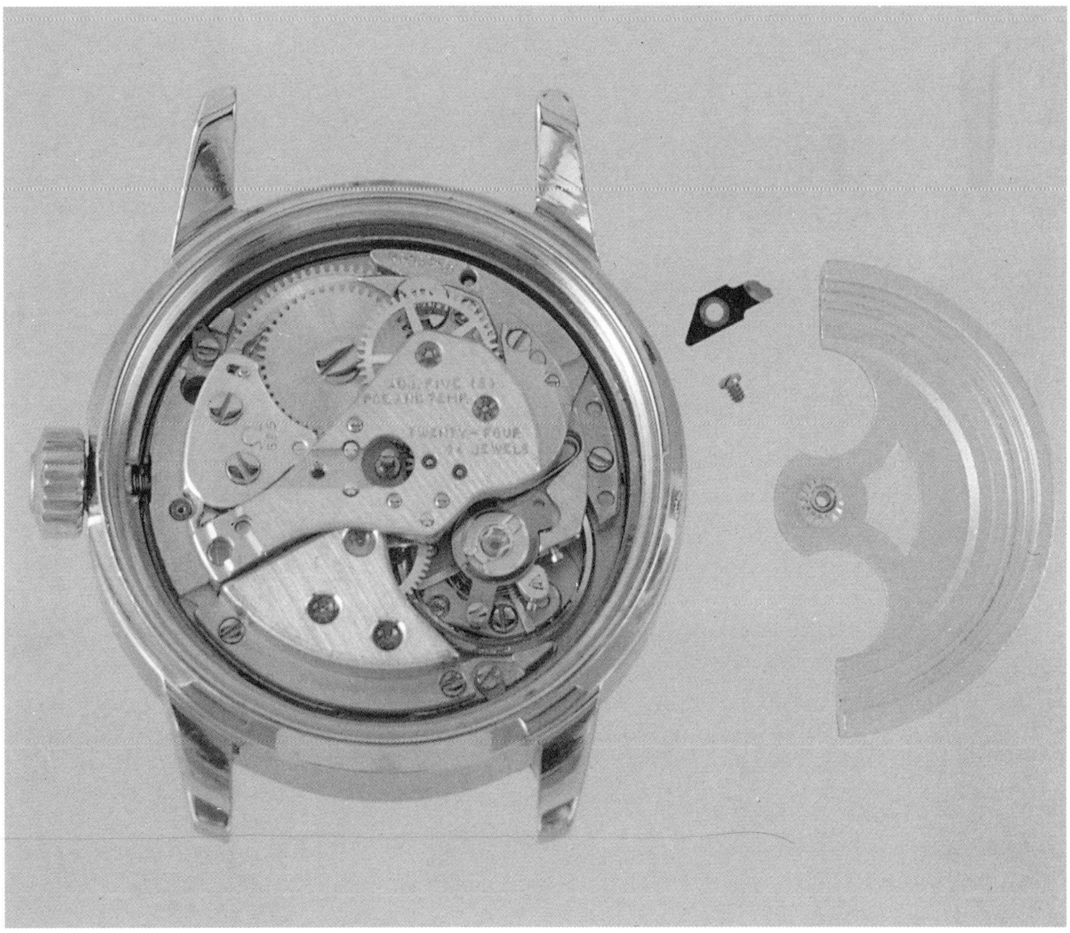

Omega Caliber 980

Automatic "Memomatic" alarm model made by Omega in 1969, with Caliber 980. Stainless steel case with screw bottom, winding crown with Omega symbol. Setting button by the 2 to stop the movement and set the alarm time to the minute. Quick date correction push-button by the 3. Rotor with ball bearings and heavy metal ring, winding in both directions via satellite-wheel switcher (at 1:00). Automatic winding of alarm mechanism by the joint barrel (at 11:00), alarm hammer strikes against sound spring (at 2:00).

Screwless Glucydur balance, Nivarox hairspring, Incabloc shock resistance, fine regulator adjustment by eccentric screw.

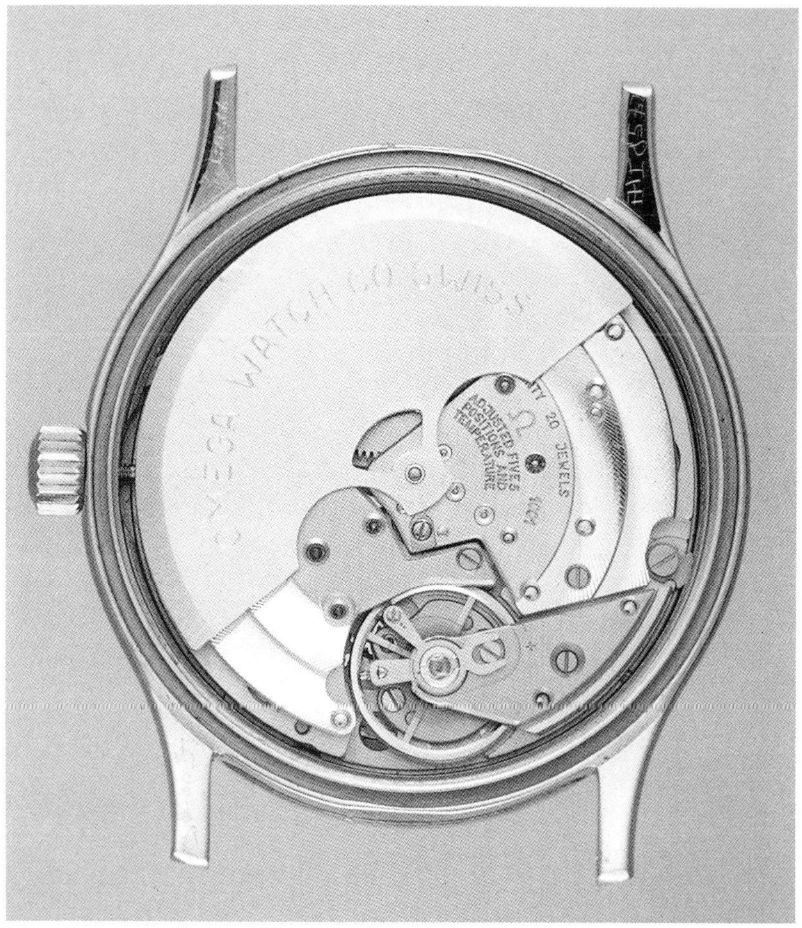

Omega Caliber 1001

Men's automatic, "Constellation" chronometer model, made by Omega in 1969, with fast beat Caliber 1001.

Stainless steel case with screw bottom (stamped observatory symbol), winding crown with Omega symbol.

Heavy metal rotor winding in both directions via satellite-wheel switcher, uncoupling of the hand-winding drive.

Screwless Glucydur balance, Nivarox hairspring, Incabloc shock resistance, fine regulator adjustment by eccentric screw.

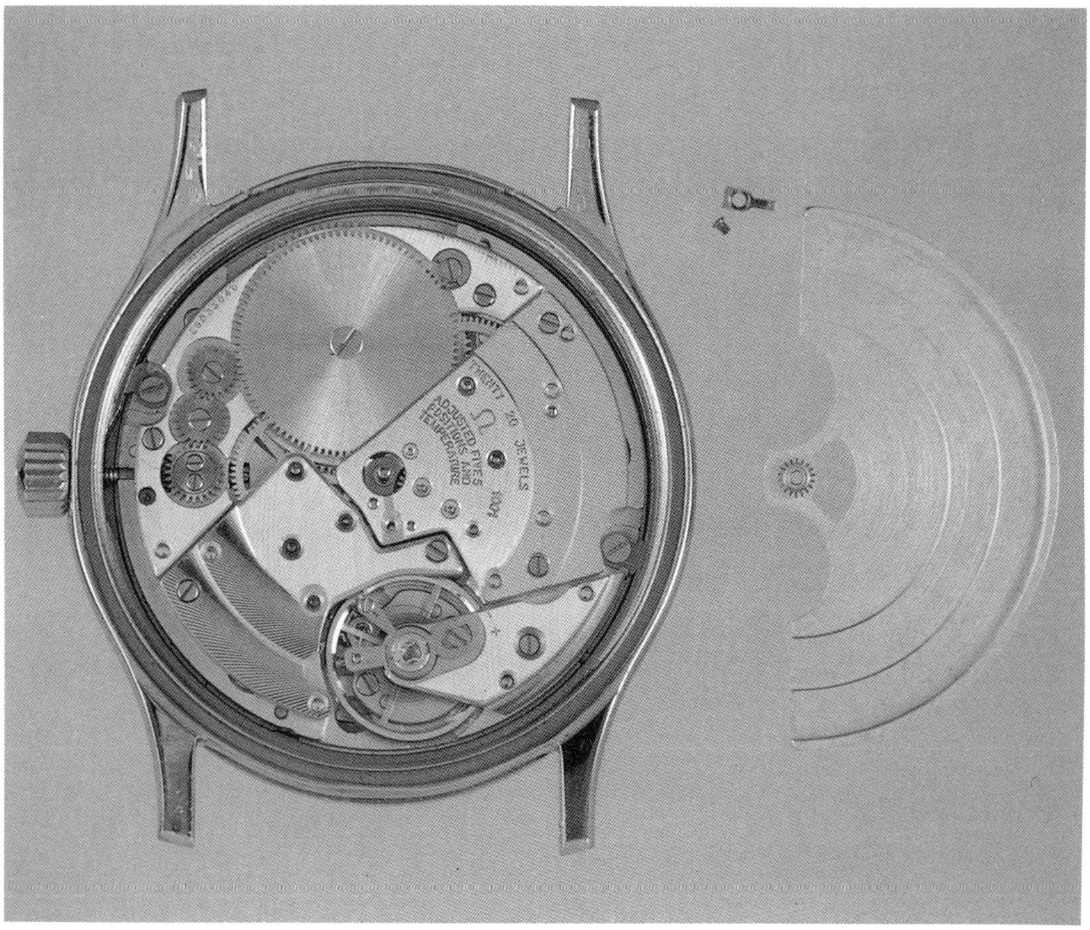

Omega Caliber 1010
Men's automatic, "Seamaster" model, made by Omega in 1972, with Caliber 1010.
Stainless steel case with screw bottom and screwed-in winding crown (with Omega emblem) and matching band. Turning bezel locking in one direction. Movement identical to Caliber 1001, but somewhat more stable.

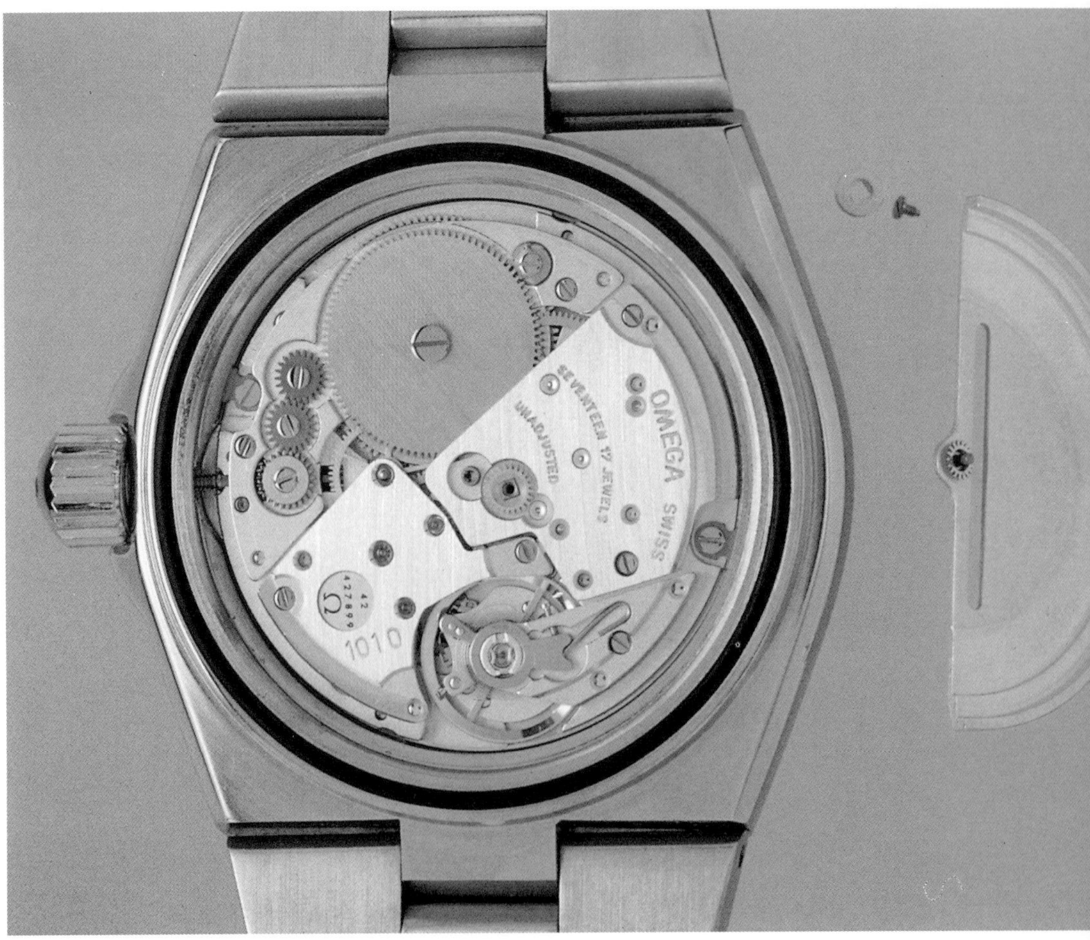

Omega Caliber 1040

Automatic chronograph, "Seamaster" model, made by Omega in 1971, with fast-swinging Caliber 1040.

Gilded case with stainless steel screw bottom, winding crown with Omega symbol.

Sweep seconds and minute-counting hands (with light markings), 12-hour counter on auxiliary dial near the 6.

Twenty-four hour indication and small second hand near the 9.

Rotor with ball bearings and heavy metal ring, winding in both directions via satellite-wheel switcher (at 11:00), chronograph mechanism with coulisse-lever switching.

Screwless Glucydur balance, Nivarox hairspring, Incabloc shock resistance, fine regulator adjustment with eccentric screw (case with clear signs of wear).

Caliber Notes

Caliber 28.10 RA PC
D = 28.1 mm, H = 4.8 mm, 17 jewels
AUTOMATIC MODULE: hammer with ruby bearings, limited by visible buffer springs, winding in one direction via click-carrier rocker on the crown wheel.
BASIC MOVEMENT: two-spoke Glucydur balance, Nivarox hairspring, 19,800 BPH, Incabloc shock resistance, small seconds hand.

Caliber 28.10 RA SC PC
D = 28.1 mm, H = 5.4 mm, 17 jewels
See Caliber 28.10 RA PC, but with indirect center seconds via double intermediate wheel.

Caliber 342
D = 28.1 mm, H = 4.8 mm, 17 jewels
See Caliber 28.10 RA PC, but with covered, loosely inserted buffer springs.

Caliber 354
D = 28.1 mm, H = 5.4 mm, 17 jewels
See Caliber 28.10 RA SC PC, but with loosely inserted, covered buffer springs.

Caliber 355
D = 28.1 mm, H = 6.1 mm, 17 jewels
See Caliber 354, but with swan's-neck fine regulation and date indication.

Caliber 455
D = 16.0 mm (7.75 lignes), H = 5.5 mm, 17 jewels
AUTOMATIC MODULE mounted on the basic movement as a complete unit, monometallic central rotor with gliding bearings, winding in both directions via a rotor disc with three teeth and three pins that contact a click-carrier rocker, the double winding click alternately moves two winding wheels in contact with each other, one reduction wheel on the locking wheel.
BASIC MOVEMENT: two-spoke Glucydur screw balance, Nivarox hairspring, 19,800 BPH, Kif-Flector shock resistance, moveable hairspring block carrier.

Caliber 471
D = 25.0 mm, H = 5.5 mm, 20 jewels
AUTOMATIC MODULE screwed to the basic movement as a complete unit, central rotor with heavy metal ring, winding in both directions via switching rocker, two reduction wheels (the first with spring click) on the double locking wheel that uncouples the hand-winding drive.
BASIC MOVEMENT: two-spoke Glucydur screw balance, Nivarox hairspring, 19,800 BPH, Incabloc shock resistance, two-part regulator hand with fine adjustment by side screw.

Caliber 501
D = 28.0 mm, H = 5.55 mm, 19 jewels
See Caliber 571, but with larger rotor and swan's-neck fine adjustment.

Caliber 561
D = 27.9 mm, H = 5.0 mm, 24 jewels
AUTOMATIC MODULE integrated into basic movement as a complete unit, central rotor with heavy metal ring, winding in both directions via springless double satellite-wheel lock (Omega patent), one reduction wheel on the double locking wheel that uncouples the hand-winding drive.
BASIC MOVEMENT: four-spoke screwless Glucydur balance, Nivarox hairspring, 19,800 BPH, Incabloc shock resistance, moveable hairspring block carrier, swan's-neck fine adjustment, date indication.

Caliber 564
D = 27.9 mm, H = 5.0 mm, 24 jewels
See Caliber 561, but with additional quick date correction.

Caliber 685
D = 17.5 mm, H = 5.05 mm, 24 jewels
AUTOMATIC MODULE partly integrated into the basic movement as a complete unit, central rotor with heavy metal ring, winding in both directions via springless double satellite-wheel lock, uncoupling of hand-winding drive by sprung pusher pinion between crown and locking wheels.
BASIC MOVEMENT: four-spoke screwless Glucydur balance, Nivarox hairspring, 21,600 BPH, Incabloc shock resistance, moveable hairspring block carrier, fine adjustment by side regulator screw, date indication.

Caliber 980
(Lemania ebauche)
D = 30.8 mm, H = 7.8 mm, 19 jewels
AUTOMATIC MODULE integrated into the basic movement under a separate potence, central rotor with ball bearings and heavy metal ring, winding in both directions via springless double satellite-wheel lock, one reduction wheel on the locking wheel (under the dial).
ALARM MECHANISM integrated into the basic movement, alarm powered by the joint barrel that is limited to one revolution for the alarm function, via the reduction wheel of the automatic drive and a second reduction wheel that is made as a planetary gear for reasons of space, to the alarm hammer and sound spring.
BASIC MOVEMENT: three-spoke screwless Glucydur balance, Nivarox hairspring, 21,600 BPH, Incabloc shock resistance, divided regulator adjustable by eccentric screw, moveable hairspring block carrier, direct center seconds, quick date correction by separate pin in the case.

Caliber 1001
D = 27.9 mm, H = 4.0 mm, 20 jewels
AUTOMATIC MODULE integrated into the basic movement, central rotor of heavy metal, winding in both directions via springless double satellite-wheel lock, one reduc-

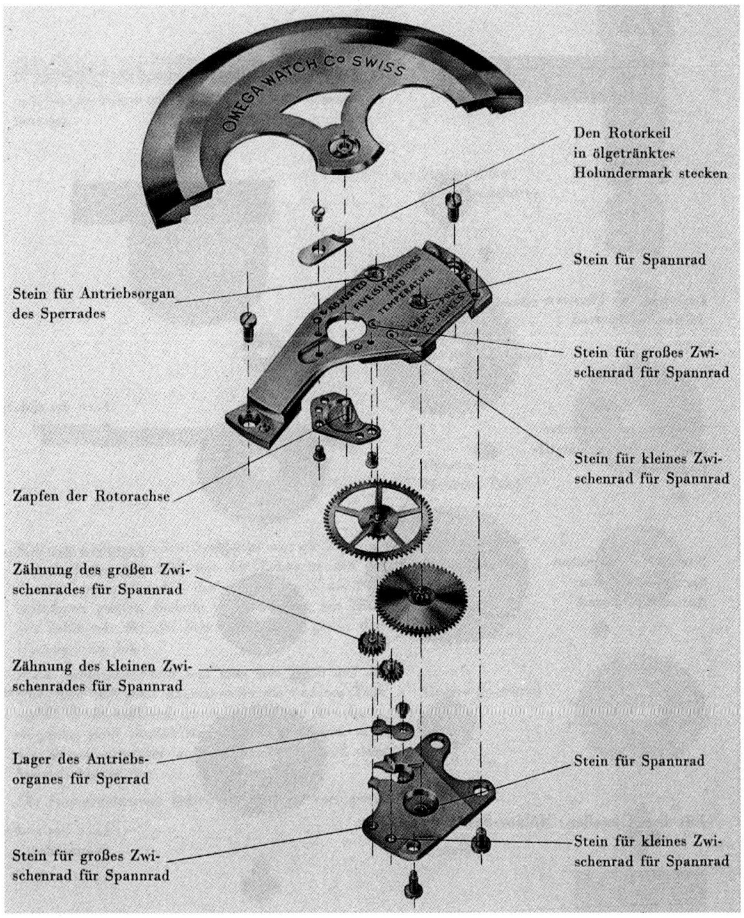

Exploded drawing of an automatic module (Omega Caliber 561): in the center is the satellite-wheel switcher, a double toothed directing lock (tension wheel), above it the single reduction wheel.

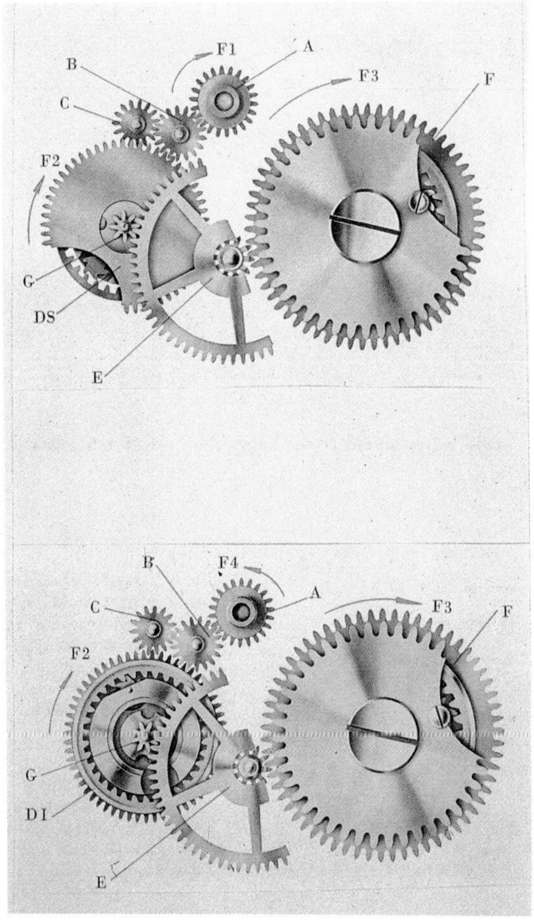

The functioning of the automatic drive (Omega Caliber 561) for both rotor directions: Double satellite-wheel lock (left) with a view of the direction-dependent special clamping teeth of the satellite wheel and the sun wheel with teeth on the inside (DS—upper, DI—lower sun wheel, A—rotor pinion, B—intermediate wheel, C—reversing wheel, E—reduction wheel, F—double locking wheel on the barrel, also serving as satellite lock).

tion wheel on the locking wheel, toothed-wheel rocker between crown and locking wheels to uncouple the hand-winding drive.
BASIC MOVEMENT: four-spoke screwless glucydur balance, Nivarox hairspring, 28,000 BPH, Incabloc shock resistance, moveable hairspring block carrier, fine regulator adjustment by eccentric screw, second stop, date indication with quick change and correction.

Caliber 1010
D = 27.9 mm, H = 4.8 mm, 17 jewels
See Caliber 1001

Caliber 1040
(Lemania ebauche)
D = 31.0 mm, H = 8.0 mm, 22 jewels
AUTOMATIC MODULE partly integrated into the basic module, central rotor with ball bearings and heavy metal ring, winding in both directions via springless double satellite-wheel lock, two reduction wheels (the first in planetary form) on the crown wheel.
Chronograph apparatus with coulisse lever switching mounted on the basic movement.
BASIC MOVEMENT: three-spoke screwless Glucydur balance, Nivarox hairspring, 28,000 BPH, Incabloc shock resistance, moveable hairspring block carrier, fine regulator adjustment with eccentric screw, date indication with quick change and correction.

Caliber List: Omega

Caliber	Diameter Lignes	mm	Height mm	Other	BPH	Jewels	Running Reserve	Notes	Year	Photo
28.10 RA PC		28,1	4,8	s	19.800	17	36		1943	xx
28.10 RA SC PC		"	5,4	SCI	"	"	"		1945	xx
30.10 RA PC		30,1	4,55	s	"	"	"		1943	
330		"	"	s	"	"	"	See 30.10 RA PC	1943	
331		"	"	s	"	"	"	Centernaire-	1947	
								Chronom. (JUB		
332		"	"	s	"	"	"	30.10 RA PC)	1949	
333		"	"	s	"	"	"		1950	
340		28,1	4,8	s	"	"	"	See 28.10 RA PC	1943	
341		"	"	s	"	"	"	Centenaire-Chronom..	1947	
								(JU 2810 RA PC)		
342		"	"	s	"	"	"		1949	xx
343		"	"	s	"	"	"		1950	
344		"	"	s	"	"	"		1953	
350		"	5,4	SCI	"	"	"	See 28.10 RA SC PC	1945	
351		"	"	SCI	"	"	"		1949	
352		"	"	SCI	"	"	"		1950	
353		"	6,1	SCI, QG	"	"	"		1951	
354		"	5,4	SCI	"	"	"		1953	xx
355		"	6,1	SCI, QG	"	"	"		1953	xx
450	7 3/4	16,0		s	19.800	17	36	Ladymatic	1955	
455	"	"	5,5	SCI	"	"	"	"	1955	xx
470	11	25,0	5,5	SCI	19.800	17	40		1955	
471	"	"	"	SCI	"	19/20	"		1955/56	xx
490		28,0	"	s	"	17	"		1956	
491		"	"	s	"	19	"		1956	
500		"	5,55	SCI	"	17	"		1956	
501		"	"	SCI	"	19	"		1956	xx
502		"	6,4	SCI, QG	"	17	"		1956	
503		"	"	SCI, QG	"	19/20	"		1956	
504		"	"	SCI, QG	"	24	"		1957	
505		"	5,55	SCI	"	24	"		1957	
550	12 1/4	27,9	4,5	SCI	19.800	17	50		1959	
551	"	"	"	SCI	"	24	"	Constellation-Chron.	1959	
552	"	"	"	SCI	"	"	"		1959	
560	"	"	5,0	SCI, QG	"	17	"		1959	
561	"	"	"	SCI	"	24	"	Constellation-Chron.	1959	xx
562	"	"	"	SCI, QG	"	"	"		1959	
563	"	"	"	SCI, QG, RDR	"	17	"		1966	
564	"	"	"	SCI, QG, RDR	"	24	"	Chronometer	1966	xx
565	"	"	"	SCI, QG, RDR	"	"	"		1966	
570		24,9	4,5	SCI	19.800	17			1959	
571		"	"	SCI	"	24			1959	

Caliber List: Omega

Caliber	Diameter Lignes mm	Height mm	Other	BPH	Jewels	Running Reserve	Notes	Year	Photo
590	27,8	5,35	SCI	19.800	17	49		1960	
591	"	"	SCI	"	20	"		1960	
592	"	"	SCI	"	17	"		1960	
593	"	"	SCI	"	20	"		1961	
660	15,0	4,85		19.800	17			1963	
661	"	"		"	24			1963	
670	17,5	"	SCI	"	17	45		1963	
671	"	"	SCI	"	24	"		1963	
672	"	"	SCI	"	"	"	Chronometer	1968	
680	"	5,0	SCI, QG	"	17	"		1963	
681	"	"	SCI, QG	"	24	"		1963	
682	"	"	SCI, QG	"	"	"	Chronometer	1968	
683	"	5,05	SCI, QG	21.600	17	"		1971	
684	"	"	SCI, QG	"	24	"		1971	
685	"	"	SCI, QG	"	"	"		1971	xx
710	25,0	3,0	SCI	19.800	17	43		1966	
711	"	"	SCI (24 Hrs.)	"	24	"		1966	
712	"	"	SCI/s	"	"	"	Chronometer	1967	
750	27,9	5,68	SCI, QG, JG, RDR	19.800	17		modif. Cal 550	1967	
751	"	"	SCI, QG, JG, RDR (24 Hrs.)	"	24		"	1967	
752	"	"	SCI, QG, JG, RDR	"	"		"	1967	
980	30,8	7,8	SCD, QG, RDR, REV	21.600	19	48 + 1	Memomatic	1969	xx
1000	27,9	4,0	SCI, QG, RDR, ST	28.800	17	50		1969	
1001	"	"	SCI, QG, RDR, ST	"	20	"	Constellation-Chron.	1969	xx
1002	"	"	SCI, QG, RDR, ST	"	"	"		1969	
1010	27,9	4,8	SCI, QG, RDR, ST	28.800	17			1972	xx
1011	"	"	SCI, QG, RDR, ST	"			Chronometer		
1012	"	"	SCI, QG, RDR, ST	"					
1020	"		SCI, QG, JG, RDR, ST	"					
1021	"		SCI, QG, JG, RDR, ST	"			Chronometer		
1022	"		SCI, QG, JG, RDR, ST	"					
1040	31,0	8,0	QG, RDR, CHR	28.000	22		Chronograf	1971	xx
1480	25,6		SCD, QG, RDR	21.600					
1481	"		SCD, QG, RDR	"					

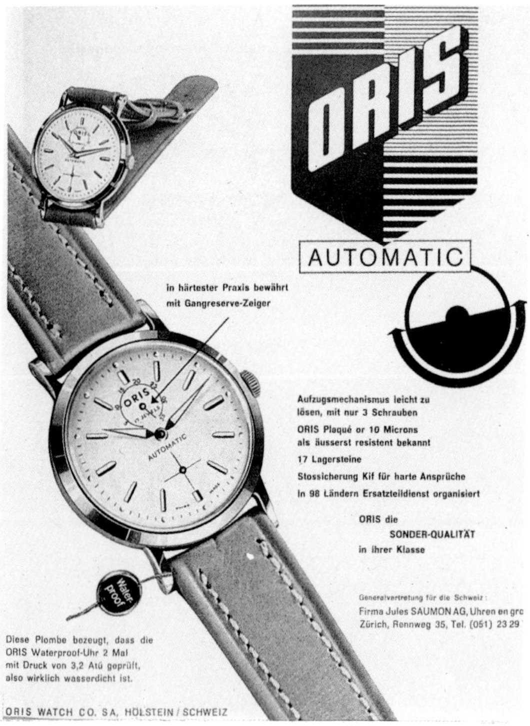

Magazine advertisement (from SUZ, 1960)

Caliber Development

Georges Christian and Paul Cattin founded the company at Hölstein in 1904; by 1938 it employed about 1000 people and produced almost exclusively pin lever watches. In 1971 the firm entered the ASUAG (Allgemeine Schweizerische Uhrenindustrie AG).

The firm's first automatic development, the Caliber 605 of 1953, was also fitted with a pin lever escapement and its central rotor wound in both directions via twin click wheels. Although the movement was relatively small, a differential drive could be installed above the barrel so the running reserve could be shown on an auxiliary dial under the 12.

After the old tradition that allowed Oris to use only pin lever escapements was abandoned, the new automatic Caliber 645 was built with a jewel-anchor escapement and kept in production until 1986. The equalization of the rotor movements, which wound in both directions, was carried out here by a centrally mounted lever switcher.

Caliber Notes

Caliber 605
D = 27 mm (12 lignes), 7 jewels
AUTOMATIC MODULE screwed to the basic movement as a complete unit, monometallic central rotor, winding in both directions via twin click wheels, two reduction wheels and one pinion arbor on the locking wheel under the dial.
BASIC MOVEMENT: two-spoke screwless nickel balance, Nivarox hairspring, 18,000 BPH, Kif shock resistance, pin lever escapement, eccentric seconds hand, differential drive over the barrel to indicate running reserve.

Caliber list: Oris

Caliber	Diameter Lignes	mm	Height mm	Other	BPH	Jewels	Running Reserve	Notes	Year	Photo
601	12	27			18.000			(Pin lever)		
605	12	27		s, RMA	18.000	7/17	35	(Pin lever)	1953	xx
645	12	27	6,25	SCD, QG	18.000	21/25	40		1966	
648	"	"	6,85	SCD, QG, JG, RDR	"	21	"		1969	

Oris Caliber 605

Men's automatic made by Oris in 1953, with Caliber 605.

Chrome plated case with aluminum screw bottom, up and down indicator on auxiliary dial under the 12.

Monometallic rotor winding in both directions via click-wheel switcher (at 5:30 and 7:00).

Ring balance, Nivarox hairspring, Kif-Protechoc shock resistance, Roskopf-type pin lever escapement (case chrome flecked).

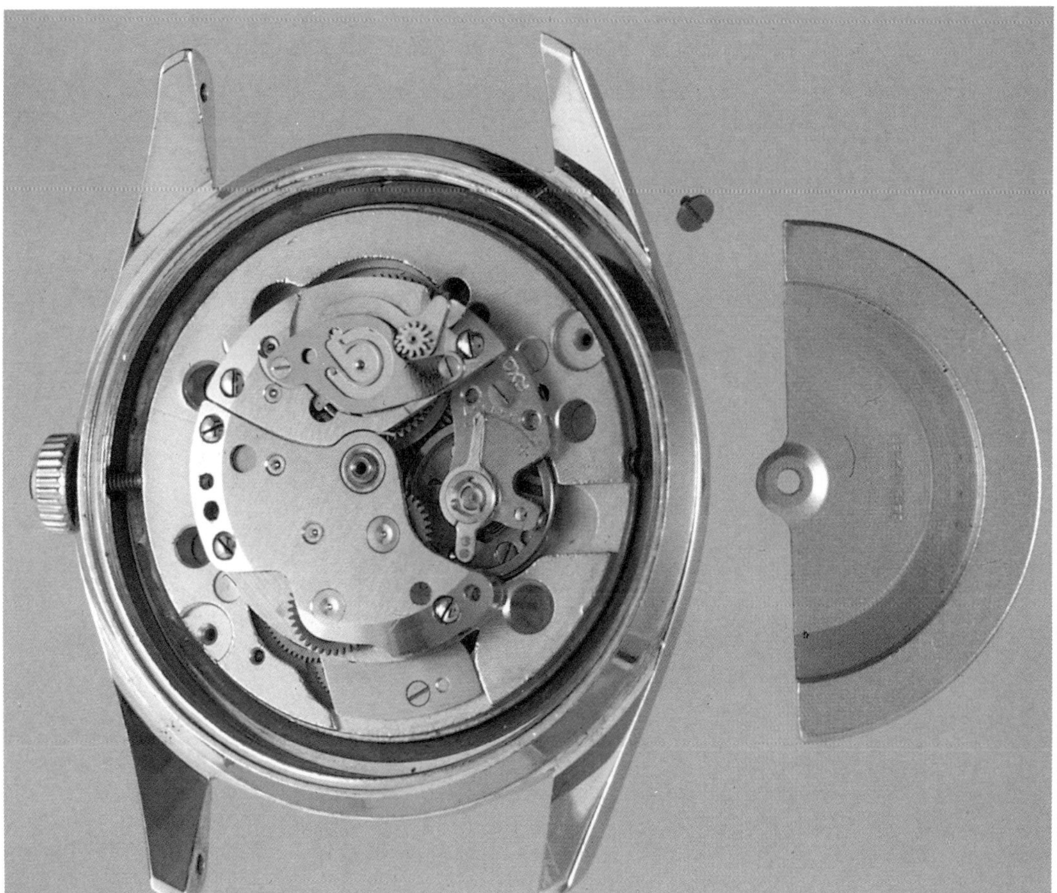

Caliber Development

In 1839 the Polish emigrant Antoine Norbert de Patek and the exiled Czech Frantisek Czapek founded the firm of "Patek, Czapek & Co." in Geneva. In 1851, when Jean Adrien Philippe joined the firm, it was renamed "Patek Philippe & Co." (PPC). In 1842 Adrien Philippe invented crown winding and also indirectly lessened the chances for further development of a self-winding pocket watch. For this new development eliminated the bothersome use of a separate winding key, which was the main argument in favor of automatic winding.

The lack of demand from the firm's very conservatively oriented customers was probably a reason for the relatively late development of the firm's first automatic caliber in 1953. With this design, Caliber 12-600 AT, the firm then tried to offer a first-class product of automatic technology. The swinging mass was a massive rotor of 18-karat gold, which provided very efficient winding of the precision movement via an elaborately built eccentric switcher. But after just a short time, the designers had to make improvements such as fastening the movement more firmly in the case, installing miniature ball bearings on the eccentric as well as a ruby jewel between the click levers. To improve the running further, the screw balance was replaced by a patented Gyromax type.

The movement design of the Caliber 27-460, introduced in 1960, was retained from its forerunner, but its height was reduced through the use of ball bearings for the rotor. In 1961 this movement was fitted with a date mechanism (Caliber 27-460 M) and as of 1962 the first automatic with a "perpetual" calendar, Caliber 27-460 Q, could be introduced, with an appropriate module added between the basic movement and the dial.

In order to satisfy the general demand for a flatter movement, PPC developed Caliber 350 in 1969. The tried and true central rotor design was discarded here, and a half-ring rotor running peripherally on ball bearings was chosen. In its open inner area, the equally high movement with integrated automatic drive was located, but the winding crown had to be moved to the back of the watch. As a model, the engineers took the Caliber 65, developed by the FHF raw-movement factory in 1959.

Unfortunately, this design proved to be very likely to break down and also very difficult to repair. The "Golden Ellipse" models were equipped with it. Only ten years later, the entire automatic mechanism was redesigned and introduced as Caliber I-350. The half-ring rotor, now made of 18-karat gold, wound in only one direction. The number of transmission elements in the reduction drive could be reduced from seven to five, but the height of the movement remained the same.

Meanwhile, a very flat raw movement by Jaeger LeCoultre was used; it was installed in modified form in the first Nautilus models from 1970 to 1980 as Caliber 28-255 without or 28-255 C with date indication.

In 1977 Patek introduced Caliber 240, one of the flattest automatic designs ever. As of 1985, as Caliber 240 Q, it became the successor to the first automatic with "perpetual" calendar, and it is still in production today. Here too, the microrotor- or planetary-rotor automatic developed by Universal and Buren in 1958 served as models, but the particularly low height was achieved by simplifying the self-winding mechanism. For unlike the models, winding was done in only one direction here and required a very effective planetary rotor, here made of 22-karat gold and running on ball bearings.

In 1981, Patek's last automatic caliber to date, numbered 335 SC, was developed. It was made flatter in 1983, as Caliber 310 SC, by modifications to the date mechanism. It was another design with a central rotor, this time of 18-karat gold, winding in only one direction.

Caliber Notes

Caliber 12-600 AT
D = 27.0 mm (12 lignes), H = 5.45 mm, 30 jewels
AUTOMATIC MODULE screwed to the basic movement under two plates, 18-karat gold central rotor with ruby bearings, winding in both directions via separate eccentric wheel, primary and subsequently switched secondary click levers, each with a click, that alternately switch the combined winding and reduction wheel further, second reduction wheel on the crown wheel.
BASIC MOVEMENT: balance, self-compensating Breguet hairspring, 19,800 BPH, Super-Shock resistance, swan's-neck fine regulation, escape wheel mount with cap jewels, eccentric seconds hand, Nerfos sliding stud.
Note: Between serial numbers 760,000 and 760,950, the following modifications, among others, were made:
1. Strengthening the rotor arbor,
2. Replacing the eccentric-wheel cam with ball bearings (1956),
3. Using ruby jewels between the primary and secondary click levers,
4. Replacing the Glucydur screw balance with a Gyromax balance,
5. Increasing the movement-attaching eyes from two to three.

Caliber 27-460
D = 27.0 mm (12 lignes), H = 4.6 mm, 37 jewels
AUTOMATIC MODULE: See Caliber 12-600 AT, but with ball bearings for the central rotor.
BASIC MOVEMENT: See Caliber 12-600 AT, but with moveable hairspring block carrier.

Notes:
1. Eloxied barrel of aluminum alloy beginning with serial number 1,115,600,

Patek Philippe Caliber 12-600 AT

Men's automatic made by Patek Philippe circa 1958, with improved form of first automatic Caliber 12-600 AT, made since 1953.

Eighteen karat gold case with press-in bottom, gold crown with PP symbol, hands and dial inlays.

Eighteen karat gold rotor with ruby bearings, winding in both directions via eccentric switcher (at 11:00).

Gyromax balance, self-compensating Breguet hairspring, Super-Shock-Resist system, fine regulator adjustment by side screw with swan's-neck spring.

Patek Philippe Caliber 27-460

Men's automatic made by Patek Philippe circa 1965, with Caliber 27-460. Eighteen karat gold case with screw bottom, gold winding crown with PP symbol.

Movement mainly identical to Caliber 12-600 AT (see page 247), but with ball bearings for rotor, Kif-Ultraflex shock resistance, no regulator.

Patek Philippe Caliber 1-350
Men's automatic made by Patek Philippe from 1979 to 1985, with Caliber 1-350. Eighteen karat gold case, flat bottom crown, gold hands and dial inlays. Eighteen karat half-ring rotor running on outside ball bearings, winding in one direction via rotor arc with inside teeth and click wheel.
Gyromax balance, self-compensating Breguet hairspring, Kif-Ultraflex shock resistance, no regulator.

Patek Philippe Caliber 28-255 C
Men's automatic, "Nautilus" model, made by Patek Philippe in 1970, with Caliber 28-255 C (exclusive raw movement by Jaeger LeCoultre).
Screwed case with integrated stainless steel or 18-karat band, waterproof to 120 meters.
Centrally mounted rotor with 21-karat gold ring and outer ring, supported by four peripheral ruby rollers, winding in both directions via switching rocker, uncoupling during hand-winding.
Gyromax balance, self-compensating flat hairspring, Kif-Ultraflex shock resistance, no regulator.

Patek Philippe Caliber 335 C
Men's automatic, "Nautilus" model, made by Patek Philippe in 1981, with Caliber 335 C.
Screwed stainless steel or 18-karat gold case with integrated band, waterproof to 120 meters.
Eighteen karat gold rotor with ball bearings, winding in one direction via click wheel.
Screwless Glucydur balance, Nivarox hairspring, Incabloc shock resistance, Triovis fine regulation.

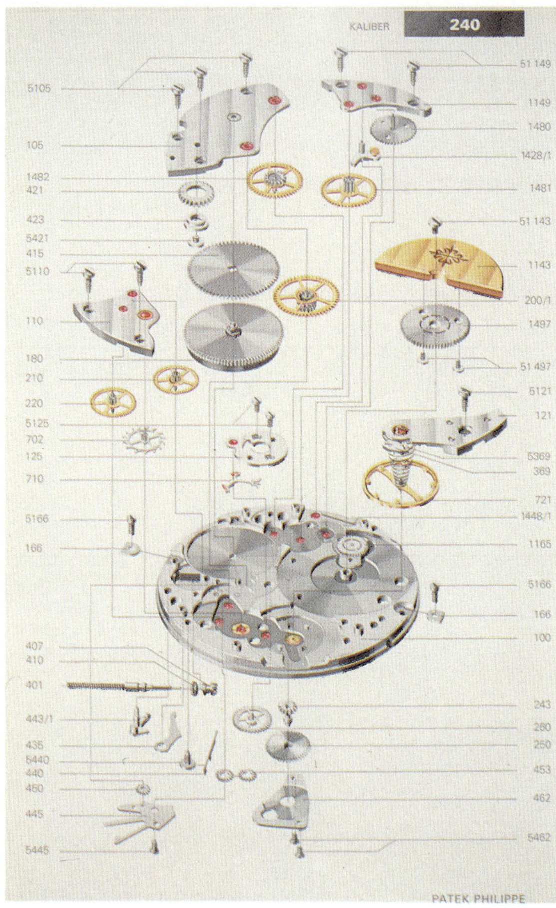

Patek Philippe Caliber 240/Q

Men's automatic with "perpetual" calendar, made by Patek Philippe as of 1985, with Caliber 240 Q.
Twenty-two karat gold microrotor (1143) with ball bearings, winding in one direction via toothed-wheel rocker (1165). Gyromax balance, self-compensating flat hairspring, Kif-Ultraflex shock resistance, no regulator (numbers in parentheses refer to the exploded view).
Calendar module (Caliber 240) between dial and basic movement.

Caliber List: PPC

Caliber	Diameter Lignes	mm	Height mm	Other	BPH	Jewels	Running Reserve	Notes	Year	Photo
12-600 AT	12	27,0	5,45	s	19.800	30			53 (-60)	xx
27-460	12	27,0	4,6	s	19.800	37			60 (-70)	xx
27-460 M	"	"	5,35	s, QG	"	"			61 (-70)	
27-460 Q	"	"		Perpetual Calendar	"				1962	
27-460 QB	"	"		Perpetual Cal. m. Schaltjahr	"				80 (-85)	
350	12 1/2	28,0	3,5	SCD	21.600	28		Ring motor, both direc	1969	
1-350	"	"	"	SCD	"	"		Ring motor, one direc	79 (-85)	xx
335 SC	11 1/2	26,0	3,45	SCI, QG, RDR	28.800	29	44	Date Mechanism	1981	xx
310 SC	"	"	3,15	SCI, QG, RDR	"	"	"	changed	1983	
28-255	12 1/2	28,0	2,55		19.800	36		Jaeger Le Coultre	70 (-80)	
28-255 C	"	"	3,15	QG	"	"		Ebauche	70 (-80)	xx
240	12	27,0	2,4		21.600	27	48	Micro-Motor	1977	xx
240 Q	13 1/4	30,0	3,75	„ewiger" Kalender	"	"	"	"	1985	

2. Number of ruby balls in rotor mount raised from 5 to 7 as of serial number 1,110,600,
3. Anchor escapement of beryllium bronze instead of steel as of serial number 1,115,000

Caliber 1-350
D = 28.0 mm (12.5 lignes), H = 3.5 mm, 28 jewels
AUTOMATIC MODULE integrated into the basic movement, 18-karat gold half-ring rotor with peripheral ball bearings, winding in one direction via rotor arc with inside teeth, click wheel, two reduction wheels and one tension wheel on the locking wheel.
BASIC MOVEMENT: two-spoke Gyromax balance with inside equalizing discs, self-compensating Breguet hairspring, 21,600 BPH, Kif shock resistance, moveable, adjustable hairspring block carrier, without regulator, direct center seconds, hand-setting and winding by bottom crown.

Caliber 28-255 C
Jeager-LeCoultre Ebauche
D = 28.0 mm (12.5 lignes), H = 3.15 mm, 36 jewels
AUTOMATIC MODULE integrated into the basic movement under the same plate, central rotor with ruby bearings, 21-karat gold ring and peripheral ring for mount support by four ruby rollers recessed in the bottom plate, winding in both directions via an intermediate wheel with ruby bearings, one small steel reduction wheel, switching rocker, two more reduction wheels (the first with spring click), one coupling wheel on the locking wheel for uncoupling during hand-winding.
BASIC MOVEMENT: two-spoke Gyromax balance, flat Nivarox hairspring, 19,800 BPH, Kif-Ultraflex shock resistance, moveable hairspring block carrier, no regulator indicator, escape wheel mounted in combined jewel bearings, no second, "flying" barrel mount, date indication with quick change.

Caliber 335 SC
D = 26.0 mm (11.5 lignes), H = 3.45 mm, 29 jewels
AUTOMATIC MODULE integrated into the basic movement, central rotor of 18-karat gold, winding in one direction via a combined winding-click and locking wheel, one reduction wheel, one intermediate wheel on the locking wheel of the barrel.
BASIC MOVEMENT: three-spoke screwless Glucydur balance, Nivarox flat hairspring, 28,800 BPH, Incabloc shock resistance, Triovis fine adjustment with moveable hairspring block carrier, escape wheel mounted between combined jewel housings, eccentrically mounted minute wheel with date indication with semi-quick change and quick correction.

Caliber 240 Q
D = 30.0 mm, H = 3.75 mm, 27 jewels
AUTOMATIC MODULE recessed in the movement out of center, 22-karat gold "Micro" rotor with ball bearings, winding in one direction via an intermediate wheel with ball bearings on a sprung rocker, three reduction wheels (the first of steel, with a spring click) on the locking wheel.
BASIC MOVEMENT: two-spoke Gyromax balance, self-compensating flat hairspring, 21,600 BPH, Kif-Ultraflex shock resistance, moveable hairspring block carrier, escape wheel mounted in combined jewel housings, no second, perpetual calendar as additional module between dial and movement.

Caliber Development

The Phenix Watch Co. S.A. was founded at Porrentruy in 1873 by Dubail, Monnin and Frossard. In 1962 it merged with the Buser, Revue and Vulcain firms, forming the MSR (Manufactures d'Horlogerie Suisse Réunies S.A.).

Phenix built its first automatic at the end of the 1930s and marketed it under its brand name "Eros". It was equipped with a hammer winding in one direction, the entire module set over the dependable hand-wound Caliber 89 like an inverted bowl. Only the regulator indicator can be reached through an opening in the basic movement, so that this design, like the Rolex Perpetual, offers additional dust protection.

For its rotor automatic (Caliber 200) of 1954, Phenix developed a simple eccentric lever drive. The strongly reducing long click lever of this design required no other element than the combined winding and reduction wheel, and along with the three-cam rotor eccentric, guaranteed very effective self-winding. The rotor ran on a central roller bearing, hence the name "Rollamatic", and was, though only to a limited extent, axially shock-resistant thanks to a spring bolt.

Phenix does not appear to have produced any other automatic movements; the listed Caliber 400 was presumably not made by Phenix.

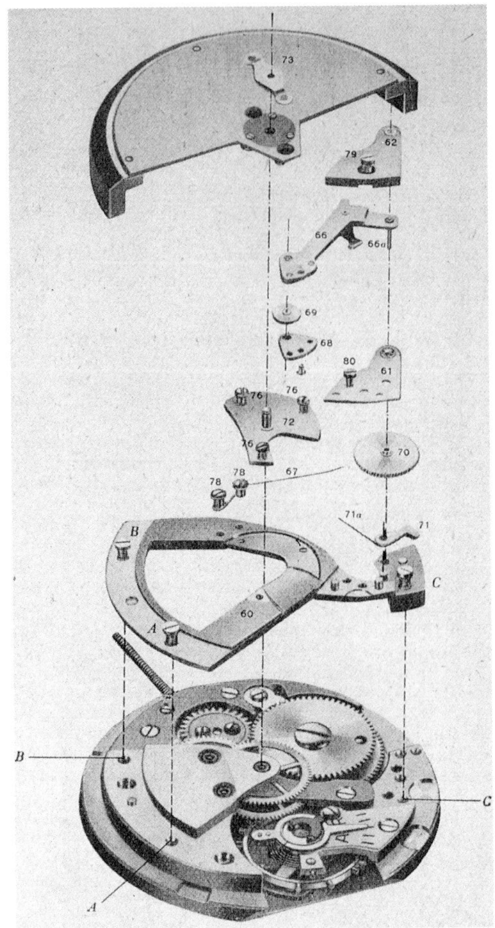

Exploded view of Phenix Caliber 200 (automatic module):
66—click lever with 70—click wheel
66a—winding click 71—locking click

Caliber Notes

Caliber (89)—Eros
D = 26 mm (11.5 lignes), 18 jewels
AUTOMATIC MODULE pressed bowl-like over the basic movement (Phenix Caliber 89) and fastened with one screw, hammer winding in one direction via hammer click, centrally mounted click wheel, two reduction wheels, the first with a spring click, on the locking wheel.
BASIC MOVEMENT (Caliber 89, 8.75 lignes, H = 3.6 mm, 15 jewels): two-spoke Glucydur screw balance, self-compensating flat hairspring, 18,000 BPH, no shock resistance, eccentric seconds hand.

Caliber 200 ("Rollador" by Nivada)
D = 26.0 mm (11.5 lignes), H = 6.35 mm, 21 jewels
AUTOMATIC MODULE screwed to the basic movement like a frame as a complete unit, central rotor with roller bearings and heavy metal ring, winding in both directions via three-cam rotor eccentric, click lever with steel roller in ruby bearings, peripherally mounted winding wheel on the locking wheel, double locking-wheel coupling to uncouple the hand-winding drive.
BASIC MOVEMENT: two-spoke Glucydur screw balance, Nivarox hairspring, 18,000 BPH, Incabloc shock resistance, direct center seconds.

Caliber List: Phenix

Caliber	Diameter Lignes	mm	Height mm	Other	BPH	Jewels	Running Reserve	Notes	Year	Photo
(89)	11 1/2	26,0		s	18.000	18		Eros	ca. 40	xx
200	11 1/2	26,0	6,35	SCD	18.000	17/21	40	Rollamatic, see	1954	xx
205	"	"		SCD, QG	"	"	"	Rollador by Nivada		
400	11 1/2	25,6	5,2	SC	18.000					

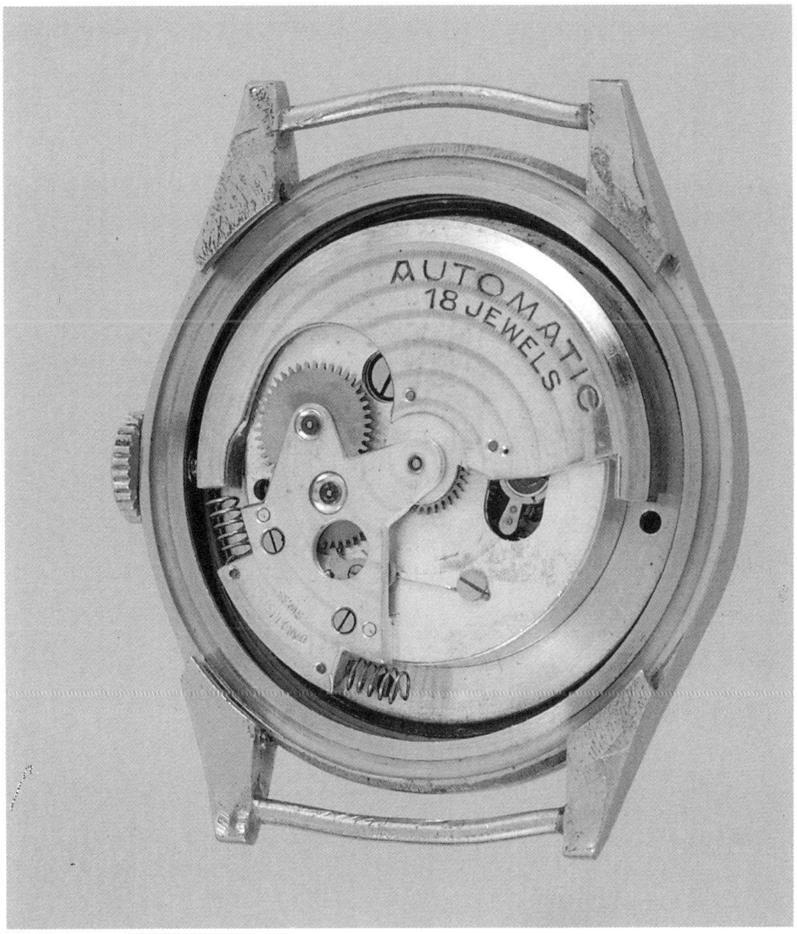

Phenix/Eros Caliber (89)
Men's automatic made by Eros circa 1940, with hand-winding Caliber 89 made by Phenix.
Chromed Staybrite case, stainless steel bottom with bayonet lock.
Hammer winding in one direction via Harwood-type automatic drive. Automatic module as a complete unit placed bowl-like over basic movement for dust protection.
Glucydur screw balance, Nivarox hairspring, no shock resistance.

Phenix Caliber 200

Men's automatic, "Rollador" model, made by Nivada circa 1955, with Rollamatic 200 raw movement made by Phenix.

Square gilded case with stainless steel press-in bowl bottom.

Rotor with ball bearings and heavy metal ring, winding in both directions via three-cam eccentric switcher.

Glucydur screw balance, Nivarox hair-spring, Incabloc shock resistance.

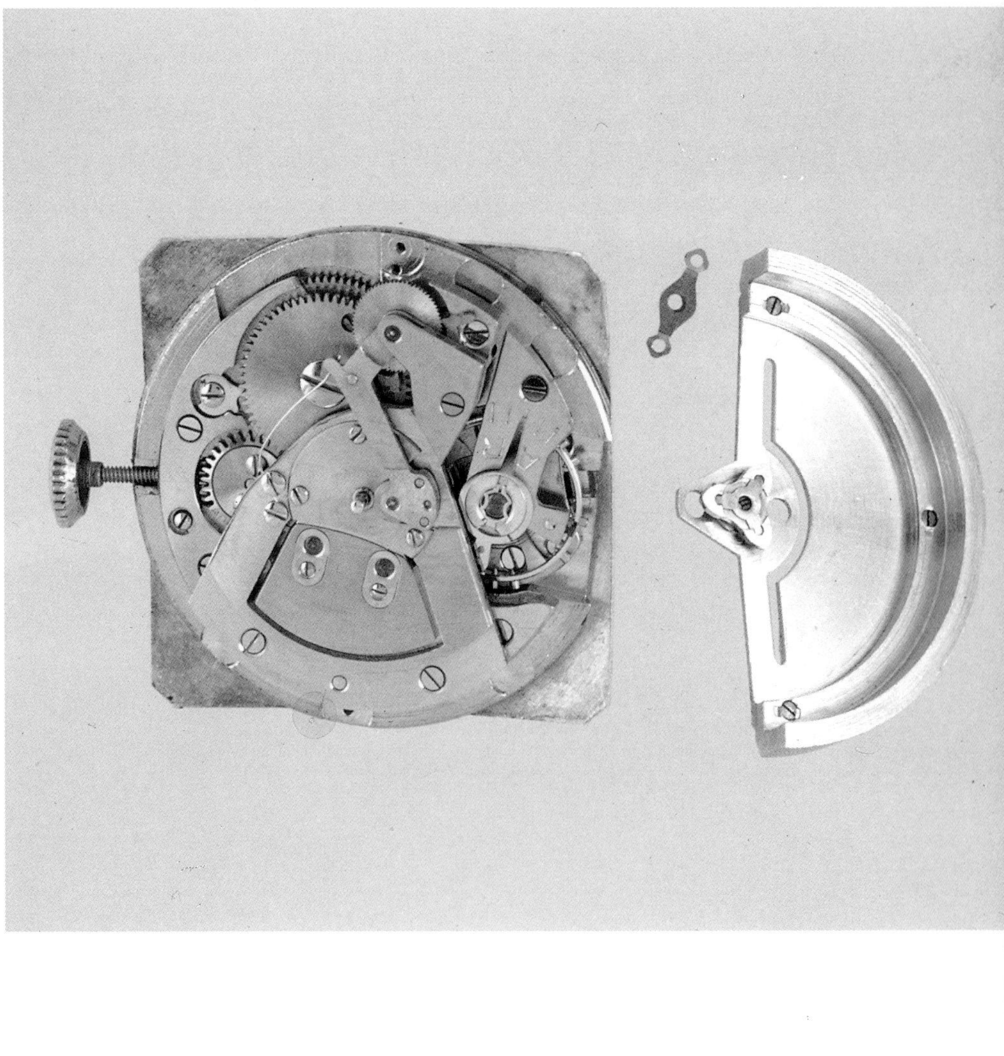

Caliber Development

The firm of Piaget S.A. was founded by Georges Piaget at La Côte-aux-Fées in 1874 and remained in family ownership until 1988 when the majority of the shares were acquired by the Cartier firm. After World War II, the founder's grandsons, Gérald and Valentin Piaget, expanded the business and made it a significant producer of ornamental watches; in 1964 they took over the firm of Baume & Mercier. While the movements continued to be made by the family firm, finishing of cases, bands and jewelry took place at the newly acquired facilities in Geneva.

In 1959 Piaget introduced an automatic that had been developed in collaboration with its neighbor firm of Complication, and that took all its Swiss competitors by surprise. This was Caliber 12 P; measuring only 2.33 mm, it ranked until 1978 as the world's flattest. Until then, the firm had no experience with the production of automatic watches, installing only movements from outside, such as those from AS, in its watch models.

Buren or Universal watches, with a small rotor recessed in the movement and mounted outside the center, served as models for this design. In Caliber 12 P, the rotor was made of 24-karat gold. Winding took place in both directions via an elaborate mechanism, with click wheels bearing specially shaped teeth serving as locks for changing the direction. The balance was mounted between precise Girocal combination jewel housings. Shock resistance was not believed to be necessary, as arbor breakage was less likely with a small, light ring balance, especially when one can proceed from the fact that these movements were made for elegant watch models not exposed to everyday use.

By buying the remaining raw movements of the Lassale Caliber 2000 in 1983, Piaget once again possessed what is still the absolutely flattest automatic design. This caliber, designated 25 P, is still used by the firm. Raw movements made by the firm of Frederic Piguet, known as Caliber P 70, have also been used.

Caliber Notes

Caliber 12 P
D = 28.1 mm (12 lignes), H = 2.30 mm, 30 jewels
AUTOMATIC MODULE with its 24-karat gold microrotor integrated into the basic movement, winding in both directions via double click wheel (tooth-directing lock) with reversing wheel, three reduction wheels in all on the locking wheel.
BASIC MOVEMENT: two-spoke screwless Glucydur balance, self-compensating flat hairspring, 19,800 BPH, Girocap combined balance mount (no shock resistance), gold anchor and escape wheel, Eloxal barrel.

Magazine advertisement (from JSH, 1949)

Caliber 25 P (2)
Lassale Ebauche (2000)
D = 20.4 mm (9 lignes), H = 2.08 mm, 9 jewels, 18 ball bearings
AUTOMATIC MODULE attached to the basic movement by two screws, as a complete unit, central rotor with ball bearings and heavy metal ring, winding in one direction via a sprung gliding pinion with direction on one side, three reduction wheels (the first with a locking spring), one intermediate wheel on the locking wheel.
BASIC MOVEMENT (H = 1.20 mm): three-spoke screwless Glucydur balance, self-compensating flat hairspring, 21,600 BPH, Kif-Ultraflex shock resistance, large bottom wheel mounted outside the center, "swimming" barrel mount by three ball bearings around its edge.

Caliber List: Piaget

Caliber	Diameter Lignes	mm	Height mm	Other	BPH	Jewels	Running Reserve	Notes	Year	Photo
12 P	12	27	2,3		19.800	30			1959	xx
25 P /2	9	20,4	2,08		21.600	9	50	Lasale 2000 with 18 ball bearings	1983	xx
P 70	11 1/2	26	2,4		18.000	35		Fr. Piguet p 70Ebauche		(xx)

Piaget Caliber 12 P

Men's automatic made by Piaget circa 1965, with Caliber 12 P, the world's flattest (from 1959 to 1978) automatic caliber.

Eighteen karat gold case with press-in bottom, gold dial and hands.

Eighteen karat gold microrotor, winding in both directions via click-wheel switcher.

Screwless Glucydur balance, self-compensating flat hairspring, no shock resistance (rotor bridge and movement ring show corrosion).

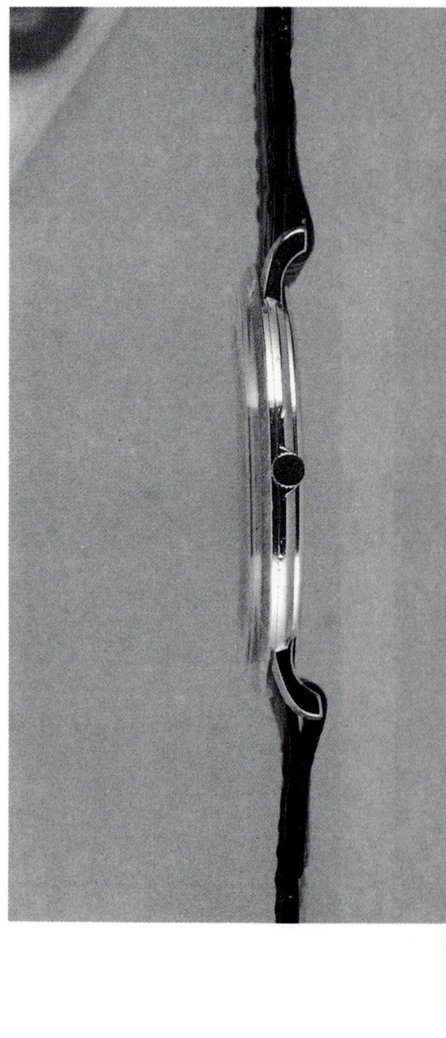

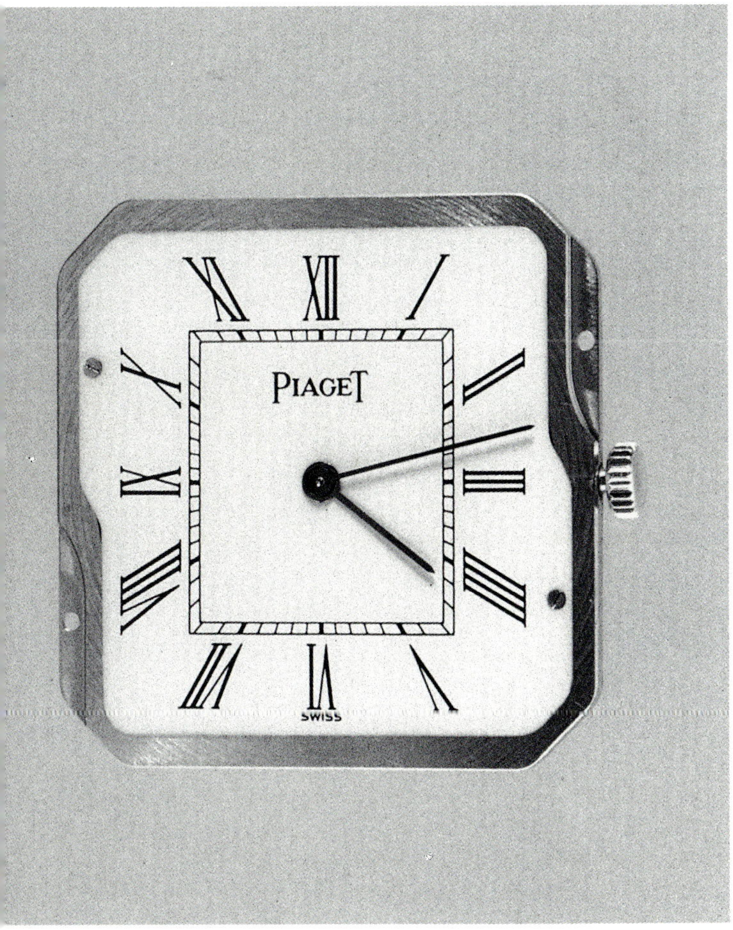

Piaget Caliber 25 P/2

Extremely flat medium-size automatic movement made by Piaget as of 1983, with Caliber 25 P/2 (Lassale raw movement 2000, see page 180).

Rotor with ball bearings and heavy metal ring, winding in one direction via gliding pinion.

Screwless Glucydur balance, self-compensating flat hairspring, Kif-Ultraflex shock resistance.

All wheels are mounted on only one side ("flying") and run on a total of 18 ball bearings.

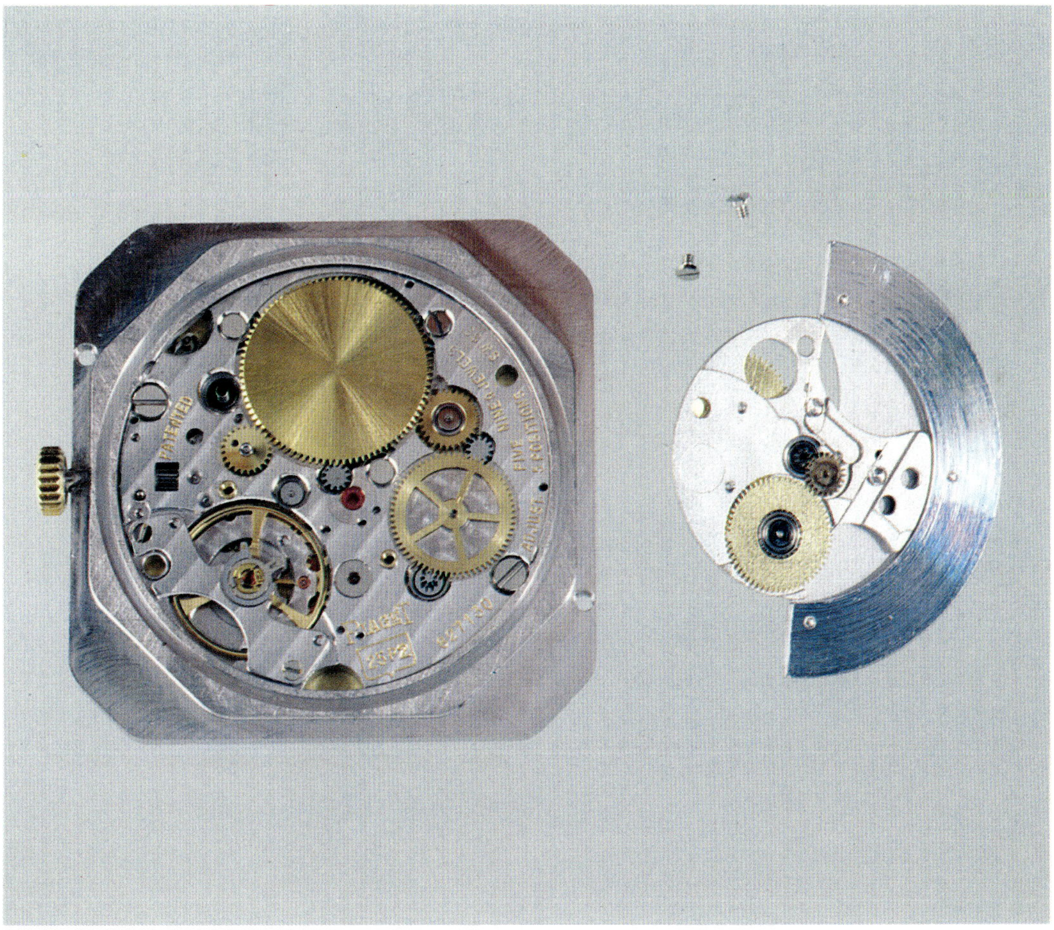

Magazine advertisement (from JSH, 1949)

Caliber Development

The firm of Pierce S.A. was founded by the Lévy brothers at Biel in 1883, and also operated in Moutier, producing reasonably priced watches with cylinder and anchor escapements, as well as a simple wrist chronograph with a special sliding coupling.

Using the firm's own 1933 patent for a swinging mass moving in a straight line on two rods, Pierce produced as of 1940 a roller automatic that was the last model with this type of winding. The basic movement was Caliber 861, with the hand-winding mechanism left off. Unlike the "Rolls" and "Wig-Wag", the whole movement did not roll here, but only a specially formed sliding weight that moved back and forth over the basic movement. The winding, in one direction, took place via a toothed rod that was attached to the swinging mass. The latter, with a maximum movement of 3 mm, was limited elastically by leaf springs attached to the case. A coil spring located on the guiding rail pushed the swinging mass back to its starting position. Pierce improved this automatic as of 1945 by adding a center seconds, a more stable attachment for the swinging mass, and the possibility of hand-winding. A locking-wheel coupling uncoupled the hand-winding mechanism during automatic winding to avoid additional friction losses. The space-saving location of the reduction drive inside the central swinging-mass cutout allowed what was then a flat movement structure. The unsatisfactory winding performance of this roller automatic in comparison with the competing hammer and rotor models did not result in its withdrawal from production until about 1954.

Caliber Notes

Caliber 861
D = 28.0 mm (12.5 lignes), H = 5.1 mm, 21 jewels
AUTOMATIC MODULE screwed to the basic movement, pointed-oval winding weight guided back and forth on two guiding rails, with the reduction drive located in its central cutout, movement distance some 3 mm, swinging mass limited by leaf springs on the case, winding in one direction via sprung toothed rod, two reduction wheels, the first with a locking click, on the upper locking wheel, uncoupling of the hand-winding mechanism during automatic winding by a locking-wheel coupling (toothed-wheel directing lock).
BASIC MOVEMENT (8.75 lignes): two-spoke nickel screw balance, self-compensating flat hairspring, 18,000 BPH, Incabloc shock resistance, 40-hour running reserve.

Caliber List: Pierce

Caliber	Diameter Lignes	mm	Height mm	Other	BPH	Jewels	Running Reserve	Notes	Year	Photo
(861)	12 1/2	28	5,1		18.000	17	36	Basic Caliber 861	ca. 40	
(861)	"	"		SCI	"	17 / 21	"	(8.75 Lignes)	ca. 45	xx

Pierce Caliber 861
Men's automatic made by Pierce circa 1947, with Caliber 861, made since 1945. Eighteen karat gold case with press-in bottom. Rolling weight moving on rails, winding in one direction via toothed rod (on rolling weight). Nickel screw balance, Nivarox hairspring, Incabloc shock resistance (pins for lower band attachment broken out).

Caliber Development

This company was founded by Louis Elysée Piguet at Le Brassus in 1858, and has limited itself to this day to the production of high-quality raw movements. The present proprietor, Jacques Piguet, is also a partner in the nearby firm of Blancpain and that firm's supplier of raw movements.

The firm's first automatic movement, Caliber P 70, was produced around 1970. The future purchasers of this caliber, such as Breguet and Corum, wanted a particularly flat design for their elegant watch models. For that reason, the placing of central mounts one above the other, which required much height, was avoided by moving the rotor mount and leaving the barrel open, without top and bottom. The rotor, as also in later models, wound in only one direction. A seconds hand and date indication were done without.

Caliber 951, completely developed by 1986, was somewhat thicker, having a center seconds and date indication, but was particularly small, with a diameter of only nine lignes and thus suitable for use in ladies' watches. The central rotor ran on ball bearings. The functioning principle of the combined winding and locking wheel, acting as a tooth-directing lock, was of interest.

The next design, Caliber 11.50, was equipped with two identical barrels, working in series, and thus attained an unusually high running reserve of 100 hours. The very complex automatic drive comprised six transmission parts in all.

The last development of 1988, Caliber 1185, was reserved for the Blancpain firm. It was a very flat automatic chronograph caliber.

Caliber Notes

Caliber P 70
D = 26.0 mm (11.5 lignes), H = 2.4 mm, 35 jewels
AUTOMATIC MODULE integrated into the basic movement, large heavy metal rotor mounted somewhat out of center, winding in one direction via one intermediate wheel, sprung lever wheel, three reduction wheels (the first with locking click) on the locking wheel, uncoupling of the hand-winding drive by a toothed-wheel rocker between the crown and locking wheels (under the dial).
BASIC MOVEMENT: two-spoke screwless Glucydur balance, self-compensating flat hairspring, 18,000 BPH, Kif-Ultraflex shock resistance, moveable hairspring block carrier, escape wheel mounted in combined Duofix jewel housings, open barrel.

Caliber 951
D = 20.2 mm (9 lignes), H = 3.2 mm, 19 jewels
AUTOMATIC MODULE integrated into the basic movement, central rotor with ball bearings and heavy metal ring, winding in one direction via a springless combined winding and locking wheel as tooth-directing lock, two reduction wheels on the locking wheel.
BASIC MOVEMENT: three-spoke screwless Glucydur balance, self-compensating flat hairspring, 21,600 BPH, Kif-Ultraflex shock resistance, moveable hairspring block carrier, direct center seconds, date indication.

Caliber 11.50
D = 25.6 mm (11.5 lignes), H = 5.25 mm, 29 jewels
AUTOMATIC MODULE integrated into the basic movement, central rotor with ball bearings and heavy metal ring, winding in one direction via one intermediate wheel, springless click wheel with locking wheel in contact (both with one tooth-directing lock), one reduction wheel, two intermediate wheels on the locking wheel of the first barrel, hand-winding mechanism uncoupled by a toothed-wheel rocker between the crown wheel and the locking wheel of the first barrel.
BASIC MOVEMENT: three-spoke screwless Glucydur balance, self-compensating flat hairspring, 21,600 BPH, Kif-Ultraflex shock resistance, Spirofin fine adjustment with moveable hairspring block carrier, 21-toothed escape wheel, direct center seconds, two identical barrels (working in series).

Caliber List: Piguet, Fr.

Caliber	Diameter Lignes	mm	Height mm	Other	BPH	Jewels	Running Reserve	Notes	Year	Photo
P 70	11 1/2	26,0	2,4		18.000	35	50		ca. 70	xx
P 71	12	27,4	"		"	"	"	Microtoothing		
951 / 952	9	20,4	3,25	SCD, QG	21.600	17	45		1986	xx
953	"	"	"	SCD	"	19	"			
1150	11 1/2	25,6	3,25	SCD, QG, RDR	21.600	29	100!	Two barrels in series		xx
1153	"	"	"	SCD	"	"	"	"		
1160	"	"	"	s, QG, RDR	"	"	"	"		
1163	"	"	"	s	"	"	"	"		
1185	11 1/2	26,0	5,5	QG, RDR, CHR	21.600	36	41	Reserved for Blancpain	1988	
9511	10 1/2	23,3	3,25	SCD, QG	21.600	19	45	Basic Caliber 951		

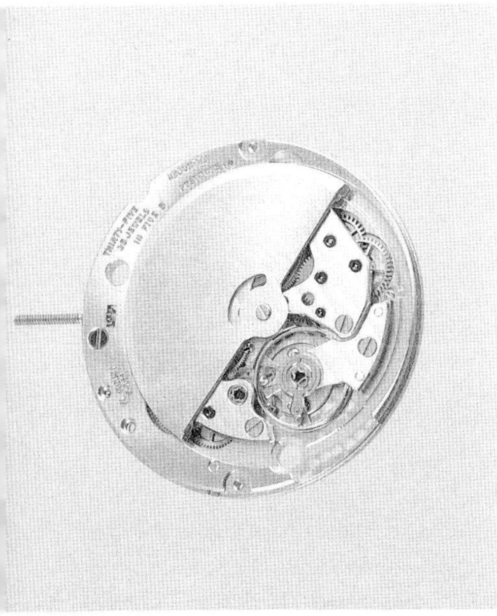

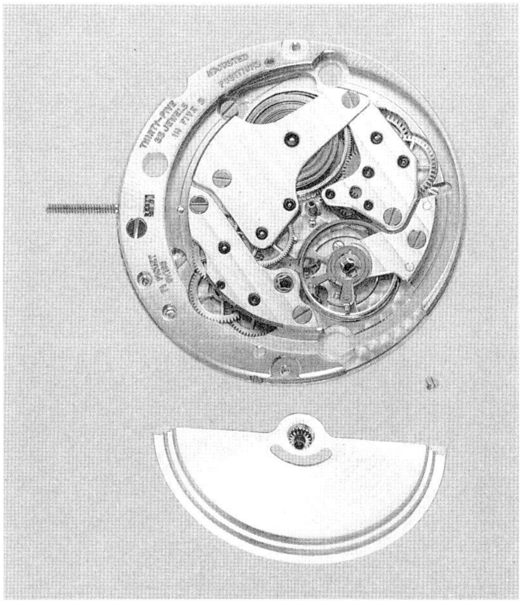

Fr. Piguet Calibers P 70, 951, 11.50

Very flat automatic Caliber P 70 for men's watches, made by Fr. Piguet circa 1970.
Heavy metal rotor mounted somewhat out of the center, winding in one direction via lever wheel.
Screwless Glucydur balance, self-compensating flat hairspring, Kif-Ultraflex shock resistance.

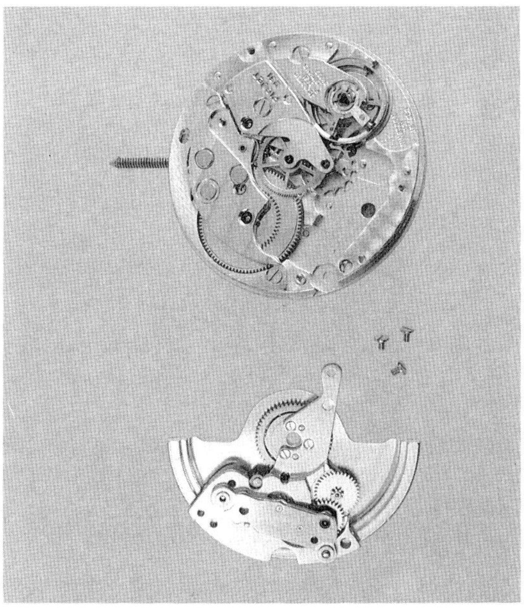

Automatic Caliber 951, medium size, made since 1986.
Heavy metal rotor with ball bearings, winding in one direction via click wheel.
Screwless Glucydur balance, self-compensating flat hairspring, Kif-Ultraflex shock resistance.

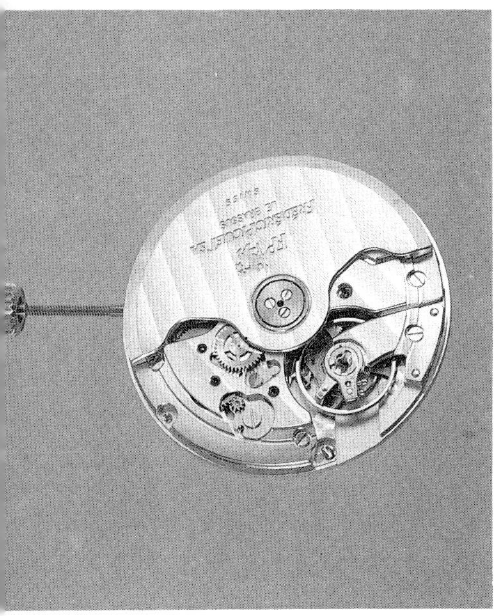

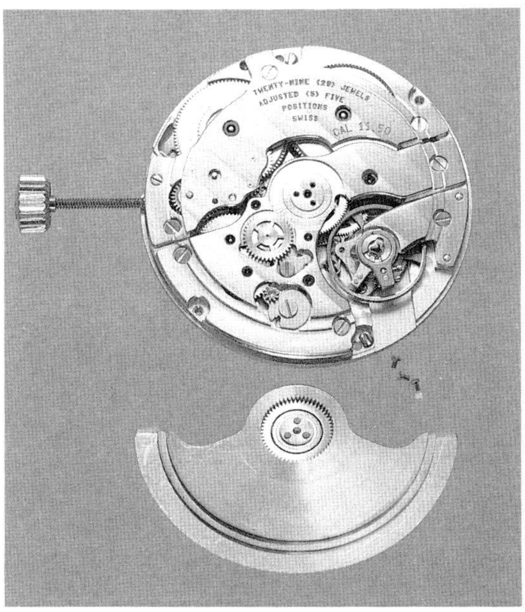

Automatic Caliber 11.50 for men's watches, circa 1988.
Heavy metal rotor with ball bearings, winding two barrels in series in one direction via click wheel, with maximum running reserve of four days.
Screwless Glucydur balance, self-compensating flat hairspring, Kif-Ultraflex shock resistance, Spirofin fine regulation.

Magazine advertisement (from JSH, 1953)

Caliber Development

The "Record Watch Co. S.A." firm was founded at Geneva in 1899. A branch facility was later opened in Tramelan. The firm's hand-wound wristwatches with simple calendar and moon-phase indication have become well-known.

Record's first automatic development, Caliber 161, was introduced in 1944. It was a hammer automatic of the Harwood type, winding in one direction, with a strikingly large reduction wheel. A crown-wheel coupling protected the reduction drive during hand-winding. This design was also fitted, somewhat later, with a center seconds (Caliber 172).

In a subsequent development of 1952, a rotor winding in both directions was used; the watches were fitted with either eccentric or center seconds, date and up-down indicator. The running reserve was shown on a subsidiary dial under the 12, with the relative position of the turning number rings to the dial being read. The elaborate differential drive could thus be dispensed with.

In this design there was little space for the basic movement so that, just as in the hammer automatic, only a relatively small caliber of 9.75 lignes could be used.

No information can be given on the calibers with higher numbers; they were presumably not made by Record.

Caliber Notes

Caliber 172
D = 25.95 mm (11.5 lignes), H = 5.95 mm, 20 jewels
AUTOMATIC MODULE screwed to the basic movement, hammer movement limited by visible buffer springs, winding in one direction via winding click on the hammer, centrally mounted winding pinion, one large combined reduction and locking wheel on the crown wheel with inside coupling to uncouple during hand-winding.
BASIC MOVEMENT (9.75 lignes): two-spoke closed screw balance, Nivarox hairspring, 18,000 BPH, Incabloc shock resistance.

Caliber List: Record

Caliber	Diameter Lignes	mm	Height mm	Other	BPH	Jewels	Running Reserve	Notes	Year	Photo
161	11 1/2	25,95	5,2	s	18.000				1944	xx
162	"	"	5,95	SCI	"	20				
171	"	"	5,2	s	"					
172	"	"	5,95	SCI	"	20				
174	11 1/2	26		s	18.000	21			1952	
174	"	"	6,35	SCI	"	"		(See advertisement)	1953	xx
174	"	"		SCI, QG	"	"		"Datofix"	1954	
174	"	"		SCI, QG, RMA	"	"		"	1954	
510	12	27	5,1	SCI	19.800	21	50		1959	
511	"	"	"		"	"	"			
515	"	"	"	SCI, QG	"	"	"			
516	"	"	"	QG	"	"	"			
520	"	"	"	SCI	"	"	"			
521	13		"	SCI	"	"	"			
525	"		"	SCI, QG	"	"	"			
526	"		"	QG	"	"	"			

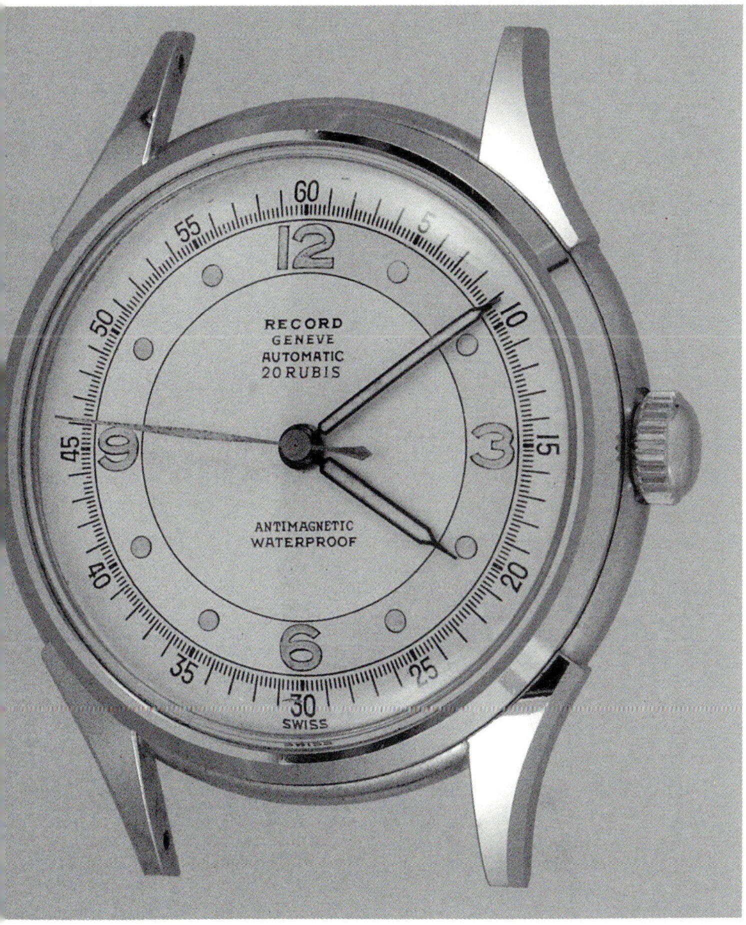

Record Caliber 172
Men's automatic made by Record circa 1946, with Caliber 172. Stainless steel case with screw bottom.
Hammer winding in one direction via Harwood-type drive, uncoupling during hand-winding.
Nickel screw balance, Nivarox hairspring, Incabloc shock resistance.

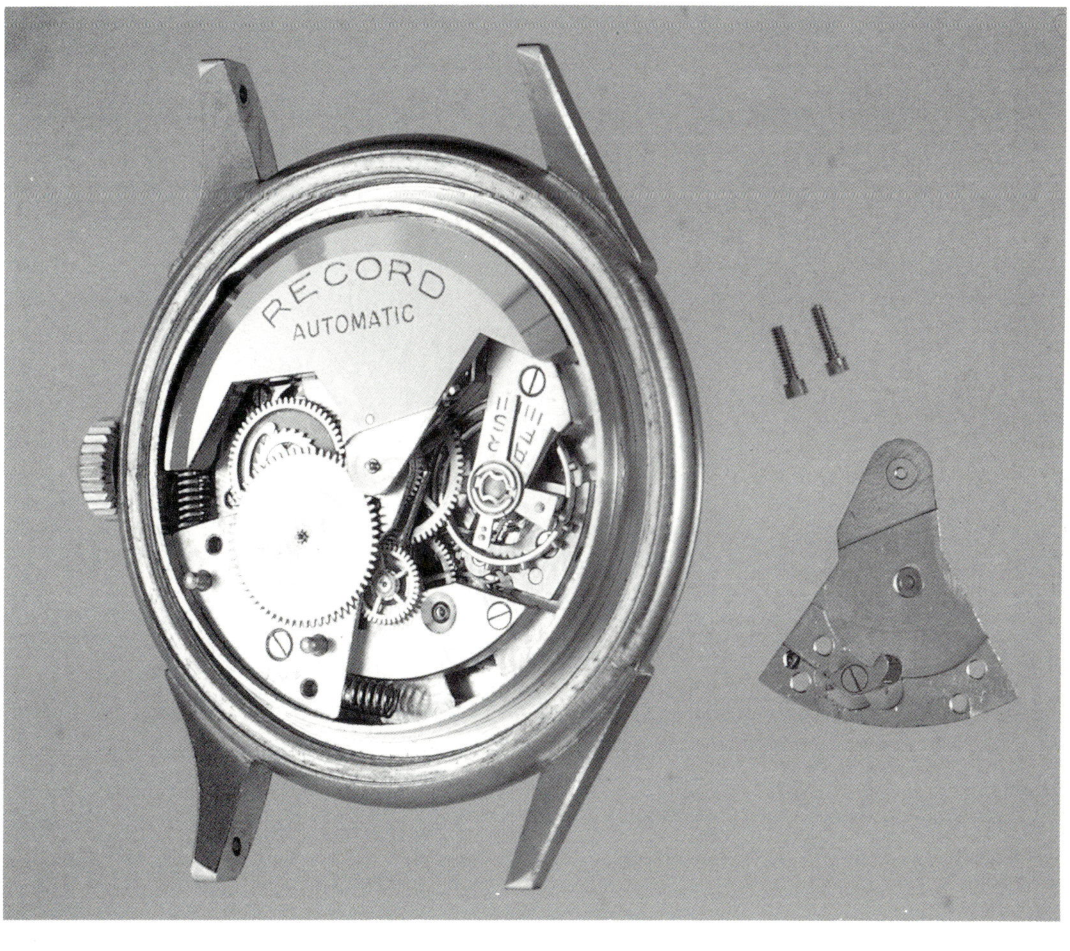

Magazine advertisement (from SUZ, a949)

Caliber Development

Gideon Thommen founded the firm of Revue-Thommen S.A. at Waldenburg in 1853, Financial conditions compelled its merger with Phenix Watch, Buser and Vulcain in 1962 to form the Manufactures d'Horlogerie Suisse Réunies S.A. (MSR).

In its first automatic, built circa 1948, Revue introduced a novelty, a hammer that wound in both directions, thus guaranteeing the watch's dependability even without much arm movement. It was Caliber 80, with its winding mechanism consisting of two click-carrier rockers moving counter to each other. According to the hammer's direction, winding was done via the click of the appropriate rocker, while the other glided freely over the teeth of the click wheel. Despite this unusual design, this movement, measuring 4.75 mm high, was the flattest of its time that had an eccentric seconds hand. Its later form, with a direct center seconds, could be made only by increasing the movement's thickness by one millimeter.

An unusual winding drive was also used in the next automatic model, the "Rotor-King" introduced in 1958. This was a coulisse-lever mechanism gliding on ruby rollers, its motion back and forth compelled by the rotor eccentric, so that winding took place in both rotor directions. The basic movement was also designed to achieve high running performance: the balance and barrel were made particularly large, and the regulator for fine adjustment by means of a pulley drive is also noteworthy.

The "Exactomatic" of 1967 was the first automatic caliber developed since the formation of the MSR conglomerate, and its design was based heavily on time-tested designs by other manufacturers. It was a movement with a heavy metal rotor mounted on ball bearings and winding in both directions via springless twin click wheels, and was also equipped with direct center seconds and a quick changing date indication.

Meanwhile the firm turned to movements from outside, using Eta movements in the 1950s and AS raw movements in the 1960s for the "Directime", in which hour, minute and second were indicated in cutouts in the dial.

Revue Caliber 80
First men's automatic made by Revue, circa 1948, with Caliber 80, 14-karat gold case with press-in bottom.
Hammer winding in both directions via double click-carrier rocker, the first of this type.
Glucydur screw balance, Nivarox hairspring, Incabloc shock resistance.

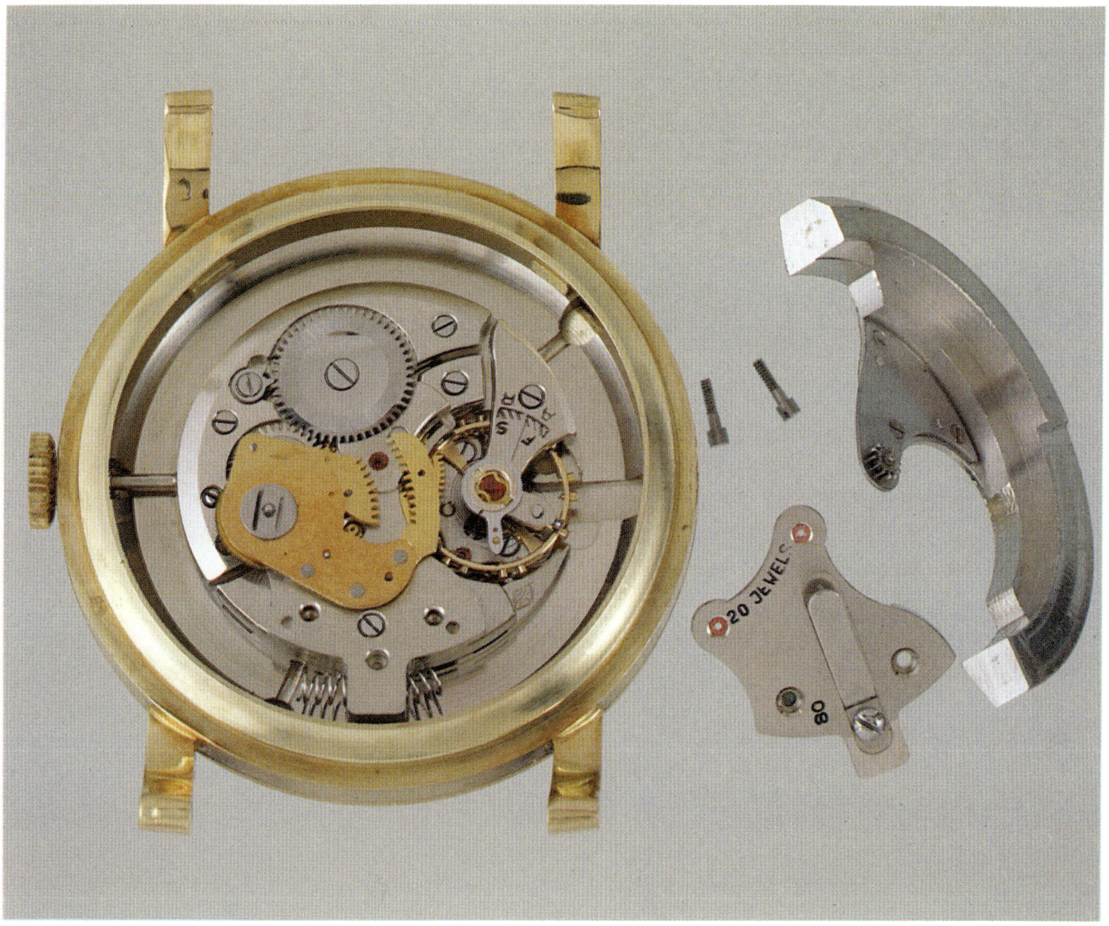

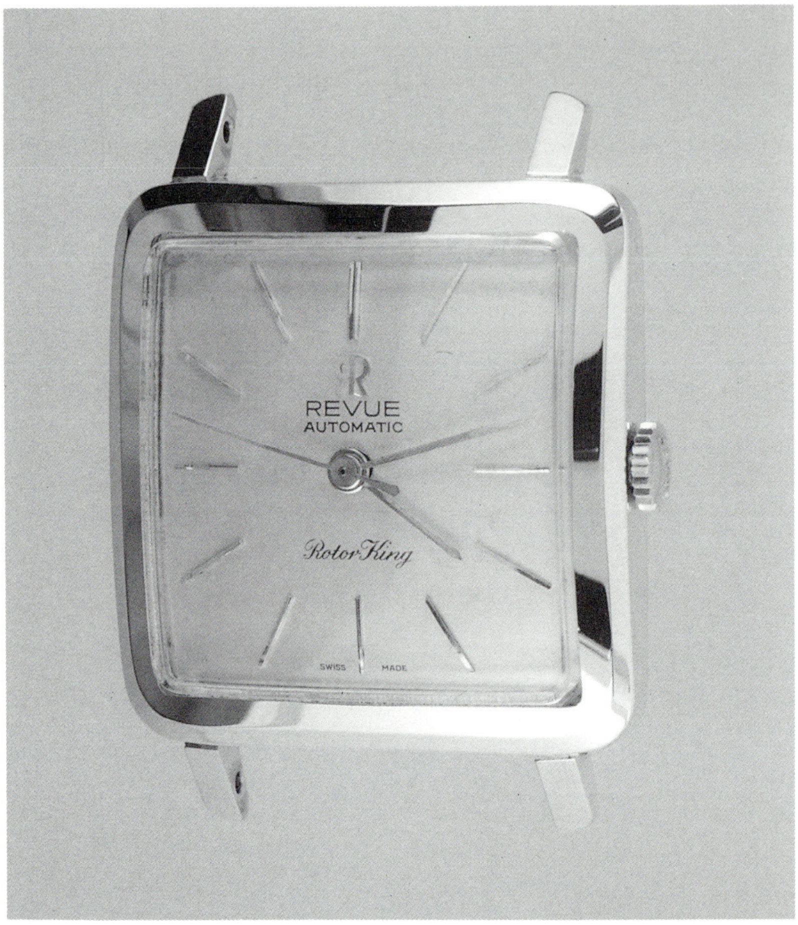

Revue Caliber 87.1

Men's automatic, "Rotor-King" model, made by Revue in 1958, with Caliber 87.1.

Square 18-karat gold case with press-in bottom, winding crown with Revue monogram.

Rotor with heavy metal ring, winding in both directions via eccentric coulisse switcher, uncoupling of the hand-winding drive.

Large Glucydur screw balance, Nivarox hairspring, Incabloc shock resistance, fine regulator adjustment by pulley disc (dial with signs of corrosion).

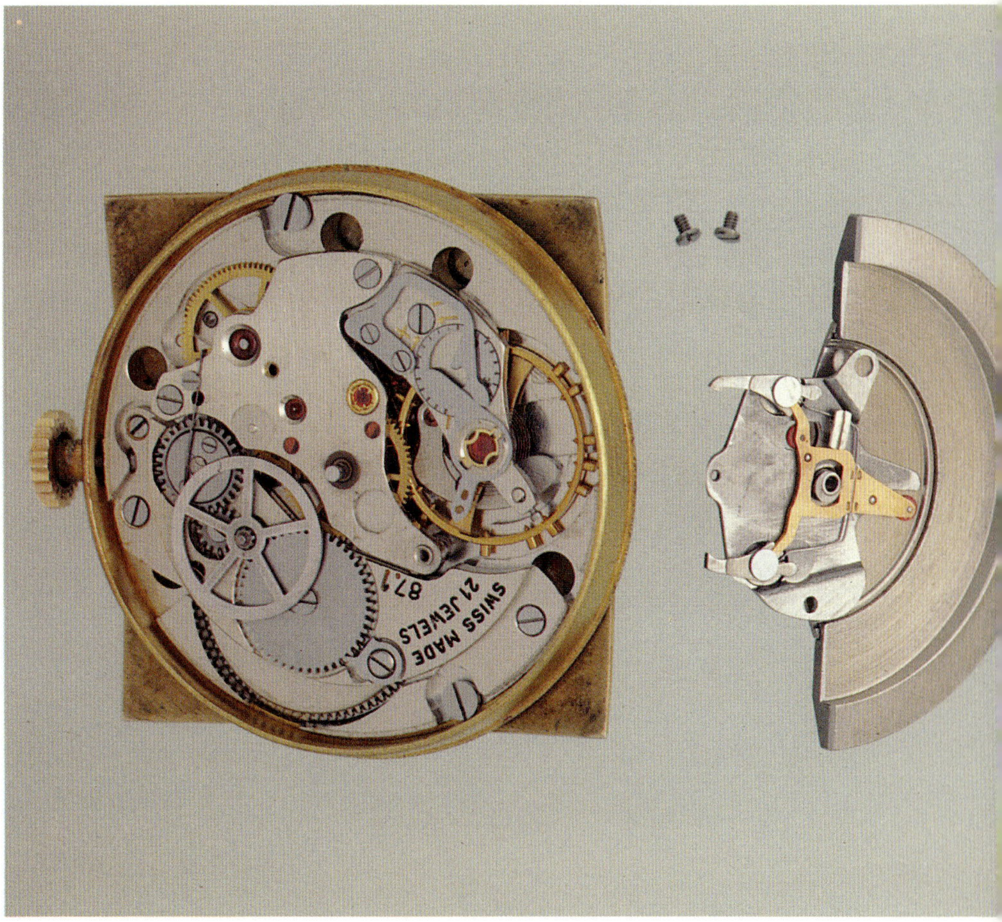

Revue/MSR Caliber T 56

Men's automatic, "Meteor IV" model, made by Revue/MSR circa 1967, with "Exactomatic" Caliber T 56.

Gilded barrel-shaped case with stainless steel screw bottom, winding crown with MSR symbol.

Heavy metal rotor with ball bearings, winding in both directions via click-wheel switcher (at 2:00).

Screwless Glucydur balance, Nivarox hairspring, Incabloc shock resistance, fine regulator adjustment by eccentric screw.

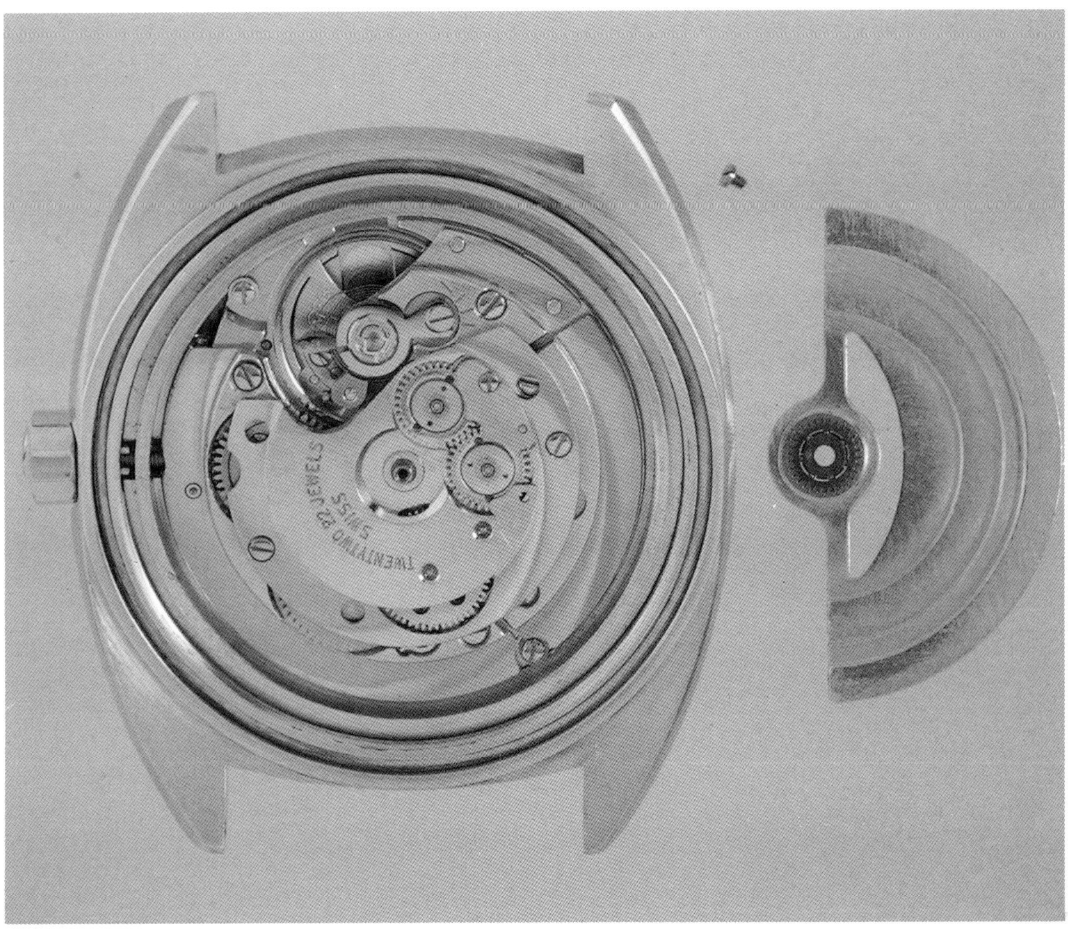

Revue Directime

Men's automatic, "Directime" model, made by Revue circa 1966, with AS raw movement (Caliber 1902).
Barrel-shaped stainless steel case with screw bottom. Digital hour, minute and second indication in dial apertures.
Rotor with heavy metal ring, winding in one direction via lever wheel (at 9:00).
Screwless Glucydur balance, Nivarox hairspring, Incabloc shock resistance.

Caliber List: Revue / MSR

Caliber	Diameter Lignes	mm	Height mm	Other	BPH	Jewels	Running Reserve	Notes	Year	Photo
80	12 1/2	28,2	4,75	s	18.000				ca. 48	
80	"	"	5,7	SCI	"	20			ca. 49	xx
87	11 1/4	25,2	4,5	SCD	18.000	21	50	Rotor-King	1958	
87	"	"	5,0	SCD, QG	"	"	"	"	1958	
87.1	"	"	4,5	SCD	"	"	"	"	1958	xx
T 1	11 1/2			SCD	18.000			Eta-Ebauche		
T 5	11 1/2			SCD, QG	19.800	22	54	Exactomatic	ca. 67	
T 53	"			SCD	"	"	"	"		
T 56	"			SCD, QG	"	"	"	"		
T 57	"			SCD, QG, JG	"	"	"	"		xx

Caliber Notes

Caliber 80
D = 28.2 mm (12.5 lignes), H = 5.7 mm, 20 jewels
AUTOMATIC MODULE screwed to the basic movement, hammer winding in both directions, limited by buffer springs, by two opposing click-carrier rockers with a combined winding and reduction wheel, on the crown wheel.
BASIC MOVEMENT (9.75 lignes): two-spoke Glucydur screw balance, Nivarox hairspring, 18,000 BPH, Incabloc shock resistance.

Caliber 87.1
D = 25.2 mm (11.25 lignes), H = 4.5 mm, 21 jewels
AUTOMATIC MODULE set atop the basic module, central rotor with heavy metal ring, winding in both directions via an eccentric on the rotor, coulisse lever on ruby rollers with two sprung lever clicks on the locking wheel that alternately activate a fine-toothed winding wheel. Uncoupling of the hand-winding mechanism by a toothed-wheel rocker between the crown wheel and the pinion of the winding wheel.
BASIC MOVEMENT: two-spoke Glucydur screw balance, Nivarox hairspring, 18,000 BPH, Incabloc shock resistance, fine regulator adjustment by pulley drive, Girocap combined mounting of the escape wheel, direct center seconds.

Caliber MSR T 56
D = 11.5 lignes, 22 jewels
AUTOMATIC MODULE screwed to the basic movement, heavy metal central rotor with ball bearings, winding in both directions via springless twin click wheels, two reduction wheels on the locking wheel.
BASIC MOVEMENT: three-spoke screwless Glucydur balance, Nivarox hairspring, 19,800 BPH, Incabloc shock resistance, moveable hairspring block carrier, fine regulator adjustment by eccentric screw, direct center seconds, date indication with quick change.

Caliber Development

Fritz Meyer founded a company at Solothurn in 1888; after a partner joined him, it was called "Meyer & Stüdeli S.A." (MST) until about 1948, when it was renamed "Roamer Watch Co. S.A.". The firm's first automatic watch (Caliber 410) was marketed at the end of the 1940s. It was a rotor automatic with a direction changer made of twin click wheels with spring clicks, and with its automatic module covering the basic movement like a bowl, so that the movement was almost completely covered. The hand-winding mechanism was uncoupled by a patented toothed-wheel rocker during automatic winding.

The firm's next development, Caliber 420 of 1952, eliminated several of the earlier design's weaknesses. The click wheels, prone to disorder because of the spring clicks, were replaced by a sturdy switching rocker, and a second toothed-wheel rocker allowed the automatic drive to be disconnected and thus protected during hand-winding. This watch, well known by the name of "Rotopower", was later equipped with date indication and called "Rotodate".

The patented waterproof Roamer case was noteworthy, consisting of a bowl-shaped bottom enclosing the movement, with the bowl-shaped glass and then the middle section pressed onto it. The winding stem was divided and fitted with a waterproof crown.

This robust but somewhat blockish watch was replaced around 1958 by the flatter Caliber 436, which ranks as the firm's most elaborate automatic design. The automatic drive was integrated into the movement, the direction changer consisted of patented free-running units on balls, and the central rotor ran on miniature ball bearings. Since all the balls were made of ruby, they added up to the astounding total of 44 functioning jewels—a sales pitch that this watch really did not need as proof of its advantages. The high quality of the self-winding mechanism allowed the watch to be simplified by leaving out the uncoupling devices between the automatic and hand-winding drives. The basic movement was also equipped with high-quality moving parts and patented fine regulation.

Unfortunately, the ruby balls in the rotor mount often broke under major shocks, so that some of them had to be replaced in Caliber 471 by steel balls, which was also done with the free-running parts.

Roamer used this caliber until 1966, using it in watches such as the renowned "Stingray", but also utilized movements made by Felsa (Caliber 441) and Eta (Caliber 474). This is especially true of the automatic models for ladies (such as Caliber 472), which were fitted exclusively with Eta movements.

The firm's last automatic design was Caliber 522 of 1972. This movement's quality of design and workmanship show the growing stress on practical production, although here too the rotor is equipped with miniature ball bearings and winds in both directions.

Caliber Notes

Caliber 420
D = 26.0 mm (11.5 lignes), H = 6.4 mm, 21 jewels
AUTOMATIC MODULE set atop the basic movement, central rotor with heavy metal ring, winding in both directions via switching rocker, two reduction wheels, the first with a spring click, on the locking wheel (under the dial), reciprocal uncoupling between the automatic and hand-winding drives by two toothed-wheel rockers.
BASIC MOVEMENT: two-spoke Glucydur screw balance, Nivarox hairspring, 18,000 BPH, Incabloc shock resistance, moveable hairspring block carrier, direct center seconds.

Caliber 436
D = 26.0 mm (11.5 lignes), H = 4.55 mm, 44 jewels
AUTOMATIC MODULE integrated into the basic movement, central rotor with ball bearings and heavy metal ring, winding in both directions via two (patented) free-running couplings, three reduction wheels in all on the locking wheel.
BASIC MOVEMENT: three-spoke screwless Glucydur balance, Nivarox hairspring, 21,600 BPH, Incabloc shock resistance, patented fine regulator adjustment by eccentric screw and moveable hairspring block carrier, direct center seconds, additional wheel between second and escape wheels, anti-corroding barrel, tension-relaxer for drive wheels.
The 44 jewels are as follows:
Basic movement: 19
Reduction drive: 8
Free-running couplings (2 x 5) 10
Rotor mount: 7

Caliber 441
Felsa Ebauche (Caliber 4002)
D = 26.0 mm (11.5 lignes), 21 jewels
AUTOMATIC MODULE set atop the basic movement, central rotor with ruby bearings and peripheral heavy metal inlay, winding in both directions via switching rocker, two reduction wheels on the crown wheel.
BASIC MOVEMENT: two-spoke Glucydur screw balance, Nivarox hairspring, 18,000 BPH, Incabloc shock resistance, direct center seconds via double intermediate wheel, date indication.

Caliber 471
See Caliber 436, but with slightly modified free-running couplings and some steel ball bearings, date indication with quick change.

Roamer Caliber 420

Men's automatic, "Rotopower" model, made by Roamer circa 1950, with Caliber 420.

Patented stainless steel bowl case with divided winding stem. Rotor with heavy metal ring, winding in both directions via switching rocker (at 4:00), complete reciprocal uncoupling of automatic and hand-winding drives.

Glucydur screw balance, Nivarox hairspring, Incabloc shock resistance (movement and case show signs of wear).

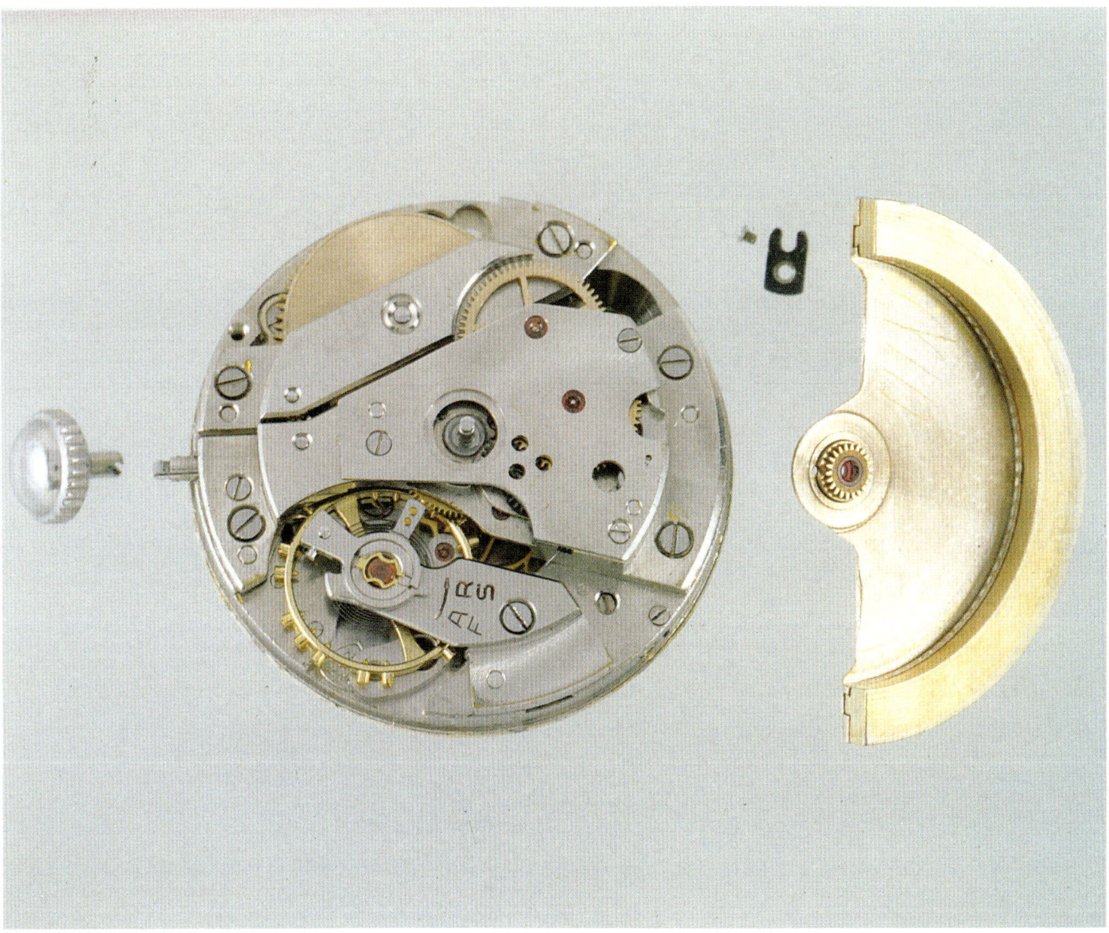

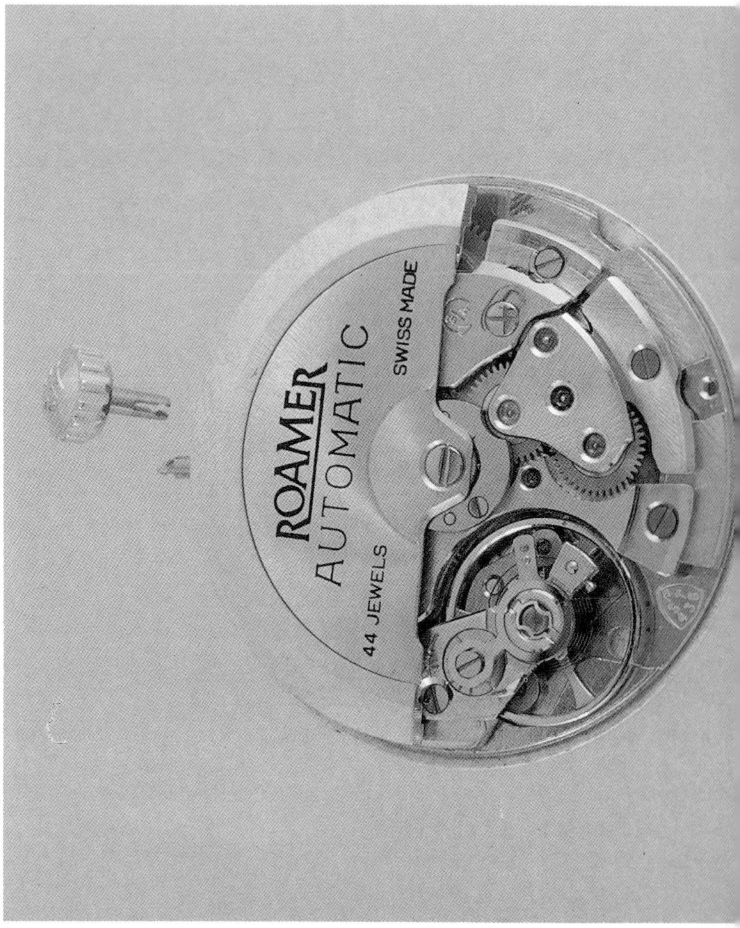

Roamer Caliber 436

Men's automatic, "Rotopower 44" model, made by Roamer circa 1958, with Caliber 436.

Gilded rounded case with stainless steel bowl bottom, divided winding stem, winding crown with RW (Roamer Watch) symbol.

Rotor with ball bearings and heavy metal ring, winding in both directions via free-running switcher (recessed in movement at 2:00).

Screwless Glucydur balance, Nivarox hairspring, Incabloc shock resistance, fine regulator adjustment by eccentric screw, 44 rubies.

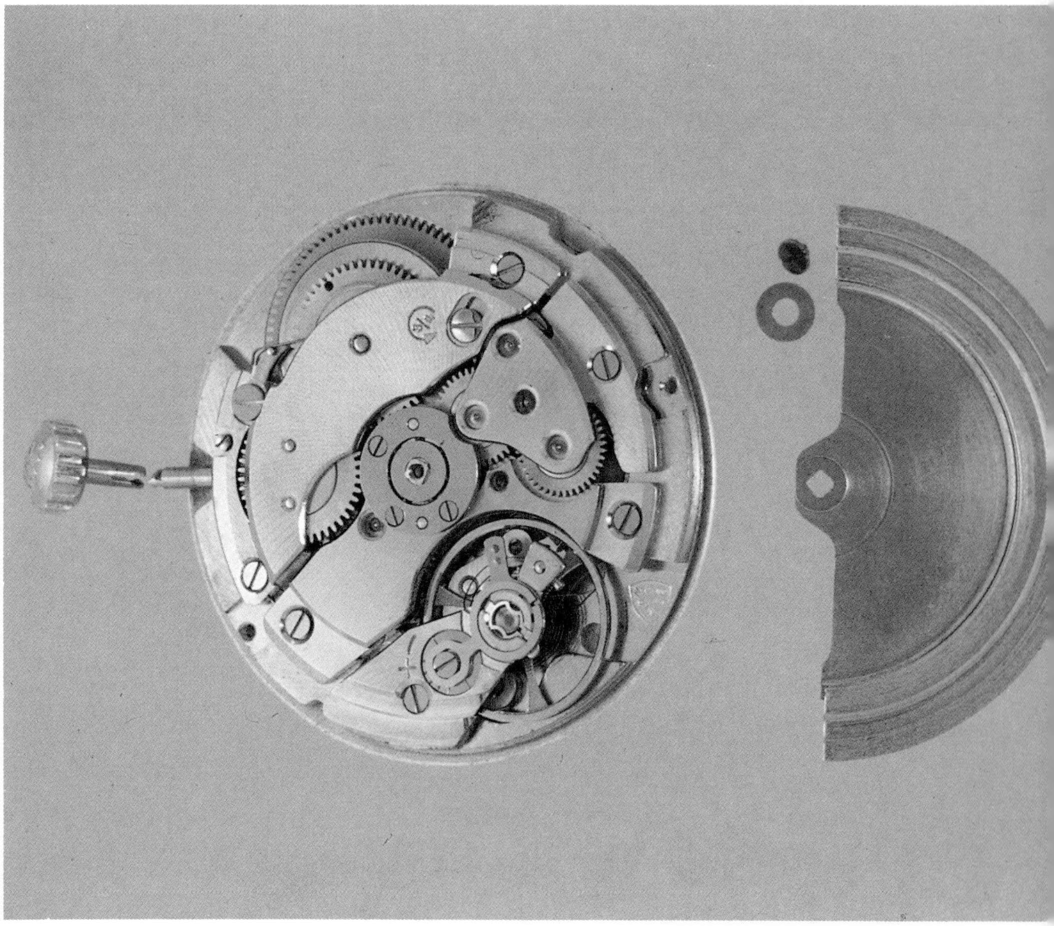

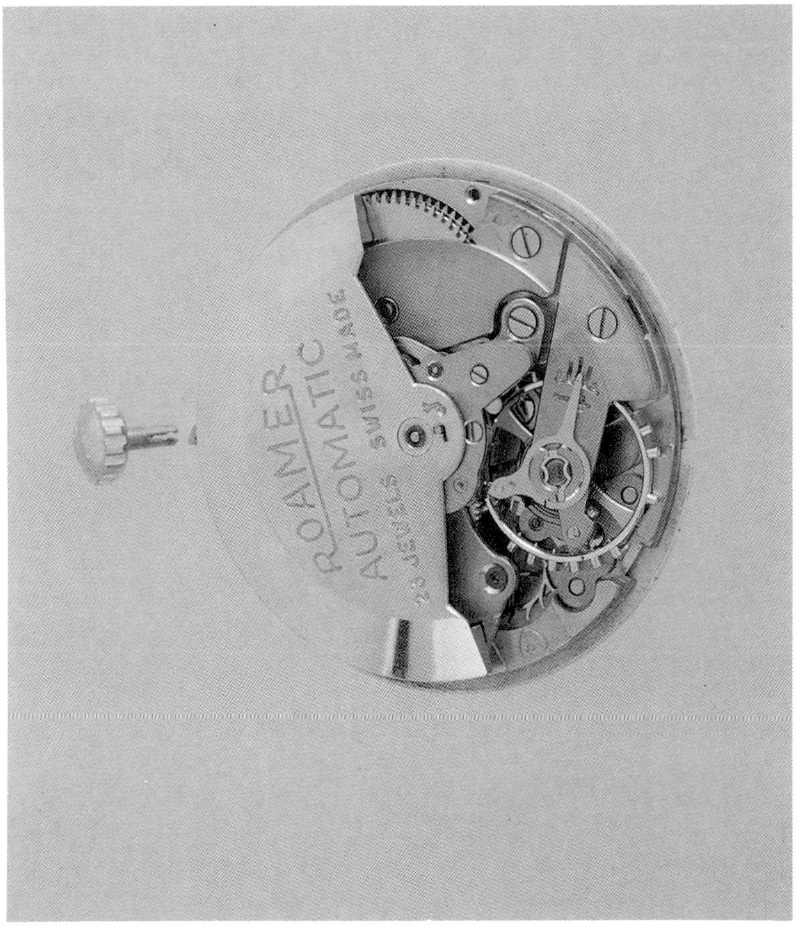

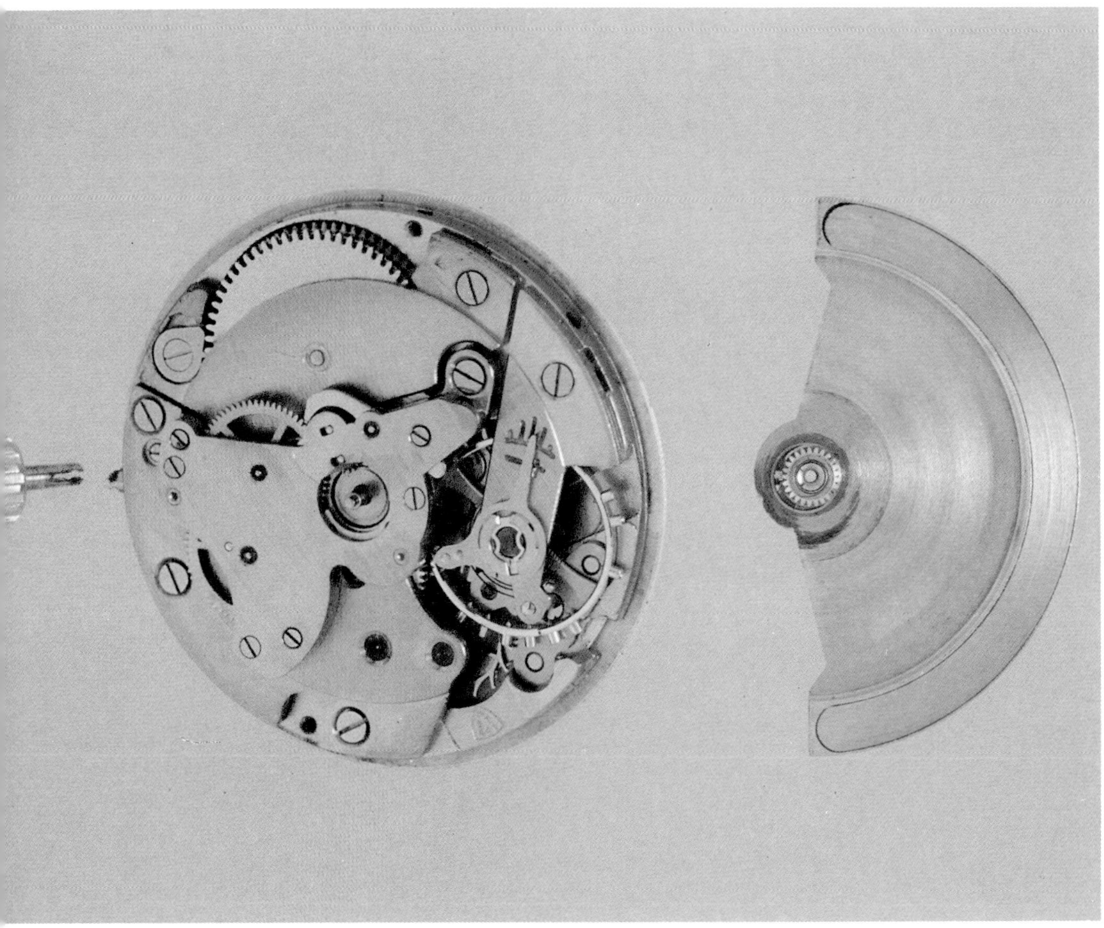

Roamer Caliber 441

*Men's automatic, "Rotodate" model, made by Roamer circa 1960 (**Caliber 441**) with Felsa raw movement.*

Patented 14-karat gold bowl case with divided winding stem, winding crown with RW symbol.

Movement identical to Felsa Caliber 4000 (see page 135).

Roamer Caliber 471

Men's automatic, "Stingray Rotodate 44" model, made by Roamer (MST) in 1959, with Caliber 471.

Gilded square case with stainless steel bowl bottom and divided winding stem, winding crown with R symbol. Movement largely identical to Caliber 436 (see page 274).

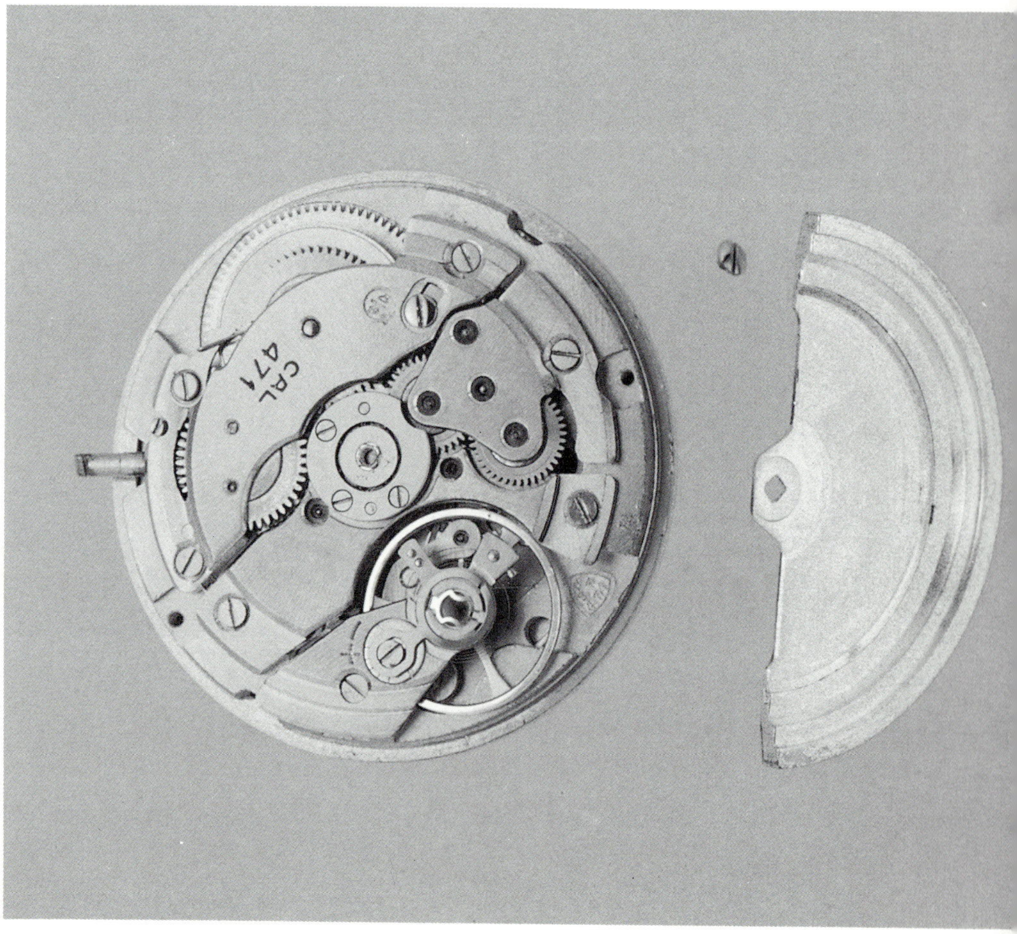

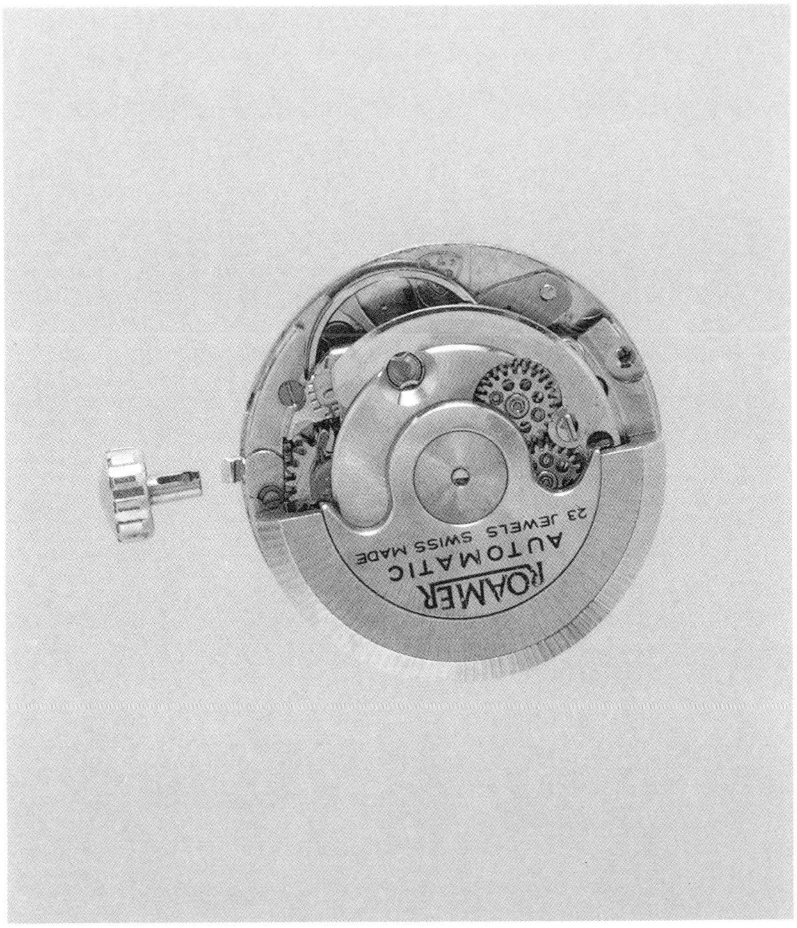

Roamer Caliber 472

Ladies' automatic, "Stingray Matic" model, made by Roamer circa 1965, with Caliber 472 (Eta raw movement).

Gilded case with stainless steel bowl bottom and divided winding stem, winding crown with R symbol.

Rotor with heavy metal ring, winding in both directions via patented click-wheel switcher (at 2 and 3;00) in upper right photo), uncoupling of hand-winding drive. Automatic module removable as a complete unit.

Screwless Glucydur balance, Nivarox hairspring, Incabloc shock resistance (case with strong signs of wear).

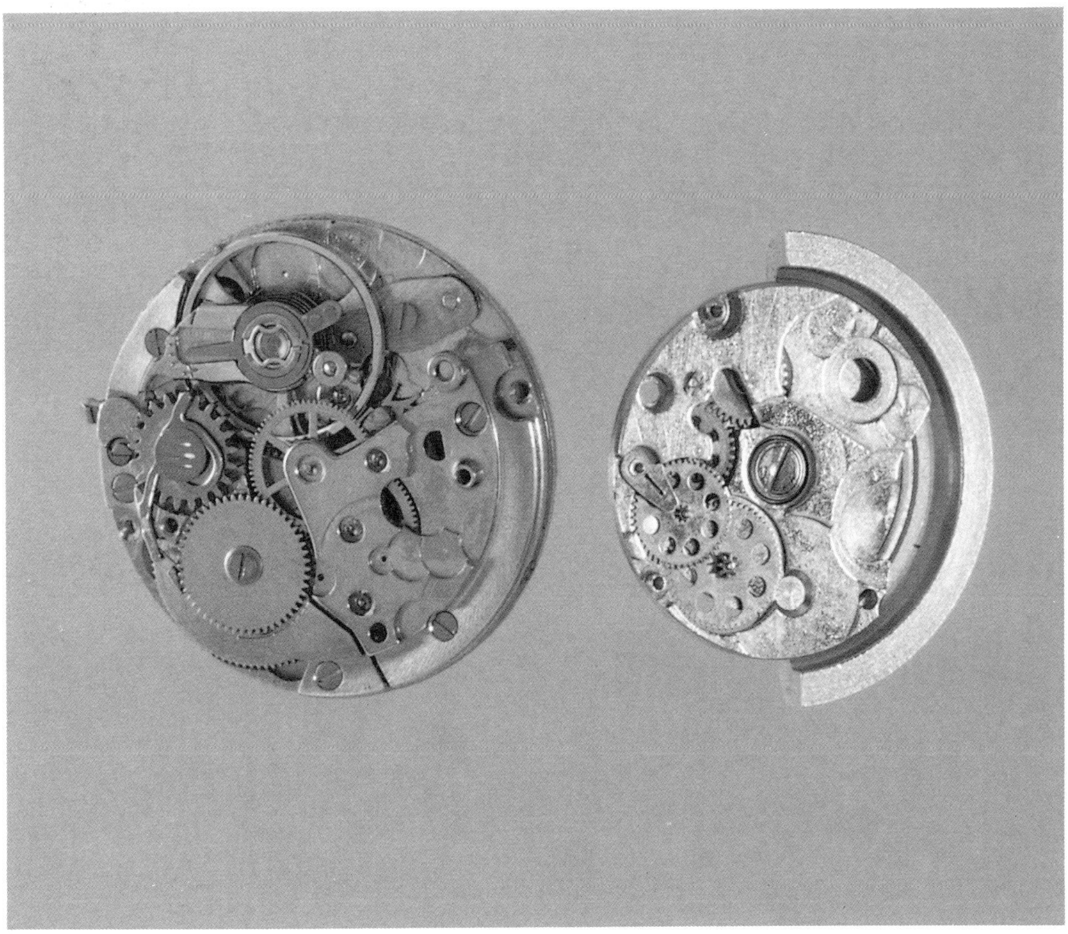

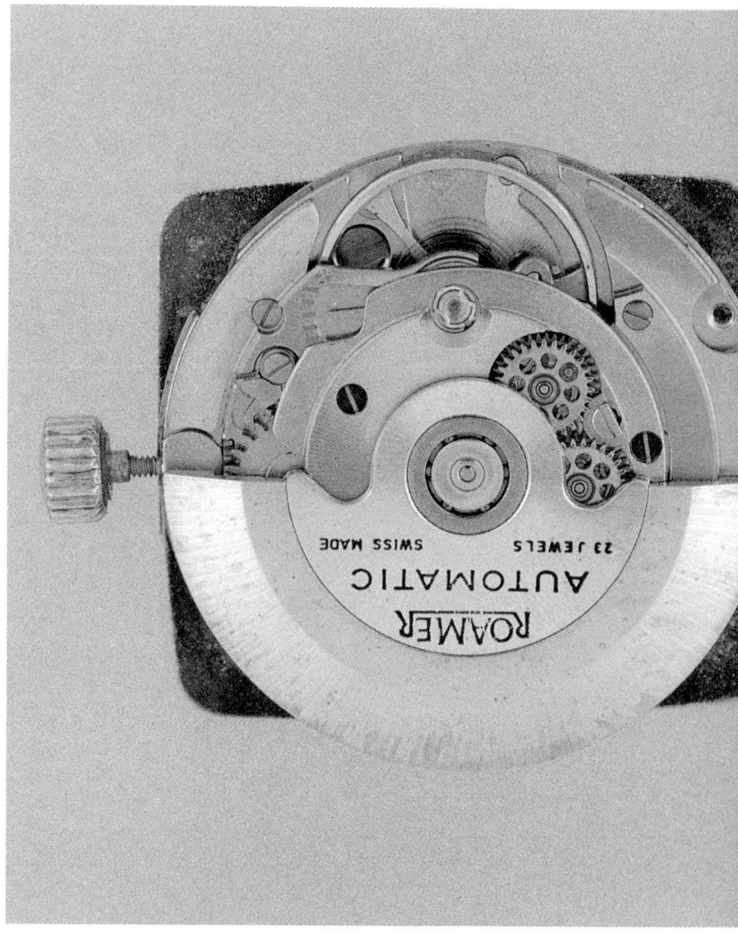

Roamer Caliber 474

Men's automatic, "Limelight" model, made by Roamer in 1966, with Caliber 474 (Eta raw movement, Caliber 2522 R).

Gilded square case with stainless steel bowl bottom.

Rotor with ball bearings and heavy metal ring, winding in both directions via click-wheel switcher (at 2:00), uncoupling of the hand-winding mechanism.

Large screwless Glucydur balance, Nivarox hairspring, Incabloc shock resistance (case and movement with signs of wear, winding crown not original).

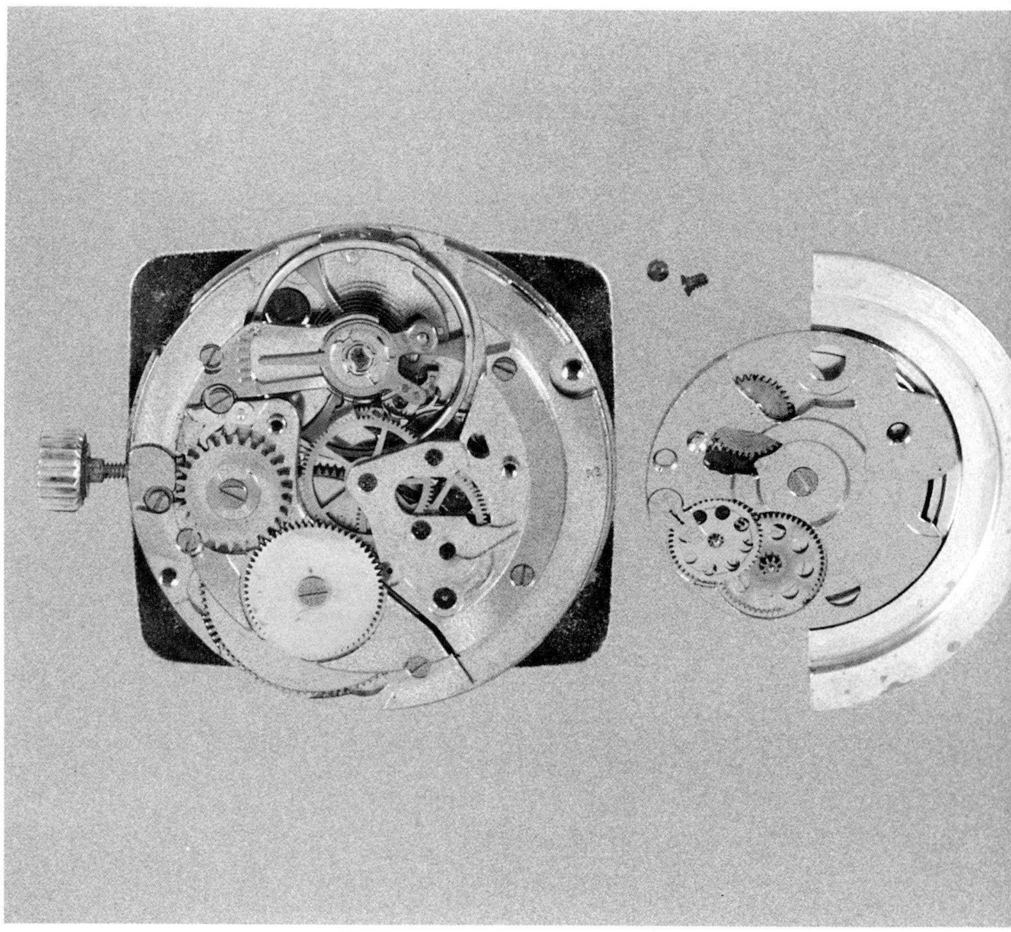

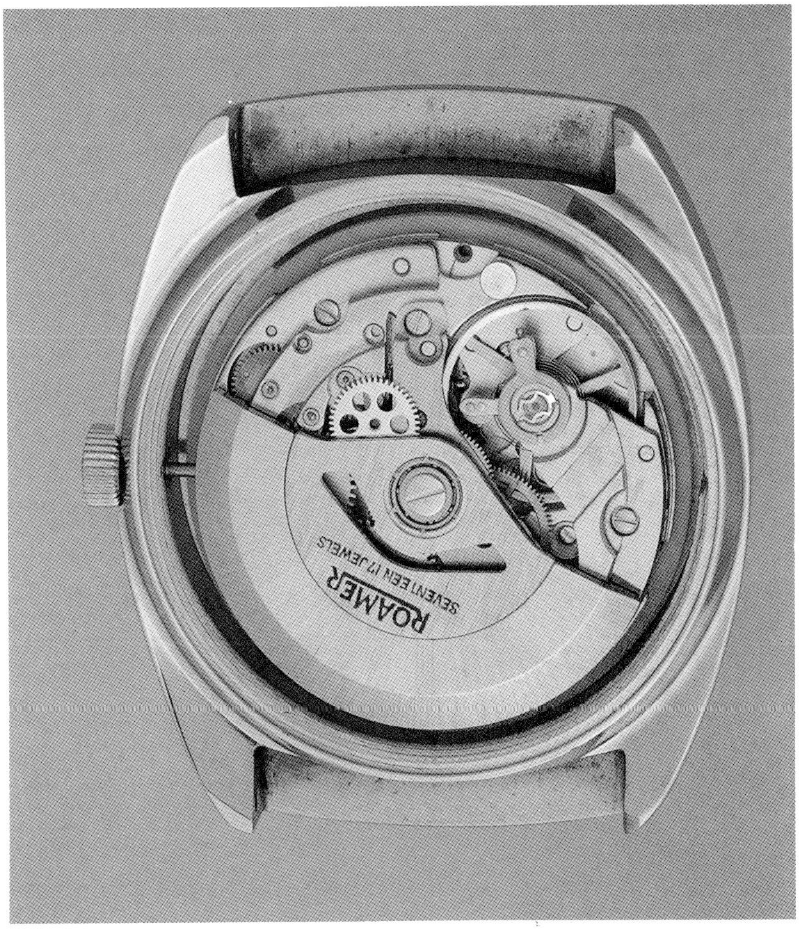

Roamer Caliber 522

Men's automatic, "Searock" model, made by Roamer in 1972, with Caliber 522.
Barrel-shaped stainless steel case with screw bottom, winding crown with R symbol.
Rotor with ball bearings and heavy metal ring, winding in both directions via click-wheel switcher.
Screwless Glucydur balance, Nivarox hairspring, Incabloc shock resistance.

Caliber List: Roamer/MST

Caliber	Diameter Lignes	mm	Height mm	Other	BPH	Jewels	Running Reserve	Notes	Year	Photo
410	11 1/2	26,0	6,3	S	18.000	17			ca. 50	
411	"	"		SCI	"	"				
420	11 1/2	26,0	6,4	SCD	18.000	21	36	Rotopower	1952	xx
421	"	"	7,3	SCD, QG	"	"	"	"		
426	7	15,8			18.000					
428				QG					1957	
436	11 1/2	26,0	4,55	SCD	21.600	44	42	Rotopower 44	ca. 58	xx
437	"	"	5,4	SCD, QG	"	"	"			
439	7 3/4	17,5		SCD	18.000					
441	11 1/2	26,0		SCD, QG	18.000	21	43	Rotodate, Felsa-Ebauche		xx
452	"	"		SCD, QG	"					
465	7 3/4	17,5		SCD, QG	18.000					
470	11 1/2	26,0	4,55	SCD	21.600	44	44		1959	
471	"	"	5,4	SCD, QG	"	44/28	"	Stingray/Vanguard 316	1959	xx
472	7 3/4	17,5		SCD	21.600	23		Stingray Matic, Eta-Ebauche		xx
473	"	"		SCD, QG	"	"				
474	11 1/2	26,0		SCD, QG	18.000	23		Limelight, Eta-Ebauche	1966	xx
477	11 1/2	26,0		SCD, QG	18.000					
478	"	"		SCD, QG, JG	21.600					
479	"	"		SCD, QG	"					
522	11 1/2	26,0	5,0	SCD, QG, RDR	21.600	17		Searock	1972	xx

Caliber 472
Eta Ebauche (Caliber 2550)
D = 17.5 mm (7.75 lignes), 23 jewels
AUTOMATIC MODULE screwed to the basic movement as a complete unit, central rotor with heavy metal ring, winding in both directions via springless twin click wheels, two reduction wheels on the locking wheel, uncoupling of hand-winding mechanism by a moveable crown wheel.
BASIC MOVEMENT: three-spoke screwless Glucydur balance, Nivarox hairspring, 21,600 BPH, Incabloc shock resistance, moveable hairspring block carrier, direct center seconds.

Caliber 474
Eta Ebauche (Caliber 2522 R)
D = 26.0 mm (11.5 lignes), 23 jewels
AUTOMATIC MODULE screwed onto the basic movement as a complete unit, central rotor with ball bearings and heavy metal ring, winding in both directions via springless twin click wheels, two reduction wheels on the locking wheel, flexibly mounted crown wheel to uncouple the hand-winding mechanism.
BASIC MOVEMENT: three-spoke screwless Glucydur balance, Nivarox hairspring, 18,000 BPH, Incabloc shock resistance, moveable hairspring block carrier, direct center seconds, date indication with quick change.

Caliber 522
D = 26.0 mm (11.5 lignes), H = 5.0 mm, 17 jewels
AUTOMATIC MODULE integrated into the basic movement, central rotor with ball bearings and heavy metal ring, winding in both directions via springless twin click wheels, two reduction wheels on the crown wheel.
BASIC MOVEMENT: three-spoke screwless nickel balance, Nivarox hairspring, 21,600 BPH, Incabloc shock resistance, moveable hairspring block carrier, direct center seconds, date indication with quick change and correction.

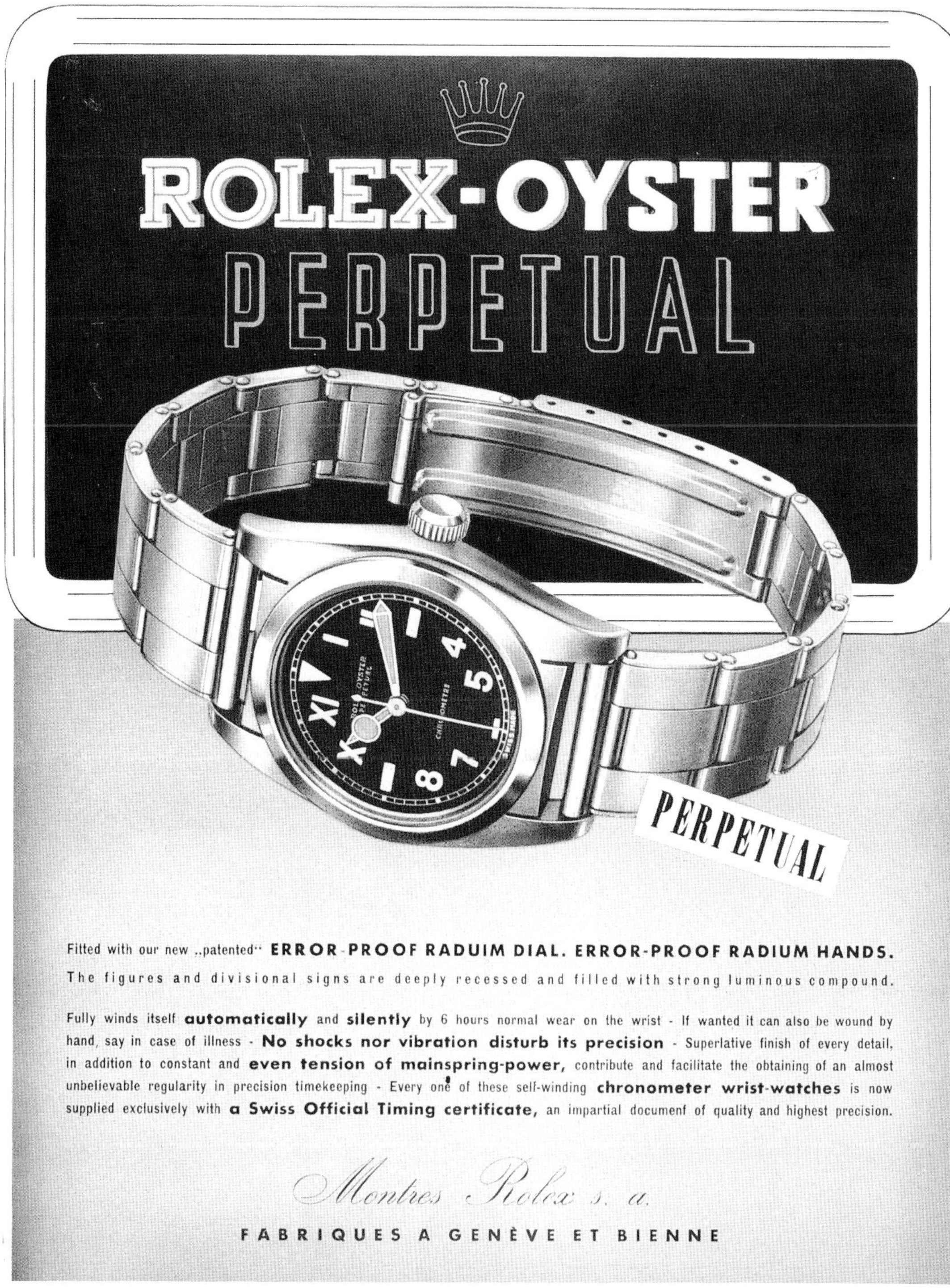

Magazine advertisement (from JSH, 1945)

Caliber Development

The German Hans Wilsdorf (1881-1960) founded the wholesale watch business of "Wilsdorf and Davis" in London in 1905, selling, among others, watches made by Jean Ageler of Biel. In 1908 Wilsdorf registered the "Rolex" brand name. In 1914 a Rolex became the first wristwatch to win a Class A chronometer certificate from the Kew Observatory in London.

As of 1920, movements produced by the Ageler S.A. firm in Biel were fitted with cases and sold by the firm of Montres Rolex S.A., now based in Geneva.

Wearing the first waterproof Rolex wristwatch, the Oyster model introduced in 1926, Mercedes Gleitze of London swam the English Channel in 1927. Rolex publicized this event with an advertising campaign of previously unheard-of extent.

Along with his chief designer Borer, Wilsdorf developed an automatic caliber (620 NA), that was introduced as the "Oyster Perpetual" in 1931 and patented a year later. For the first time in a wristwatch, a rotor functioned as a swinging mass winding in one direction. During automatic winding, a new design automatically uncoupled the hand-winding mechanism, thus increasing its degree of efficiency. As a result, a running reserve of 35 hours built up after an average wearing time of six hours. A sliding stud in the barrel protected the balance from damage resulting from overwinding the mainspring. The complete automatic unit was located in a bowl-like frame that was screwed onto the basic movement, covering it like a hat and making it waterproof. Only a relatively small hand-wound movement was used, but it was built with precision and equipped with appropriate fine adjustment.

Caliber List: Rolex

Caliber	Diameter		Height	Other	BPH	Jewels	Running Reserve	Notes	Year	Photo
	Lignes	mm	mm							
420 PA	8 1/2	19,4			18.000	18		On Caliber 400	1941	
520 AR	10 3/4	24,0			18.000	17		On Caliber 500	1936	
530 AR	"	"		SCI	"	"		"	1936	
620 NA	11 3/4	26,4	7,5	s	18.000	17	40	On Caliber 600	1931	xx
630 NA	"	"		SCI	"	18	"	"		xx
635 (A 260)	"	"	6,5		"		"	"		
640	"	"		QG, JG, PH	"		"	"		
645 (A 260)	"	"		SCI	"	18+2	"	"	1950	xx
720 (A 295)	13	29,5			18.000	18		On Caliber 700	1940	
730	"	"		SCI	"	18+2		"	1945	
740	"	"		SCI, QG	"	18		", first	1945	
745	"	"		SCI, QG	"	18+2		" automatic with date	1952	
765 (A 296)	"	"		SCI, QG	"	"		"	1950	
775	"	"		SCI	"	"		"	1950	
780	"	"		QG, JG, PH	"	18		"	1952	
1030	12 1/2	28,1	5,85	SCI	18.000	25			1950	xx
1035	"	"		SCI, QG	"					
1036	"	"		SCI, QG (GMT)	"					
1040	"	"		SCI	"					
1055	"	"		SCI, QG	"			250,000th	1954	
1065	"	"		SCI, QG (M, GMT)	"			Chronometer	1955	
1066	"	"		SCI, QG (M, GMT)	"				1957	
1080	"	"		SCI	"					
1120	8 3/4	19,7			19.800	27			1953	
1130	"	"		SCI	"	26			1955	
1135	"	"		SCI, QG	"	"			1955	
1160	"	"	5,5	SCI	"				1964	
1161	"	"		SCI	"				1964	
1165	"	"		SCI, QG	"				1965	
1166	"	"		SCI, QG	"				1967	
1520	12 1/2	28,1	5,75	SCI	19.800				1963	
1525	"	"	6,3	SCI, QG	"				1965	
1530	"	"	5,75	SCI	18.000				1957	
1535	"	"	6,3	SCI, QG	"				1957	
1555	"	"	7,03	SCI, QG, JG	"				1959	
1556	"	"	"	SCI, QG, JG	19.800				1965	
1560	"	"	5,75	SCI	18.000				1965	
1565	"	"	6,3	SCI, QG	"				1959	
1565 GMT	"	"	6,47	SCI, QG (GMT)	"				1962	
1570	"	"	5,75	SCI	19.800	25			1965	xx
1575	"	"	6,3	SCI, QG					1965	
1575 GMT	"	"	6,47	SCI, QG, ST (GMT)		26				
1580	"	"		SCI, ST (M)					1963	
2030	8 3/4	19,7	5,4	SCI, ST	28.800	28			1970	
2035	"	"	5,83	SCI, QG, ST	"	"			1970	

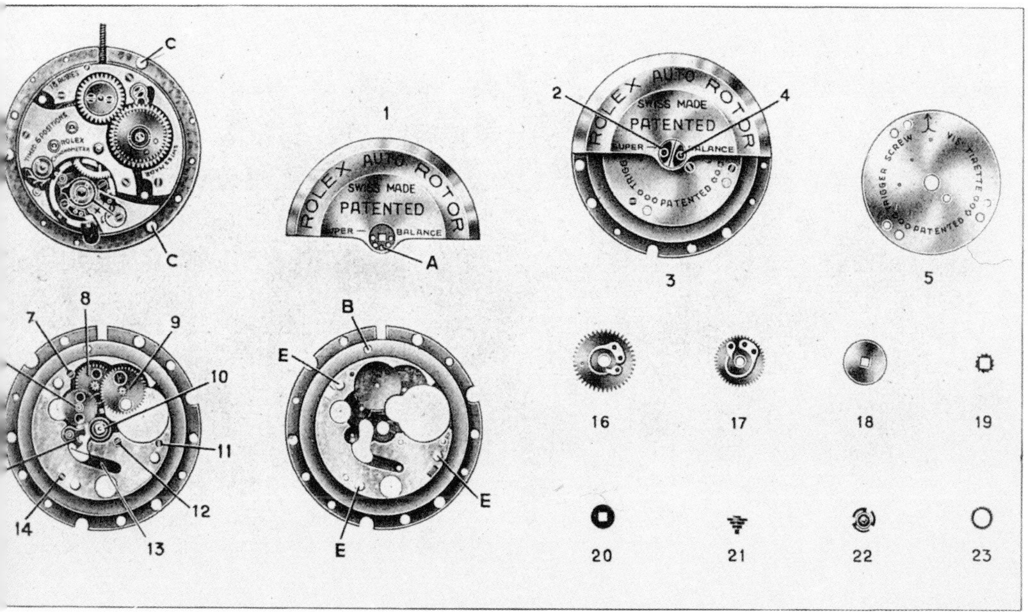

Illustrated parts catalog for Rolex Caliber 620 NA: the automatic drive, after removal of the upper plate (5), is shown at lower left.

By combining several Rolex innovations, all the essential requirements of a modern wristwatch were provided in the "Oyster Perpetual Chronometer", which also offered the qualities of sturdiness, wearing comfort and precision. Thus Rolex was able to develop an automatic concept that pointed the way to the future and thus lay the cornerstone for the firm's success, which has lasted to this day.

On the basis of this first rotor automatic, movements were built until 1950 with only minor modifications, their sizes and caliber numbers always based on the basic movement they used:

Caliber 420 Basic caliber 400 (7.75 lignes)
Calibers 520, 530 Basic caliber 500 (8.75 lignes)
Calibers 620-645 Basic caliber 600 (9.75 lignes)
Calibers 720-780 Basic caliber 700 (10.5 lignes)

Over the years, additional equipment was added to this series of calibers, including shock resistance for the balance arbor, an indirect center seconds and, in 1945, the first automatic with date indication, Caliber 740.
Only by changing the design of the locking-wheel coupling could the height of the Caliber 745 movement be reduced by one millimeter. The obvious thickness of the previous Rolex movements necessarily required a thick case shape with the arched screwed-in bottom that was nicknamed "Bubble Back".

Only in 1950, with Caliber 1030, was a new automatic design introduced, its advantages including two-way winding that set it apart from its forerunners. Twin wheels with springless anchor clicks were used for switching and made of a red-eloxied, low-wear aluminum alloy.

The central rotor now turned above the big basic movement that filled the entire space inside the case and, with a larger balance and barrel installed, resulted in further improvement to the chronometer quality. With it, the Rolex firm was able to sell its 250,000th wrist chronometer in 1954.

In the ensuing years, among other things the designers also improved the basic movement by installing a screwless ring balance with a faster beat.

This time-tested automatic design was replaced only in 1970 by the medium-sized Caliber 2030, and by the larger Caliber 3035 in 1977.

While the automatic module of these two watches was taken over from their predecessors without any major changes, Rolex equipped the basic movements according to the most modern state of precision technology. All the wheels were made of beryllium bronze and with microgearing. This plus the fast-turning barrel guaranteed a particularly steady transmission of energy to the fast-beat balance assembly, which worked with a Clinergic 21 anchor escapement made by FAR (Fabriques d'Assortiments Réunies). The lack of a regulator is noteworthy; here it was replaced by two pairs of patented Microstella regulatory nuts on the inside of the balance ring. An additional screw in the foot of the balance potence allowed axial balance play to be eliminated. The entire construction was essentially modified with the introduction of a direct center second and a minute wheel mounted out of the center.

Without a doubt, Rolex deserves the credit of having made the most significant contribution to the general development and acceptance of the automatic wristwatch.

Caliber List: Rolex

Caliber	Diameter Lignes	mm	Height mm	Other	BPH	Jewels	Running Reserve	Notes	Year	Photo
2130	8 3/4	19,7	5,4	SCI, ST	28.800	29			1984	
2135	"	"	5,83	SCI, QG, RDR, ST	"					
3035	12 1/2	28,1	6,35	SCD, QG, RDR, ST	28.800	27	50		1977	xx
3055	"	"	7,11	SCD, QG, JG, RDR, ST	"	"	"			
3075	"	"	6,35	SCD, QG, RDR, ST (GMT)	"	"	"		1983	
3085	"	"	6,8	SCD, QG, ST (GMT)						
3135	12 1/2	28,1	6,0	SCD, QG, RDR, ST	28.800	31		With balance bridge	ca. 88	

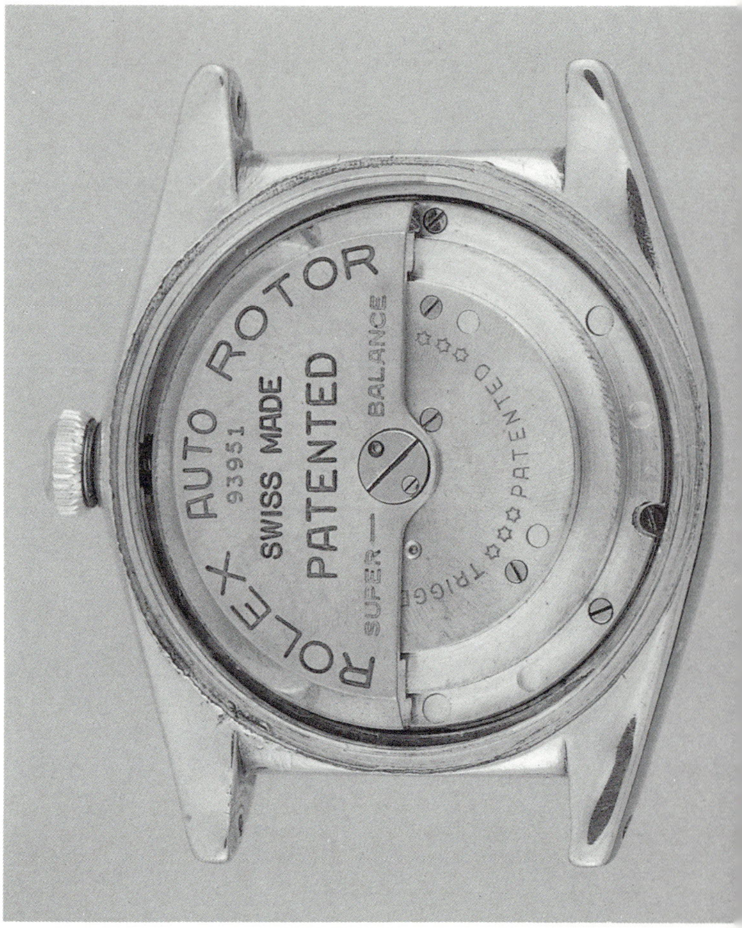

Rolex Caliber 620 NA
Men's automatic, "Oyster Perpetual" model, made by Rolex circa 1933, with Caliber 620 NA, the world's first rotor caliber, made since 1931.
Gilded stainless steel case with screw bottom, screwed winding crown with Rolex monogram ("bubble back").
Brass rotor winding in one direction via Breguet coupling, complete uncoupling between automatic and hand-winding drives. Automatic module as a complete unit fully covering the precision movement.
"Super Balance" screw balance, Breguet hairspring, no shock resistance (dial is freshened up, case regilded, winding crown not original, movement with signs of wear, swan's-neck spring missing).

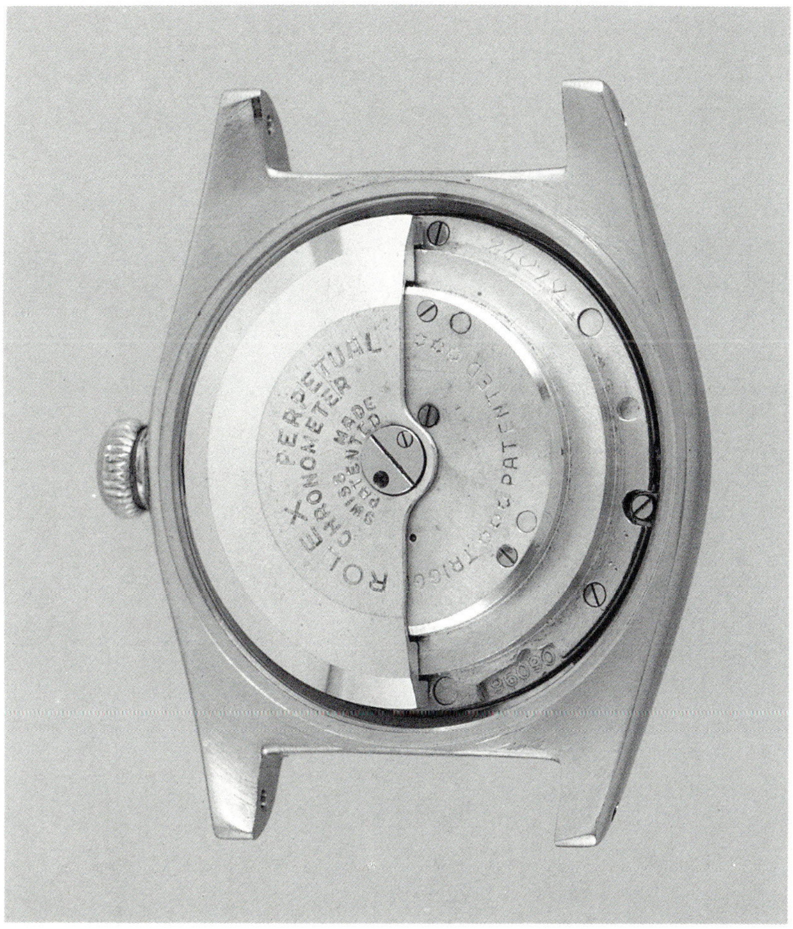

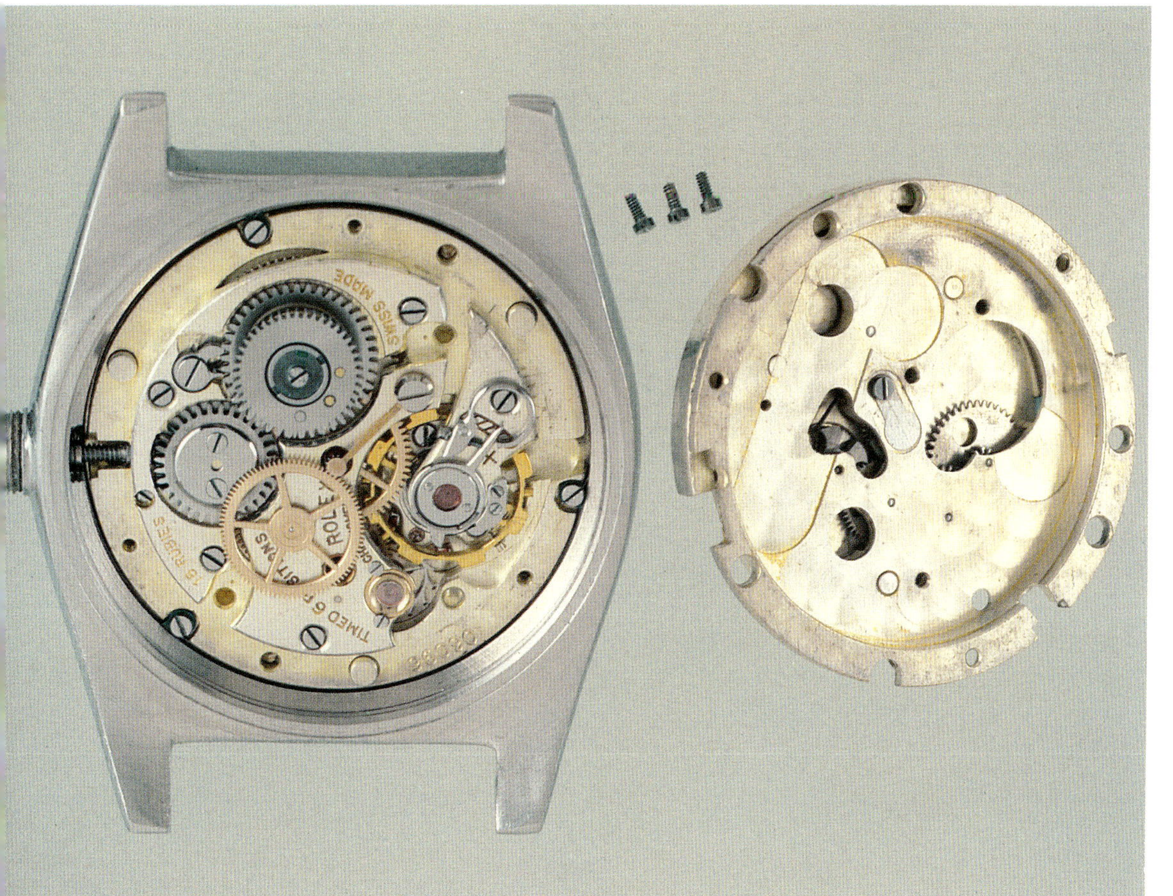

Rolex Caliber 630 NA
Men's automatic, "Oyster Perpetual" model, made by Rolex circa 1938 as "Super Precision" chronometer, with Caliber 630 NA. Stainless steel case with screw bottom and screwed crown with Rolex monogram ("bubble back"). Movement identical to Caliber 620 NA, but with center seconds (seconds hand not original, surface of movement slightly corroded).

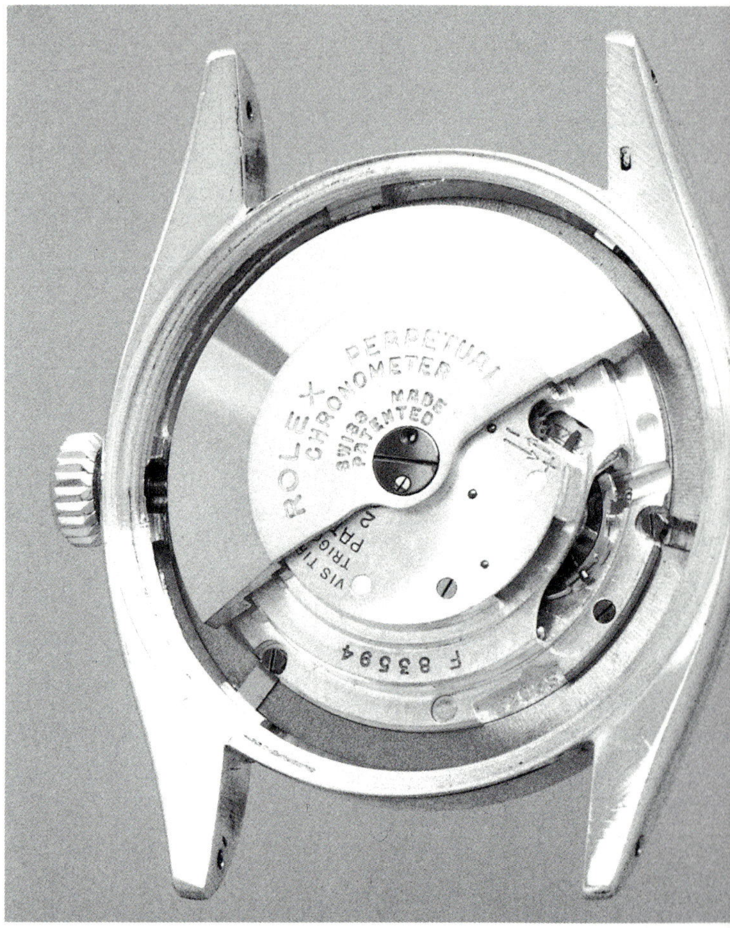

Rolex Caliber 645

Men's automatic, "Oyster Perpetual" chronometer model, made by Rolex circa 1950, with Caliber 645. Eighteen karat gold case with screw bottom and screwed winding crown (with crown symbol). Movement essential identical to Caliber 630, but made flatter and equipped with Monorex shock resistance (dial inlays and hands corroded).

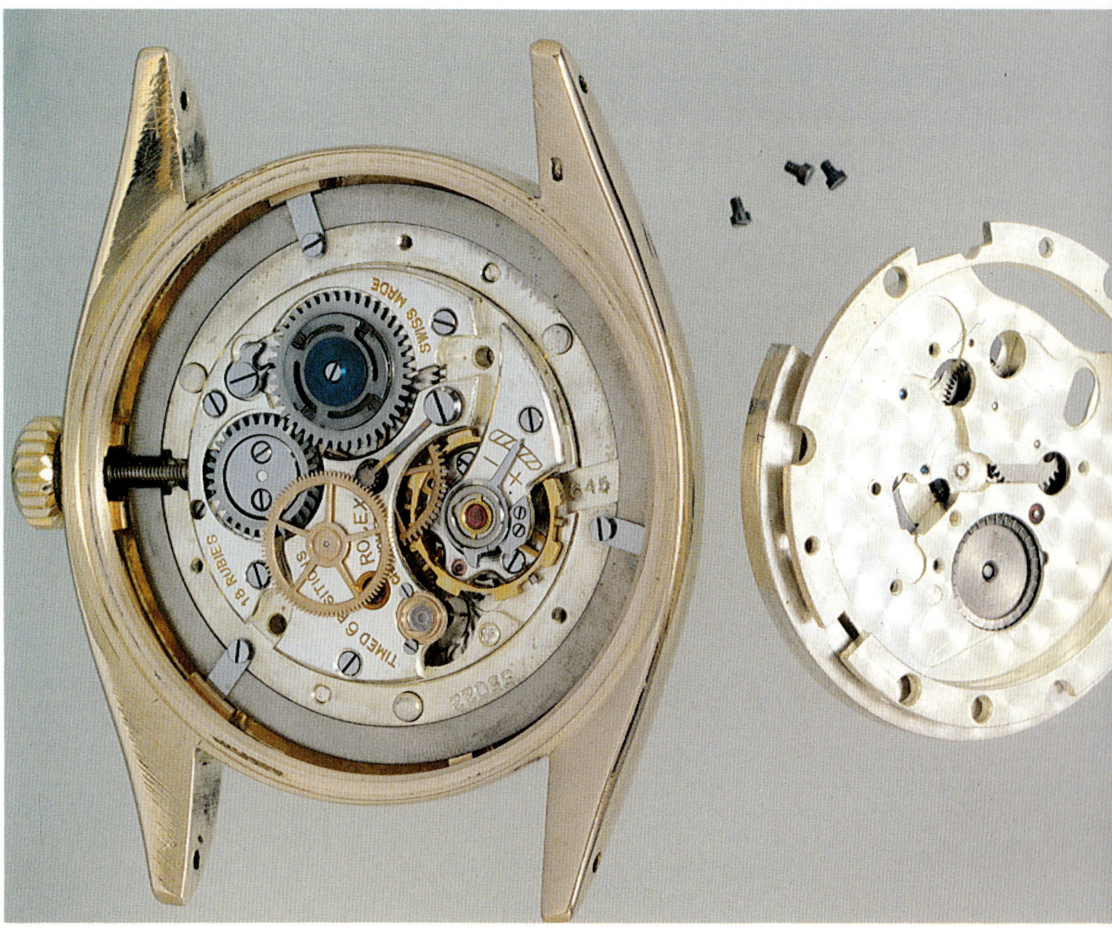

Rolex Caliber 1030
Men's automatic, "Oyster Perpetual" chronometer model, made by Rolex circa 1957, with Caliber 1030.
Stainless steel case with screw bottom and screwed winding crown (with crown symbol).
Rotor with heavy metal ring, winding in both directions via click-wheel switcher, uncoupling of the hand-winding drive.
Glucydur screw balance, self-compensating Breguet hairspring, Kif-Flector shock resistance (dial discolored, hand, movement, etc. corroded).

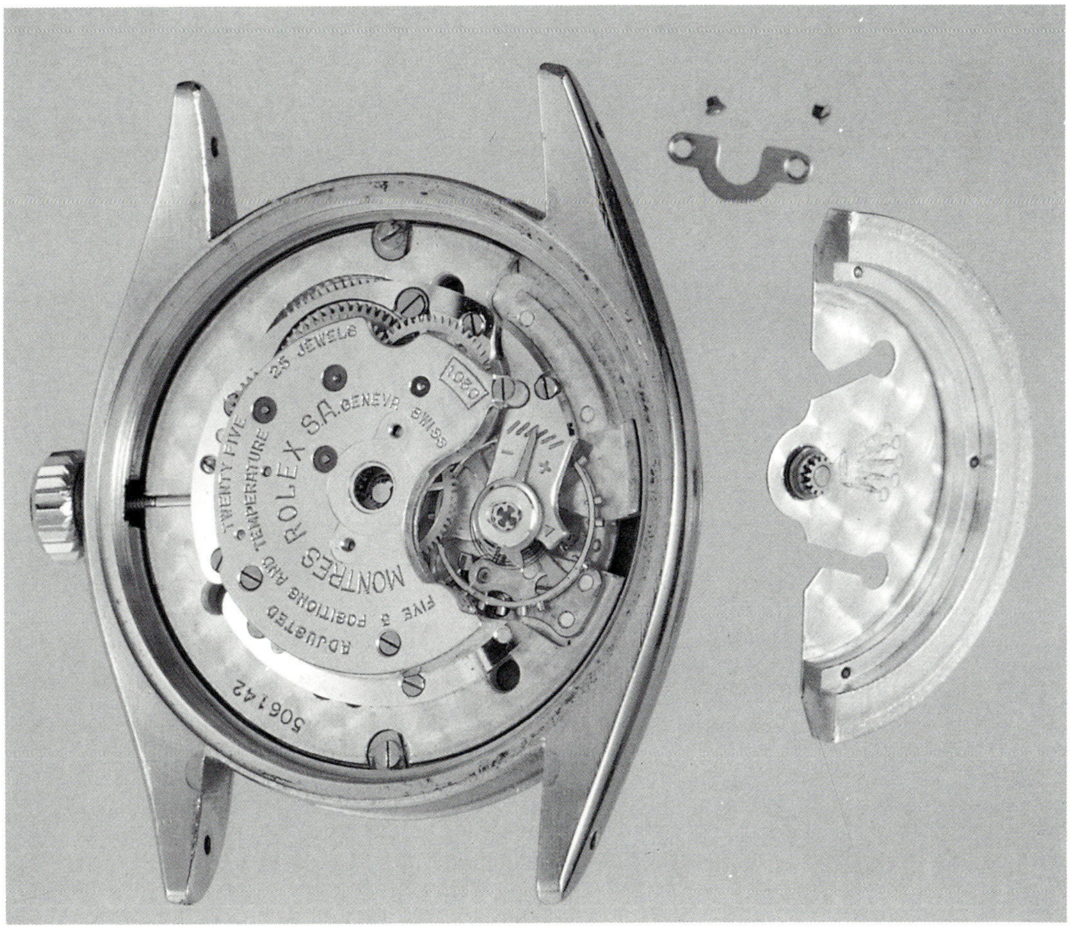

Rolex Caliber 1575 GMT

Men's automatic, "Oyster Perpetual Explorer II" chronometer model, made by Rolex circa 1970, with Caliber 1575 GMT (basic caliber 1570). Stainless steel case with screw bottom and screwed winding crown (with crown symbol), 24-hour indication by a 4th central hand.

Rotor with heavy metal ring, winding in both directions via click-wheel switcher (eloxied red), automatic module removable as a complete unit. Glucydur screw balance, self-compensating Breguet hairspring, Kif-Ultraflex shock resistance, no regulator.

Rolex Caliber 2030

Ladies' automatic, "Oyster Perpetual" model, made by Rolex circa 1975, with Caliber 2030.
Stainless steel case with screw bottom, gold bezel and winding crown.
Rotor with heavy metal ring, winding in both directions via click-wheel switcher (eloxied red), uncoupling of the hand-winding drive. Automatic module removable as a complete unit.
Screwless Glucydur balance, self-compensating flat hairspring, Kif-Ultraflex shock resistance, Spirofin fine adjustment (dial and hands corroded, some movement screws rusty, signs of wear on case and movement).

Rolex Caliber 3035

Men's automatic, "Oyster Perpetual" chronometer model, made by Rolex circa 1982, with "Datejust" Caliber 3035. Stainless steel case with screw bottom and screwed winding crown (with crown symbol). Rotor with heavy metal ring, winding in both directions via click-wheel switcher (eloxied red), uncoupling of hand-winding drive. Automatic module removable as a complete unit. Glucydur ring balance with two pairs of Microstella weight screws, self-compensating Breguet hairspring, Kif-Ultraflex shock resistance, no regulator.

Caliber Notes

Caliber 620 NA
D = 26.4 mm (11.75 lignes), H = 7.5 mm, 17 jewels
AUTOMATIC MODULE set onto the basic movement as a complete unit, like a cap, monometallic central rotor, winding in one direction via a centrally mounted Breguet wheel with click spring, three reduction wheels, the first with a spring click, on the double locking wheel, which is built as a click coupling for reciprocal uncoupling between the automatic and hand-winding drives.
BASIC MOVEMENT (Caliber 600, D = 9.75 lignes): two-spoke monometallic screw balance (Super-Balance), Breguet hairspring, 18,000 BPH, no shock resistance, swan's-neck fine regulation of the regulator, escape wheel mounted between cap jewels, eccentric seconds hand.

Caliber 630 NA
D = 26.4 mm (11.75 lignes), 18 jewels
See Caliber 620 NA, but with indirect center seconds.

Caliber 645
D = 26.4 mm (11.75 lignes), 20 jewels
Movement as Caliber 630 NA, but with larger rotor, cutouts in automatic plate for regulator adjustment, monorex shock resistance and Breguet coupling on the locking wheel.

Caliber 1030
D = 28.1 mm (12.5 lignes), H = 5.85 mm, 25 jewels
AUTOMATIC MODULE screwed onto the basic movement as a complete unit, central rotor with heavy metal ring, winding in both directions via springless twin click wheels of red-eloxied aluminum alloy, two reduction wheels on the double locking wheel (to uncouple the hand-winding mechanism).
BASIC MOVEMENT: two-spoke Glucydur screw balance, self-compensating Breguet hairspring, 18,000 BPH, Kif-Flector shock resistance, escape wheel mounted in combined Duofix jewel housing.

Caliber 1575 GTM
D = 28.1 mm (12.5 lignes), H = 6.47 mm, 26 jewels
AUTOMATIC MODULE screwed to the basic movement as a complete unit, central rotor with heavy metal ring, winding in both directions via springless twin click wheels of red-eloxied aluminum alloy, one reduction wheel on the locking wheel.
BASIC MOVEMENT: two-spoke Glucydur screw balance, self-compensating Breguet hairspring, 19,800 BPH, Kif-Ultraflex shock resistance, stoppable-moveable hairspring block carrier without hairspring key, escape wheel mounted in combined Duofix jewel housing, second stop, center 24-hour hand, date indication.

Caliber 2030
D = 19.7 mm (8.75 lignes, H = 5.4 mm, 28 jewels
AUTOMATIC MODULE screwed to the basic movement as a complete unit, central rotor with heavy metal ring, winding in both directions via springless twin click wheels of red-eloxied aluminum alloy, one reduction wheel on the locking wheel, uncoupling of the hand-winding drive between the crown and locking wheels.
BASIC MOVEMENT: four-spoke screwless Glucydur balance, self-compensating flat hairspring with Greiner hairspring roller, 28,800 BPH, Kif-Ultraflex shock resistance, Spirofin fine adjustment with two ruby regulator pins and moveable hairspring block carrier, Clinergic 21 anchor escapement, escape wheel mounted in combined Duofix jewel housing, date indication with instant change.

Caliber 3035
D = 28.1 mm (12.5 lignes), H = 6.25 mm, 27 jewels
AUTOMATIC MODULE screwed to the basic movement as a complete unit, central rotor with heavy metal ring, winding in both directions via springless twin click wheels of red-eloxied aluminum alloy, one reduction wheel on the locking wheel, uncoupling of the hand-winding mechanism by an alternating pinion between the crown and locking wheels.
BASIC MOVEMENT: four-spoked Glucydur ring balance with two pairs of Microstella balance screws. self-compensating Breguet hairspring, 28,800 BPH, Kif-Ultraflex shock resistance, moveable hairspring block carrier, high play of the balance regulated by a potence screw, Clinergic 21 anchor escapement, escape wheel mounted in combined Duofix jewel housing, direct center seconds, fast-turning barrel (n = 10), second stop, quick date change and correction.

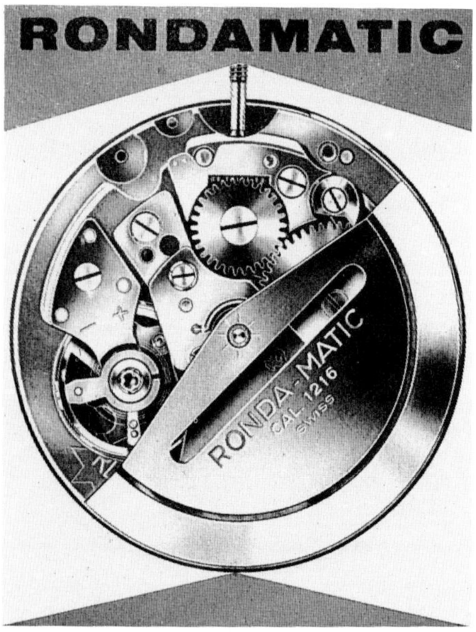

Magazine advertisement (from The Swiss Watch, 1972)

Caliber Development

William Mosset founded the firm of Ronda S.A. at Lausen in 1946 and began to produce individual watch components. In 1961 he was given permission by the Swiss Department of Commerce to produce complete pin lever movements of his own. After the Pin Lever Convention was repealed in 1966, Ronda also supplied raw movements with jewel-anchor escapements and modern construction. The usually utilized central rotor and the microrotor installed in Caliber 2539 D wound in only one direction and made a simple and sturdy automatic module possible.

Caliber Notes

Caliber 1239-21 (RAX)
D = 27 mm (12 lignes), 17 jewels
AUTOMATIC MODULE screwed to the basic movement, central rotor with heavy metal ring and sliding stud, winding in one direction via springless gliding pinion, two reduction wheels on the locking wheel.
BASIC MOVEMENT: two-spoke screwless nickel balance, Nivarox hairspring, 21,600 BPH, Kif-Protechoc shock resistance, moveable hairspring block carrier, jewel-anchor escapement, direct center second, date indication with quick change and correction, weekday indication.

Caliber List: Ronda

Caliber	Diameter Lignes	mm	Height mm	Other	BPH	Jewels	Running Reserve	Notes	Year	Photo
Hook-Anchor Caliber (RL):										
1216-21	12	27			18.000					
1216-21	"	"	5,5	SCD	21.600				ca. 72	
1218-21	"	"		SCD, QG, RDR	"				"	
1219-21	"	"		SCD, QG, JG, RDR	"				"	
1226	12	27		REV	21.600				ca. 72	
1226-21	"	"		REV, SCD (Digital)	"					
9018	8 3/4	20		SCD, QG	21.600	17				
Jewel-Anchor Caliber (RAX):										
1236-21	12	27			21.600					
1237-21	"	"		CLD	"					
1238-21	"	"		CLD	"					
1239-21	"	"		SCD, QG, JG	"	17				xx
1246-21	12	27		SCD (Digital)	21.600					
2539 D	11 1/2	26	4,25	SCD, QG, JG, RDR	21.600	22	40	(Micro-Rotor)	1986	

Ronda Caliber 1239-21

Men's automatic of the "Sicura" brand, circa 1975, with Ronda Caliber 1239-21 (RAX).

Barrel-shaped gilded case with stainless steel screw bottom.

Rotor with heavy metal ring, winding in one direction via gliding pinion (at 12:00).

Screwless Glucydur balance, Nivarox hairspring, Kif-Protechoc shock resistance.

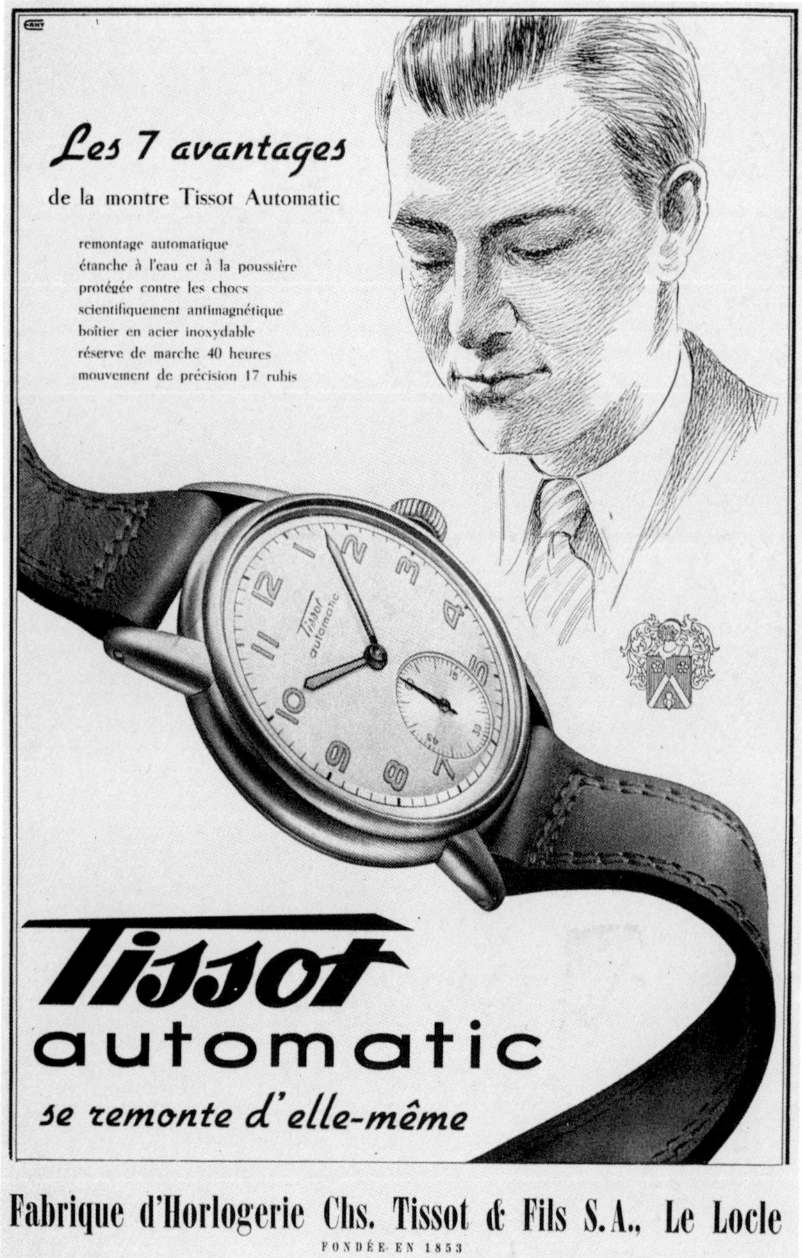

Magazine advertisement (from JSH, 1945).

Caliber Development

Charles F. Tissot founded the firm at Le Locle in 1853, and in 1930 it merged with Omega and other firms to form the SSIH (Holding Société Suisse pour l'Industrie Horlogère), after having cooperated with Omega for several years. The firm's further development has been influenced by Omega; Tissot's own caliber developments complete their offerings in the realm of reasonably priced watches.

Thus the first Tissot automatic caliber, made in 1944, shows marked similarities with those of Omega. It was a hammer automatic with one-sided winding in the typical manner, with the caliber number 28-1.

For the firm's 100th anniversary, the "Navigator" world-time watch was introduced; it was fitted with a slightly modified form of this caliber (Caliber 28.5 N-21). Here a simple mechanism controlled by the crown indicates the local time without requiring the world-time scale on the dial to be turned (see illustration).

The firm's one new development is the rotor automatic introduced in 1953 and winding in both directions via a centrally mounted switching rocker typical of Tissot watches. The subsequent caliber designs were based on this movement and show only minor changes in construction.

Caliber Notes

Caliber 28.5-1
D = 28.5 mm (12.5 lignes), H = 5.05 mm, 17 jewels
AUTOMATIC MODULE screwed to the basic movement, hammer with ruby bearings, limited by buffer springs, winding in one direction via click-carrier rocker, combined winding and reduction wheel on the crown wheel.

Advertisement for the "Navigator" automatic world-time watch made by Tissot in 1953. Pushing the button by the 2 uncouples the world-time circle from the hand-setting mechanism.
Exploded view of Tissot Caliber 28.5 R 621 (for lubrication).
BASIC MOVEMENT: two-spoke nickel screw balance, Nivarox hairspring, 19,800 BPH, Incabloc shock resistance, eccentric seconds hand.

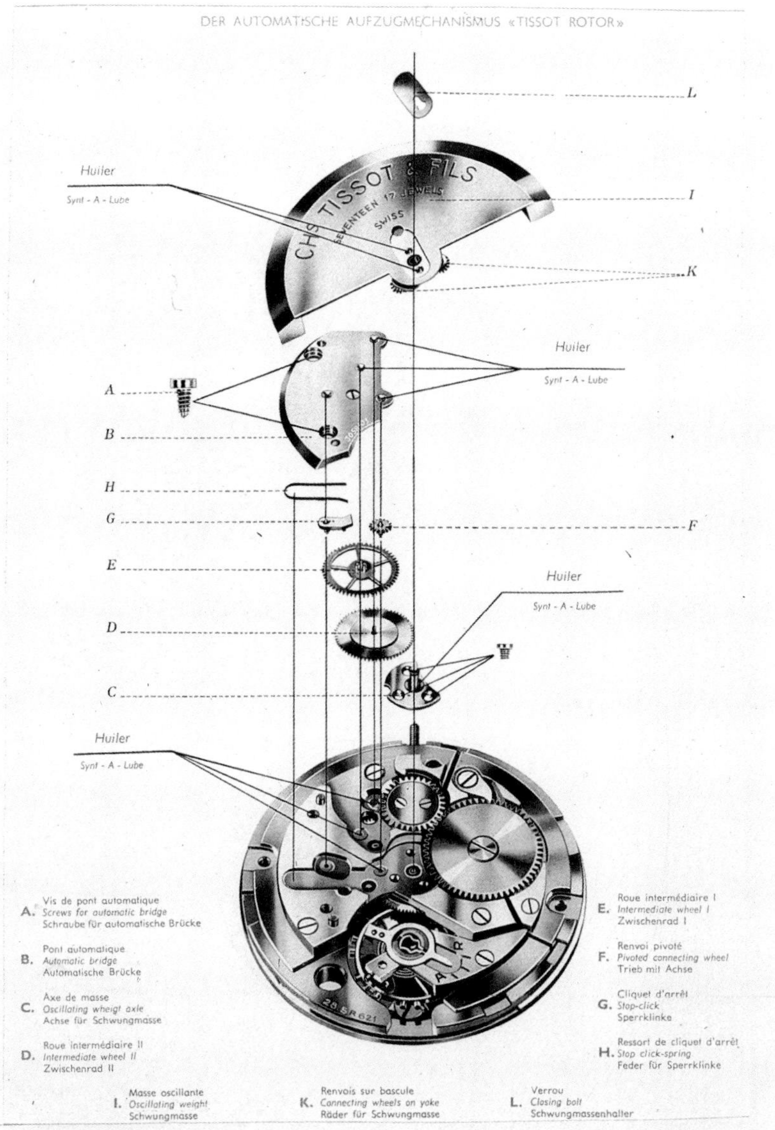

Tissot Caliber 28.5-1
Men's automatic, Model T 400, made by Tissot circa 1950, with Caliber 28.5-1. Square 14-karat gold case with press-in bowl bottom.
Hammer winding in one direction via click-carrier rocker (at 8:00).
Nickel screw balance, Nivarox hairspring, Incabloc shock resistance.

Caliber 28.5 R-21
D = 28.5 mm (12.5 lignes), H = 6.05 mm, 17 jewels.
AUTOMATIC MODULE screwed to the basic movement, monometallic central rotor, winding in both directions via a centrally mounted switching rocker with a small intermediate wheel, two reduction wheels, the first with a spring click, on the crown wheel, uncoupling of the automatic drive during hand-winding by a coupling in the second reduction wheel.
BASIC MOVEMENT: two-spoke nickel screw balance, Nivarox hairspring, 19,800 BPH, Incabloc shock resistance, direct center seconds.

Caliber 784-2
D = 26.0 mm (11.5 lignes), 21 jewels
AUTOMATIC MODULE screwed to the basic movement as a complete unit, central rotor with heavy metal ring, winding in both directions via centrally mounted switching rocker with small intermediate wheel, two reduction wheels, the first with a spring click, on the crown wheel.
BASIC MOVEMENT: three-spoke screwless nickel balance, Nivarox hairspring, 18,000 BPH, Incabloc shock resistance, moveable hairspring block carrier, fine regulator adjustment by eccentric screw, direct center seconds, date indication with quick change.

Caliber 2481
D = 26.0 mm (11.5 lignes), H = 5.2 mm, 21 jewels
Movement as Caliber 784-2, but with 21,600 BPH, date indication with quick change and correction.

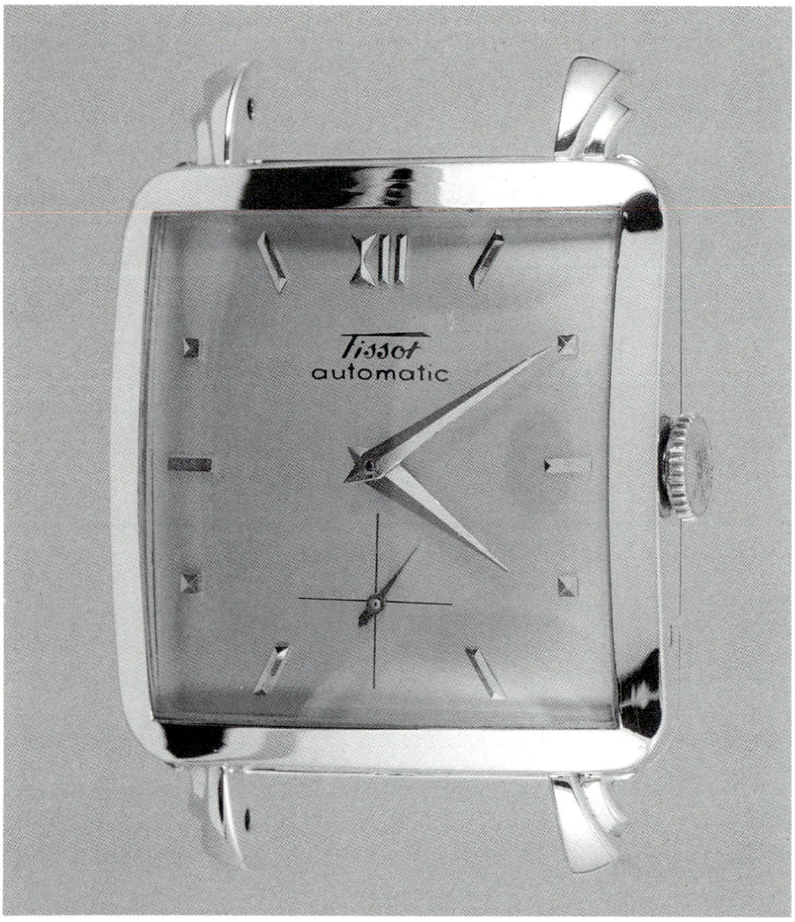

Tissot Caliber 28.5-1

Men's automatic, Model T 400, made by Tissot circa 1950, with Caliber 28.5-1. Square 14-karat gold case with press-in bowl bottom.

Hammer winding in one direction via click-carrier rocker (at 8:00).

Nickel screw balance, Nivarox hair-spring, Incabloc shock resistance.

Tissot Caliber 28.5R-21

Men's automatic, "Seastar" model, made by Tissot in 1953, with Caliber 28.5R-21.
Stainless steel case with screw bottom, winding crown with Tissot monogram.
Rotor with heavy metal ring, winding in both directions via central switching rocker (under the rotor), uncoupling during hand-winding.
Nickel screw balance, Nivarox hairspring, Incabloc shock resistance.

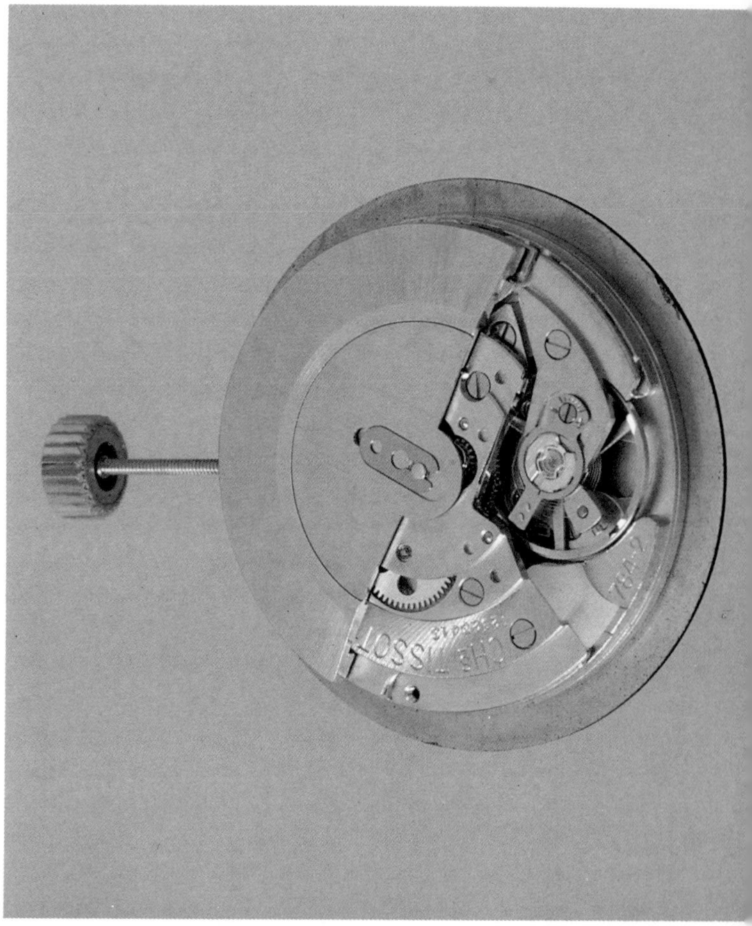

Tissot Caliber 784-2

Men's automatic, "Sideral S" model, made by Tissot circa 1970, with Caliber 784-2.

Plastic case (one-piece) and band, removal of movement possible only after removing the glass and dial, winding crown with T symbol.

Rotor with heavy metal ring, winding in both directions via central switching rocker (under the rotor).

Screwless nickel balance, Nivarox hairspring, Incabloc shock resistance, fine regulator adjustment by eccentric screw.

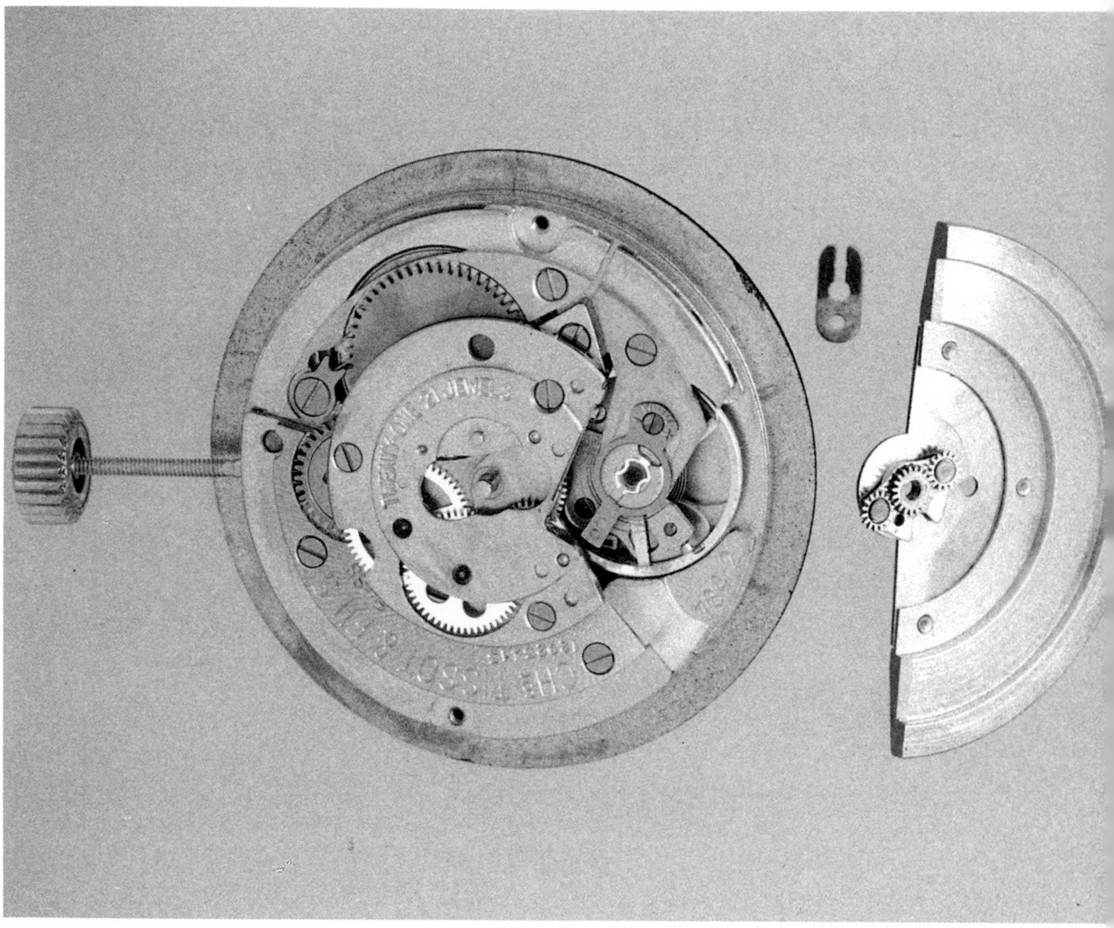

Tissot Caliber 2481
Men's automatic, "Seastar" model, made by Tissot circa 1975, with Caliber 2481.
Barrel-shaped gilded case with stainless steel screw bottom, winding crown with T symbol.
Movement mainly identical to Caliber 784-2 (see page 298).

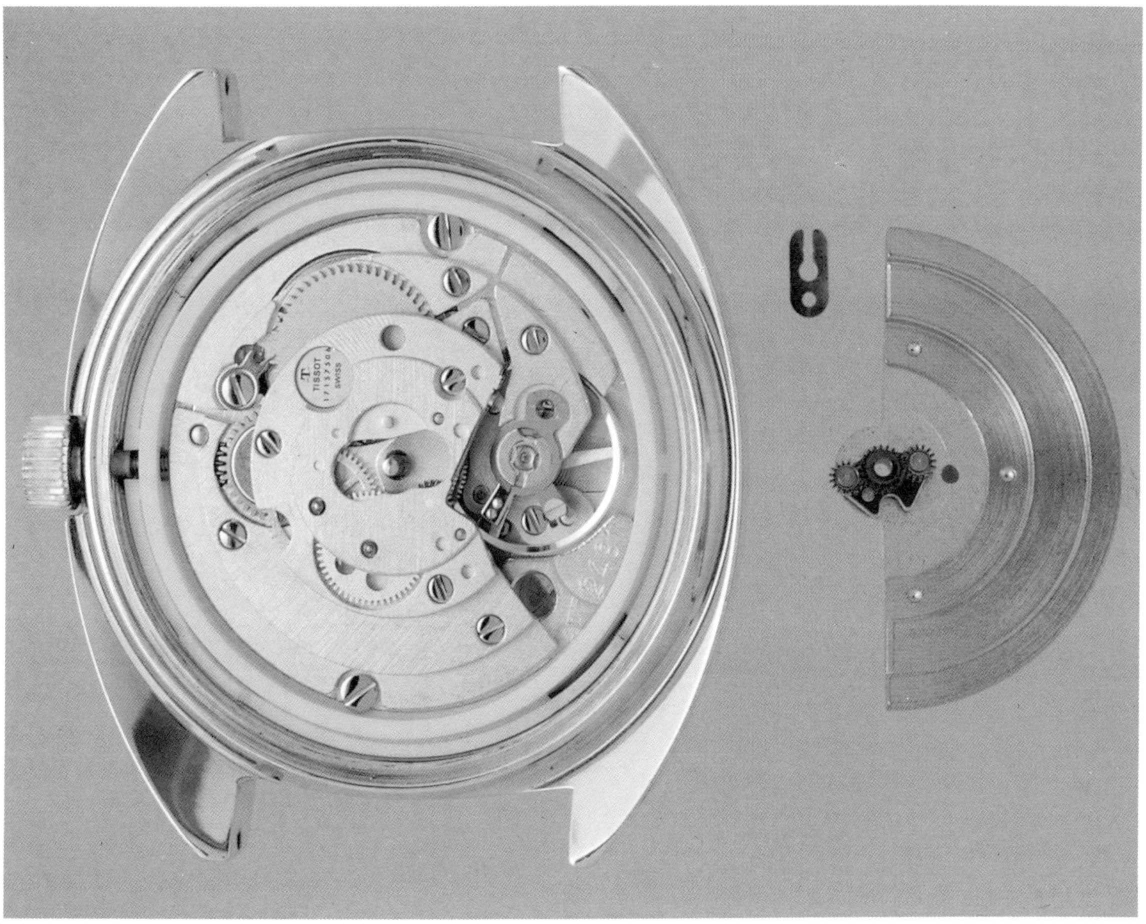

Caliber List: Tissot

Caliber	Diameter Lignes	mm	Height mm	Other	BPH	Jewels	Running Reserve	Notes	Year	Photo
17.2 R-21	7 3/4	17,2	5,3	SCD	18.000	17		Eta-Ebauche		
28-1	12 1/2	28,0	5,05	s	19.800	17	40		1944	
28-21	"	"	6,0	SCI	"	"	"			
28.5-1	"	28,5	5,05	s	"	"	"		1950	xx
28.5-21	"	"	6,0	SCI	"	"	"		1952	
28.5 N-21	"	"	6,6	SCI, 24 h	"	"	"	Navigator	1951	
28.5 R-21	12 1/2	28,5	6,05	SCD	19.800	17	40	Centenary	1953	xx
28.5 R-621	"	"	6,45	SCD, QG	"	"	"		1954	
31-1	13 1/2	31,0	5,05	s	19.800	17	40			
31-21	"	"	6,0	SCI	"	"	"			
717	7 3/4	17,2	5,3	SCD	18.000	21		Eta-Ebauche		
718	"	"	5,6	SCD, QG	"	"		"		
743	"	"		SCD	21.600					
744	"	"		SCD, QG	"					
760	"	"		SCD	"					
783/-1/-2	11 1/2	26	5,2	SCD	18.000	21			ca. 60	
784	"	"	"	SCD	"	"				
784-1/-2	"	"		SCD, QG/ST	"	"			1970	xx
785	"	"		SCD	"					
786	"	"		SCD, QG	"	"				
788 T 12	"	"		SCD, QG 24 h	"			Navigator		
794	11 1/2	26		SCD, QG, JG	18.000			Visoday		
823	12 1/2	28,5	6,05	SCD, QG	19.800	17	40	See Cal 28.5 R-21		
824	"	"	6,45	SCD, QG	"	"	"	See Cal 285 R-621		
2321	7 3/4	17,2	4,8	SCD, QG	28.800	17		Eta- Ebauche		
2322	"	"	"	SCD, QG	21.600	"		"		
2341	"	"	5,35	SCD, QG, JG	28.800	"		"		
2342	"	"	"	SCD, QG, JG	21.600	"		"		
2381	"	"	4,8	SCD, QG	28.800	21		"		
2382	"	"	"	SCD, QG	21.600	"		"		
2391	"	"	5,35	SCD, QG, JG	28.800	"		"		
2392	"	"	"	SCD, QG, JG	21.600	"		"		
2481	11 1/2	26	5,2	SCD, QG	21.600	21			ca. 75	xx
2531	"	"		SCD, QG	21.600					
2571	"	"		SCD, QG, JG						
2780	"	"	6,6						1977	

Caliber Development

The Rolex firm sold reasonably priced watches under the brand name of "Tudor", using dependable raw movements of Ebauches S.A., at first made by AS and later by Eta, in its Oyster cases.

An exception was the "Rotor-Prince" Caliber 390 developed by Rolex around 1950 especially for the Tudor brand. It was a rugged and rather high automatic movement with a central rotor winding in both directions. Its components were markedly similar to those of the contemporary Rolex Caliber 1030, including the shape of the rotor, the Maltese spring to stop the rotor, and the sprung discs of the twin click wheels (those of the older Perpetual movements).

Caliber Notes

Caliber 390
D = 28.5 mm (12.5 lignes), H = 6.25 mm, 17 jewels
AUTOMATIC MODULE screwed to the basic movement, central rotor with heavy metal ring and sliding mount, winding in both directions via twin click wheels with sprung discs, two reduction wheels on the locking wheel.
BASIC MOVEMENT: two-spoke Glucydur screw balance, Nivarox hairspring, 18,000 BPH, Kif-Trior(alt) shock resistance, direct center seconds.

Caliber List: Tudor

Caliber	Diameter Lignes	mm	Height mm	Other	BPH	Jewels	Running Reserve	Notes	Year	Photo
390	12 1/2	28,5	6,25	SCD	18.000	17		„Rotor-Prince"	ca. 50	xx
395	"	"		SCD, QG	"	"				
1250	11	24,8		SCD	18.000			AS 1250 with Breguet	ca. 47	
1410	9 1/4	20,4	5,35	SCI	18.000	17		AS 1320 Hairspring		
1411	10 1/2	23,3	5,35	SCI	18.000	17		AS 1364		
1412	9 3/4	20,4		SCI, QG	18.000	17		AS		
1413	10 1/2	23,3		SCI, QG	18.000	17		AS		
1566	7	16,3	5,31		18.000	17/25		AS 1477		
1895	13 3/4	31,5	5,25	SCD, QG, JG	21.600	25		AS 1895	1969	
2446	7 3/4	17,5	5,3	SCD	18.000	21		Eta 2365		
2447	"	"	5,6	SCD, QG	"	"		Eta		
2461	11 1/2	26,0	5,2	SCD	18.000	25		Eta 2459		
2462	"	"	"	SCD, QG	"	"		Eta 2460		
2483	"	"	"	SCD	"	"		Eta	1967	
2484	"	"	"	SCD, QG	"	"		Eta	1967	
2554	7 3/4	17,5	4,5	SCD	21.600	21		Eta 2550	1967	
2555	"	"	"	SCD, QG	"	"		Eta	1967	
2650	7 3/4	17,5	4,8	SCD	21.600	21		Eta 2650	1972	
2651	"	"	"	SCD, QG	"	"		Eta	1982	
2670	"	"	"	SCD	28.800	25		Eta	1977	
2671	"	"	"	SCD, QG	"	"		Eta	1975	
2772	11 1/2	26,0	5,2	SCD, QG	21.600	25		Eta 2772	1969	
2776	"	"	"	SCD	"	"		Eta 2776	1974	
2783	"	"	"	SCD, QG	"	"		Eta 2776	1981	
2784	"	"	"	SCD, QG	28.800	"		Eta 2784	1973	
2824-1	11 1/2	25,6	4,6	SCD, QG	28.800	25		Eta	1980	
2834-1	13	29	5,05	SCD, QG, JG	"	"		Eta	1980	

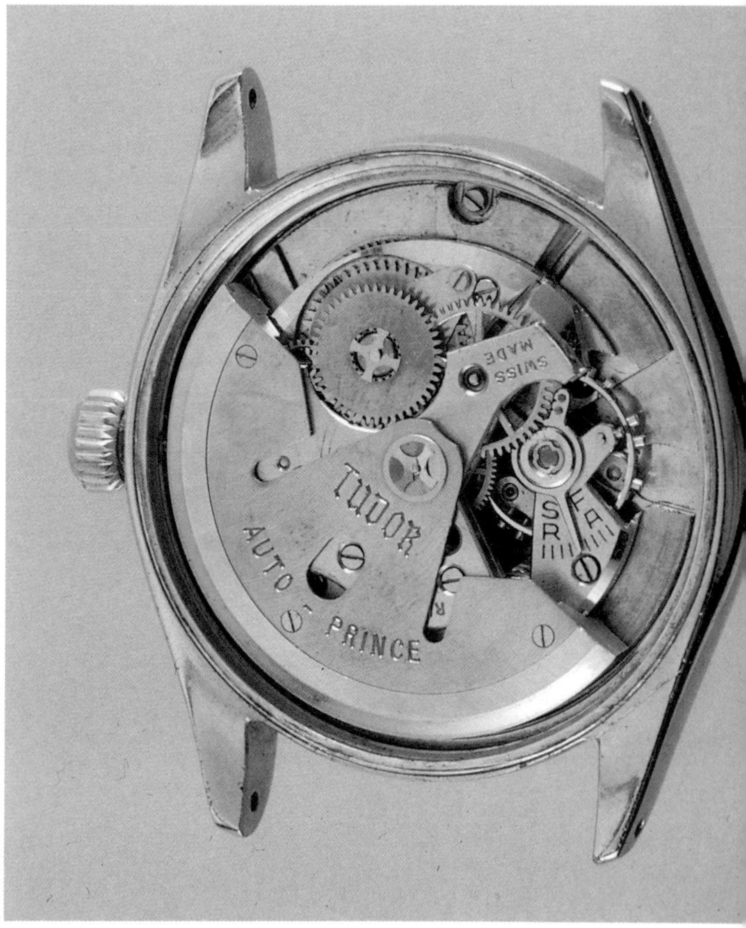

Tissot Caliber 390

Men's automatic, "Oyster Prince" model, "Tudor" brand, made by Rolex with Caliber 390.

Stainless steel case with screw bottom and screwed winding crown with Tudor symbol.

Rotor with heavy metal ring, winding in both directions via click-wheel switcher.

Glucydur screw balance, Nivarox hairspring, Kif-Trior shock resistance (dial discolored, hands, inlays and movement corroded).

Caliber Development

G. Perret and L. Berthoud founded the "Universal" firm at Le Locle in 1884, later moving the company's headquarters to Geneva.

The firm's first automatic development of 1947 had a conventionally built hammer winding in one direction and was fitted somewhat later with a center seconds, or with date indication as the "Monodatic".

Then, in 1958, Universal introduced the "Microtor", a completely new type of automatic (Caliber 215). Here the small rotor was completely recessed in the basic movement, allowing free access to the other components, now located side by side, such as the reduction drive, barrel, wheel train and balance. The entire height of the movement was thus available for the individual components; the separate elements such as the wheels, barrel and escapement could thus be mounted more securely than when the flat components were mounted one above the other with a central rotor.

In order to compensate somewhat for the loss of performance provided by the small rotor, the swinging mass was greater and made completely of heavy metal. Winding took place in both directions, via a switching rocker which always had to be located under the rotor, where it was hard to reach. The reduction drive was made with a greater ratio, so as to wind effectively with even the slightest arm movements. During automatic winding, the hand-winding mechanism was uncoupled by a locking-wheel coupling to decrease friction losses.

The simultaneous development of the Buren firm's very similarly built planetary-rotor caliber is regarded as coincidence to this day. In any case, Universal deserves the credit of having produced the flattest automatic caliber of its time with a center seconds (4.1 mm), without having to make compromises in terms of running precision.

Universal's well-known Polerouter models, which served as the official pilots' watches of the Scandinavian airline, SAS, and were used when it opened its North Pole route from Copenhagen to New York on November 15, 1954, were later equipped with this movement as well.

Between 1963 and 1965 Universal found itself compelled to use raw movements made by Movado under the somewhat imprecise name of "Super-Microtor", for the rotor used in them was a normal large central type.

The further development of the "Microtor" led in 1965 to the particularly flat Caliber 66. The movement height, reduced to only 2.5 mm, was achieved by the following means: Second indication was eliminated—the swinging mass was flatter but its diameter was larger so that it now spread over the center of the movement and thus required the minute wheel to be mounted out of the center. The rotor and connected switching rocker were fitted with a total of four miniature ball bearings, and the hand-winding mechanism was now located on the dial side.

The Universal "White Shadow" and "Golden Shadow" models, with steel or gold cases, were fitted with this movement.

Magazine advertisement (from JSH, 1948)

Caliber Notes

Caliber 138 C
D = 28.2 mm (12.25 lignes), H = 6.0 mm, 17 jewels
AUTOMATIC MODULE screwed to the basic movement, hammer with ruby bearings, winding in one direction via click-carrier rocker, combined winding and reduction wheel on the crown wheel.
BASIC MOVEMENT (9.75 lignes): two-spoke Glucydur screw bearing, Nivarox hairspring, 18,000 BPH, Incabloc shock resistance, eccentric seconds hand, date indication.

Caliber 139
D = 28.2 mm (12.25 lignes), H = 5.55 mm, 17 jewels
See Caliber 138 C, but with indirect center seconds.

Caliber 215-9
D = 28 mm (12.25 lignes), H = 4.1 mm, 28 jewels
AUTOMATIC MODULE including small heavy metal rotor mounted out of the center, integrated into the basic movement, winding in both directions via switching rocker, two reduction wheels (the first with spring click) on the locking wheel, complete reciprocal uncoupling of automatic and hand-winding drives by double locking wheel with integrated click disc.
BASIC MOVEMENT: two-spoke screwless Glucydur balance, Nivarox hairspring (Kl. I), 18,000 BPH, Incabloc shock resistance, moveable hairspring block carrier, escape wheel mounted in combined Girocap jewel housing.

Caliber 256 C (Movado Ebauche)
D = 25.3 mm (11.25 lignes), H = 3.9 mm, 28 jewels
AUTOMATIC MODULE integrated into the basic movement as a complete unit, central rotor with ball bearings and heavy metal ring, winding in both directions via switch-

Universal Caliber 138 C

Men's automatic, "Monodatic" model, made by Universal in 1951, with Caliber 138 C.

Stainless steel case with screw bottom, winding crown with U symbol. Hammer winding in one direction via click-carrier rocker (at 10:00).

Glucydur screw balance, Nivarox hairspring, Incabloc shock resistance.

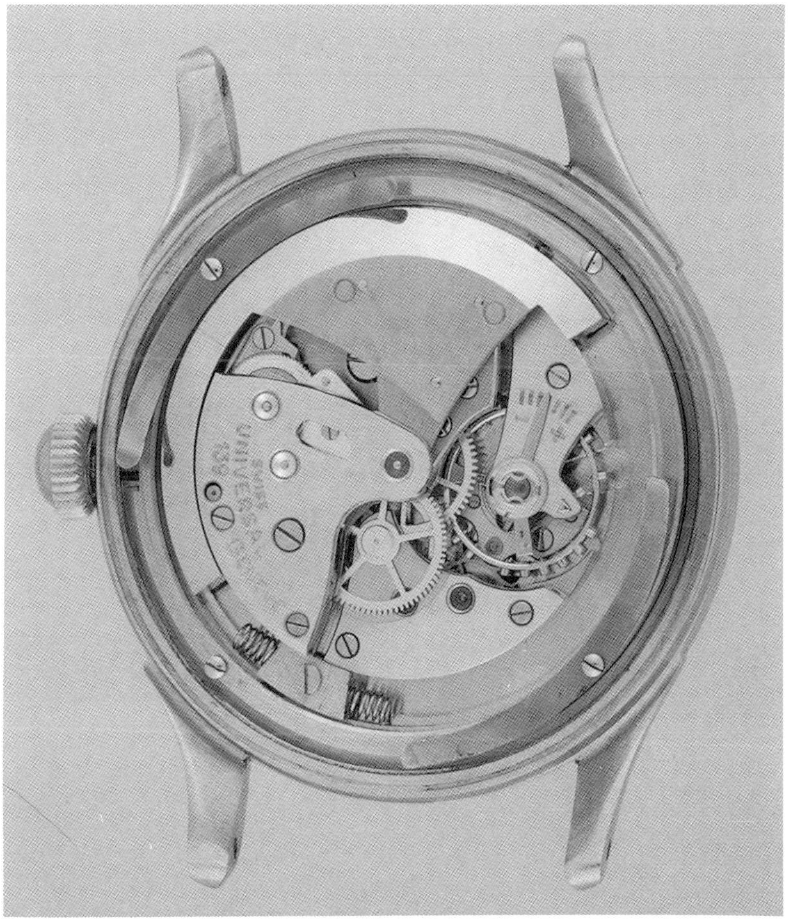

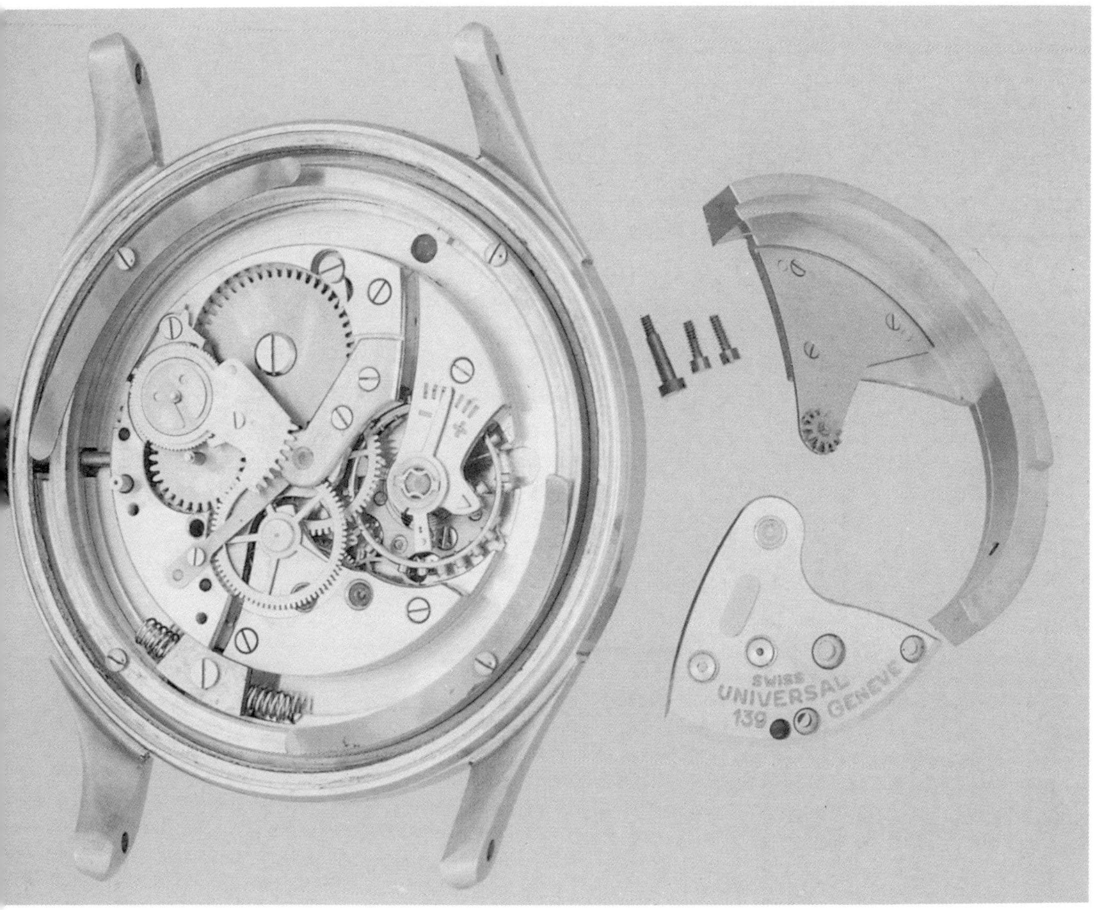

Universal Caliber 139

Men's automatic made by Universal in 1948, with Caliber 139. Stainless steel case with screw bottom.

Movement identical to Caliber 138 C (see page 304), but with center seconds. Leaf springs on movement ring for elastic mounting of movement.

Universal Caliber 215-9
Men's automatic, "Polerouter Jet" model, made by Universal circa 1960, with "Microtor" Caliber 215-9. Stainless steel case with screw bottom, winding crown with U symbol. Microtor made of heavy metal, recessed in movement, winding in both directions via switching rocker (under the blued rotor plate). Complete reciprocal uncoupling of automatic and hand-winding drives. Screwless Glucydur balance, Nivarox hairspring, Incabloc shock resistance.

Universal Caliber 256 C

Men's automatic made by Universal circa 1965, with "Super-Microtor" Caliber 256 C (Movado raw movement). Eighteen karat gold case with screw bottom, gold winding crown with U symbol.

Movement largely identical to Movado Caliber 395 E (see page 221), but with Spirofix fine adjustment.

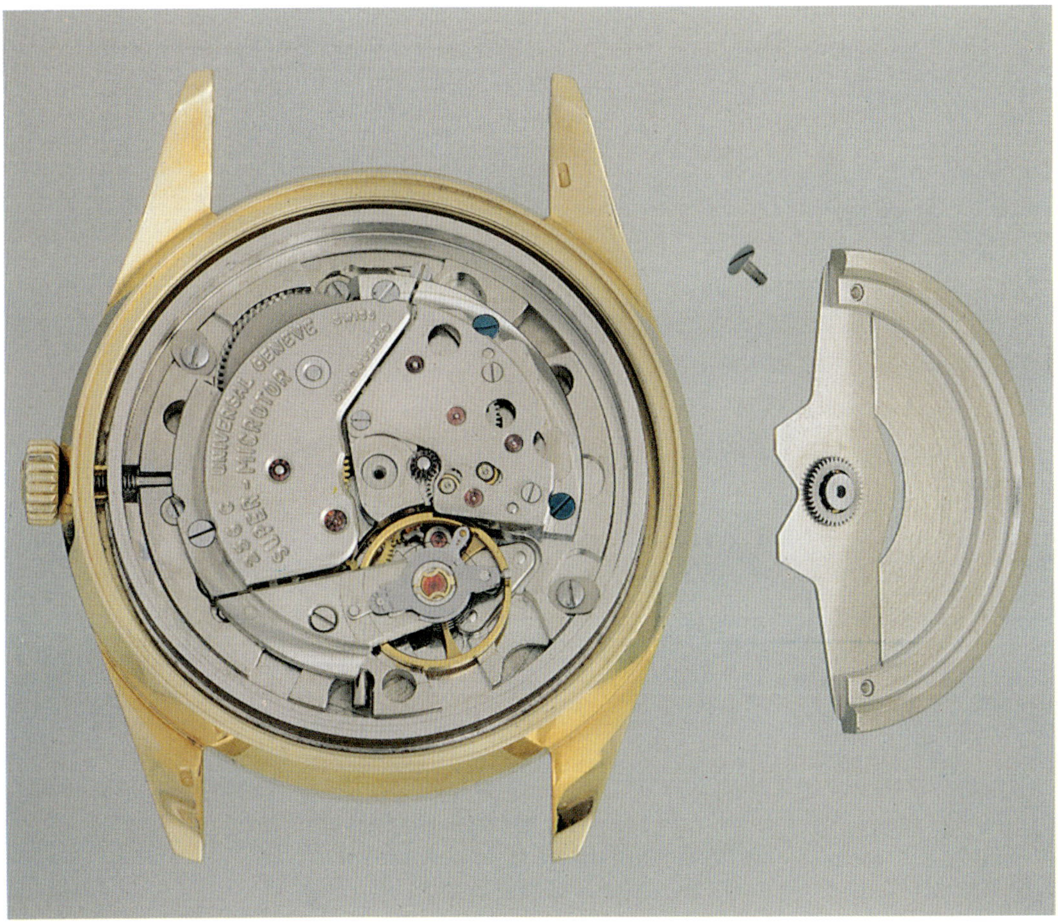

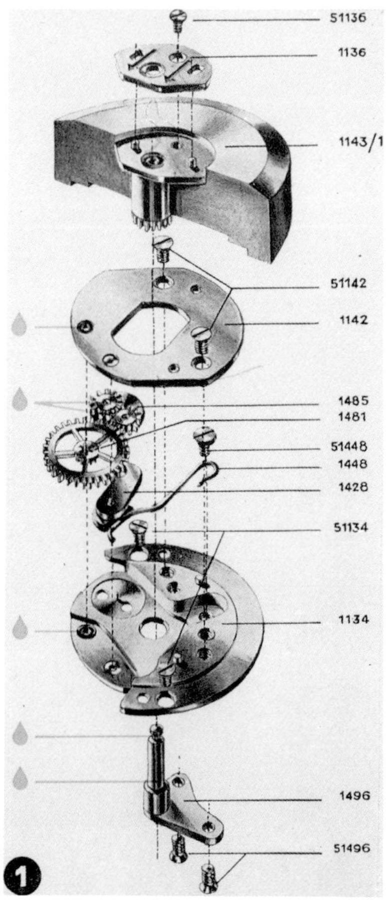

1

Universal Caliber 2-66

Men's very flat automatic, "White Shadow" model, made by Universal circa 1975, with Caliber 2-66.

Stainless steel case with press-in bottom, winding crown with U symbol. Heavy metal microrotor with ball bearings recessed in the movement, winding in both directions via switching rocker with ball bearings (at 1:00), complete reciprocal uncoupling of automatic and hand-winding drives.

Screwless Glucydur balance, Nivarox hairspring, Incabloc shock resistance.

Exploded view of Universal Caliber 215-9 (automatic module):

1143/1 microrotor

1428 locking click

1481 combined locking and reduction wheel

1485 switching rocker

Caliber List: Universal

Caliber	Diameter Lignes	mm	Height mm	Other	BPH	Jewels	Running Reserve	Notes	Year	Photo
24	7 1/4	16,2	4,25		19.800		40			
1-24	"	"	4,75	SCI	"		42			
25	"	"	"	SCI	"		"			
1-25	"	"	"	SCI	"		"			
66	12 1/4	27,6	2,5		19.800	25	48		1965	
1-66	"	"	"		"	"	"			
2-66	"	"	"		"	"	"	White-Shadow		xx
1-67	"	"		QG	"					
68	12 1/2		4,1	SCI	18.000			Microtor		
1-68	"		"	SCI	"			"		
69	"		5,15	SCI, QG	"			"		
1-69	"		"	SCI, QG	"			"	ca. 60	xx
71	13	30	6,1	SCD, QG						
75/76	7 3/4	17,2	5,3	SCD	18.000	25		Eta-Ebauche		
138	12 1/2	28,2	4,85	s	18.000	17	32		1947	
138 SS	"	"	5,55	SCI	"	"	"			
138 C	"	"	6,0	QG	"	"	"	Monodatic	1951	xx
139	"	"	5,55	SCI	"	"	40		1948	xx
215	12 1/2	28	4,1	SCI	18.000	28	60	Microtor	1958	
215.1	"	"	5,15	SCI, QG	"	"	"	"		
215.2	"	"	4,7	SCI, QG	"	"	"	"		
215.07	"	"	4,1	SCI	"	17	"	"		
215.17/ .27	"	"	4,7	SCI, QG	"	28		"		
215.9 / .97	"	"	4,1	SCI	"	28/17		"		
218	"			SCI	"			"		
218-2/-27	"			SCI, QG	"			"		
218-9/-97	"			SCI	"	28		"		
255 /A	11 1/4	25,3	3,9	SCD	21.600	28	50	Super-Microtor (Movado-Ebauche)	1963	
256 /A/B/C	"	"	"	SCD	"	"	"	"		xx
275	11 3/4			SCD, QG	"			"		

ing rocker with ruby bearings, two intermediate wheels in all, two reduction wheels
(the first with spring click with ruby bearing) on the locking wheel (under the dial).
BASIC MOVEMENT: three-spoke screwless Glucydur balance, Nivarox hairspring, 21,600 BPH, Incabloc shock resistance, Spirofix fine adjustment with moveable hairspring block carrier, direct center second, eloxied aluminum barrel (Antikorodial 15 B). Rocker mechanism for hand-winding and play-free hand-setting via double toothed wheels.

Caliber 2-66
D = 27.6 mm (12.25 lignes), H = 2.5 mm, 25 jewels
AUTOMATIC MODULE, including small heavy metal rotor mounted on ball bearings outside the center, integrated into the basic movement, winding in both directions via switching rocker (three ball bearings), three reduction wheels (the first with spring click) on the locking wheel, which is located under the dial with the hand-winding mechanism, uncoupling of automatic and hand-winding.
BASIC MOVEMENT: three-spoke screwless Glucydur balance, Nivarox hairspring, 19,800 BPH, Incabloc shock resistance, moveable hairspring block carrier, anchor wheel mounted in Girocap combined jewel housing, no seconds hand.

Caliber Development

Jean-Marc Vacheron founded the company at Geneva in 1755. In 1819 François Constantin became a business partner in the firm. Georges-Auguste Leschot, who became its technical director in 1839, introduced machine production of parts to the watch industry for the first time. In 1880 the firm registered the Maltese cross as its trade mark. The firm's first wristwatch was made in 1910.

A result of the firm's cooperation with Jaeger LeCoultre that began in 1938 was the firm's future strategy of not developing the raw movements for its automatic wristwatches itself, but to order them from Jaeger LeCoultre.

The first VC automatic watch was introduced around 1950. The movement used was that already installed in Jaeger LeCoultre's "Powermatic" several years before, with hammer winding in one direction; it was used almost unchanged as Caliber 477.

Rotor winding first appeared in VC watches around 1955. It used the somewhat larger Jaeger LeCoultre Caliber 493 of 1951, which was reserved especially for VC and later also for Audemars Piguet. This movement (VC Caliber 498 without and 499 with center second) had a central rotor running on four steel balls in a covered position under the central section and winding in both directions with the help of a sturdy switching rocker. To protect the automatic drive, it was uncoupled by a crown-wheel coupling during hand-winding.

In the following years, this movement was used by VC almost unchanged, though with new caliber numbers. Only around 1959 was it, now known as Caliber 1072, improved significantly: instead of four steel balls, the rotor ran on four ruby rollers that were now mounted visibly in cutouts of the central section. The rotor weight was made of 18-karat gold. The locking click in the reduction drive, a source of trouble, was located under a separate plate, where it was easy to service. About a year later its screw balance, fitted with a Breguet hairspring, was replaced by a Gyromax balance for even better running precision as Caliber 1072/1.

A very flat automatic design, Caliber 1120, was introduced in 1967, to appear later in VC models with an added perpetual calendar and moon-phase device. A peripherally mounted ring of beryllium, running on four ruby rollers recessed in the bottom plate, served to guide the centrally mounted rotor. Here too, the sensitive automatic drive and its switching rocker were uncoupled by a coupling during hand-winding. Jaeger LeCoultre reserved this raw movement not only for VC, but also for Audemars Piguet and Patek Philippe.

Caliber Notes

Caliber 477/1
Jaeger LeCoultre Ebauche (Caliber 476)
See page 162, but with self-compensating Breguet hairspring.

Caliber P 1019/2
Jaeger LeCoultre Ebauche
D = 29.0 mm (13 lignes), H = 5.4 mm, 21 jewels
AUTOMATIC MODULE screwed to the basic movement as a complete unit (three blued screws), central rotor with heavy metal ring running paracentrally on four steel balls, winding in both directions via switching rocker, two reduction wheels, the first with a locking click, on the double crown wheel, crown-wheel coupling for uncoupling during hand-winding.
BASIC MOVEMENT: two-spoke Glucydur screw balance, self-compensating Breguet hairspring, 18,000 BPH, JLC shock resistance, swan's-neck spring for fine regulator adjustment.

Caliber K 1072
Jaeger LeCoultre Ebauche
D = 29.5 mm (13 lignes), H = 5.4 mm, 29 jewels
See Caliber P 1019/2, but with 18-karat gold rotor ring, steel balls replaced by four visible ruby rollers in cutouts of the rotor's central section, locking click on the first reduction wheel under a separate plate, Kif-Flector shock resistance, date indication.

Caliber K 1120
Jaeger LeCoultre Ebauche
D = 26 mm (12.5 lignes), H = 2.45 mm, 36 jewels
AUTOMATIC MODULE integrated in the basic movement under a shared plate, central rotor with ruby bearings, 21-karat gold ring and peripheral beryllium ring gliding on four ruby rollers, winding in both directions via a steel intermediate wheel mounted on ball bearings, the first reduction wheel, the switching rocker, two more reduction wheels (the first with a locking click), another coupling wheel on the locking wheel for uncoupling during hand-winding.
BASIC MOVEMENT: two-spoke Gyromax balance, Nivarox hairspring, 19,800 BPH, Kif-Ultraflex shock resistance, moveable hairspring block carrier without a hairspring regulator, anchor wheel mounted in combined jewel housings, no second, barrel with one-sided "floating" mount.

Vacheron Constantin Caliber 477/1
Men's automatic made by Vacheron Constantin circa 1950, with Caliber 477/1 (raw movement by Jaeger LeCoultre). Eighteen karat gold case with screw bottom, gold hands and dial inlays. Movement largely identical to Jaeger LeCoultre Caliber 476 (see also Caliber 481 on page 162, winding crown not original).

Caliber List: Vacheron Constantin

Caliber	Diameter Lignes	mm	Height mm	Other	BPH	Jewels	Running Reserve	Notes	Year	Photo
477 /1	13	29,56	6,0	SCI	18.000	17		J. Le Coultre Ebauche	ca. 50	xx
498	13	29,0	5,1		18.000			J. Le Coultre Ebauche	ca. 55	
499	"	"	5,4	SCI	"			"		
K 1008	"	29,5	"	SCI	"	19		See Caliber 499	ca. 57	
P1019 /1 /2	"	"	"	SCI	"	21		See Caliber 499		xx
K1072 /1	"	"		SCI, QG	"	29		See Caliber 499	ca. 59	xx
K 1120	12 1/2	26	2,45		19.800	36	50	Ebauche J. Le Coultre	1967	xx
K 1121	"	"	6,05	QG	"	"	"	"		
K 1122	"	"	3,4	SCI, QG (?)	"		"	"		

Vacheron Constantin Caliber P 1019/2

Men's automatic made by Vacheron Constantin circa 1957, with Caliber P 1019/2 (exclusive raw movement by Jaeger LeCoultre).

Eighteen karat gold case with screw bottom, gold winding crown with Maltese cross, hands and dial inlays.

Rotor with heavy metal ring, supported by four balls, winding in both directions via switching rocker, uncoupling during hand winding.

Glucydur screw balance, self-compensating Breguet hairspring, JLC shock resistance, fine regulator adjustment by side screw with swan's-neck spring.

Vacheron Constantin Caliber K 1072

Men's automatic made by Vacheron Constantin circa 1965, with Caliber K 1072 (exclusive raw movement by Jaeger LeCoultre).
Eighteen karat white gold case with screw bottom, white gold winding crown with Maltese cross, hands and dial inlays.
Rotor with 18-karat gold ring, supported by four visible ruby rollers. Movement otherwise identical to Caliber P 1019/2 (see page 312), but with Kif-Flector shock resistance and separate locking-click plate (at 12:00).

Vacheron Constantin Caliber K 1120

Men's very flat automatic made by Vacheron Constantin in 1967, with Caliber K 1120 (exclusive raw movement by Jaeger LeCoultre).

Eighteen karat gold case with press-in bottom, gold hands and dial inlays. Centrally mounted rotor with 21-karat gold ring and outer ring, supported by four peripheral ruby rollers (at 1, 4, 7 and 10:00), winding in both directions via switching rocker (at 2:30), uncoupling during hand-winding.

Gyromax balance, self-compensating flat hairspring, Kif-Ultraflex shock resistance, no regulator (winding crown not original).

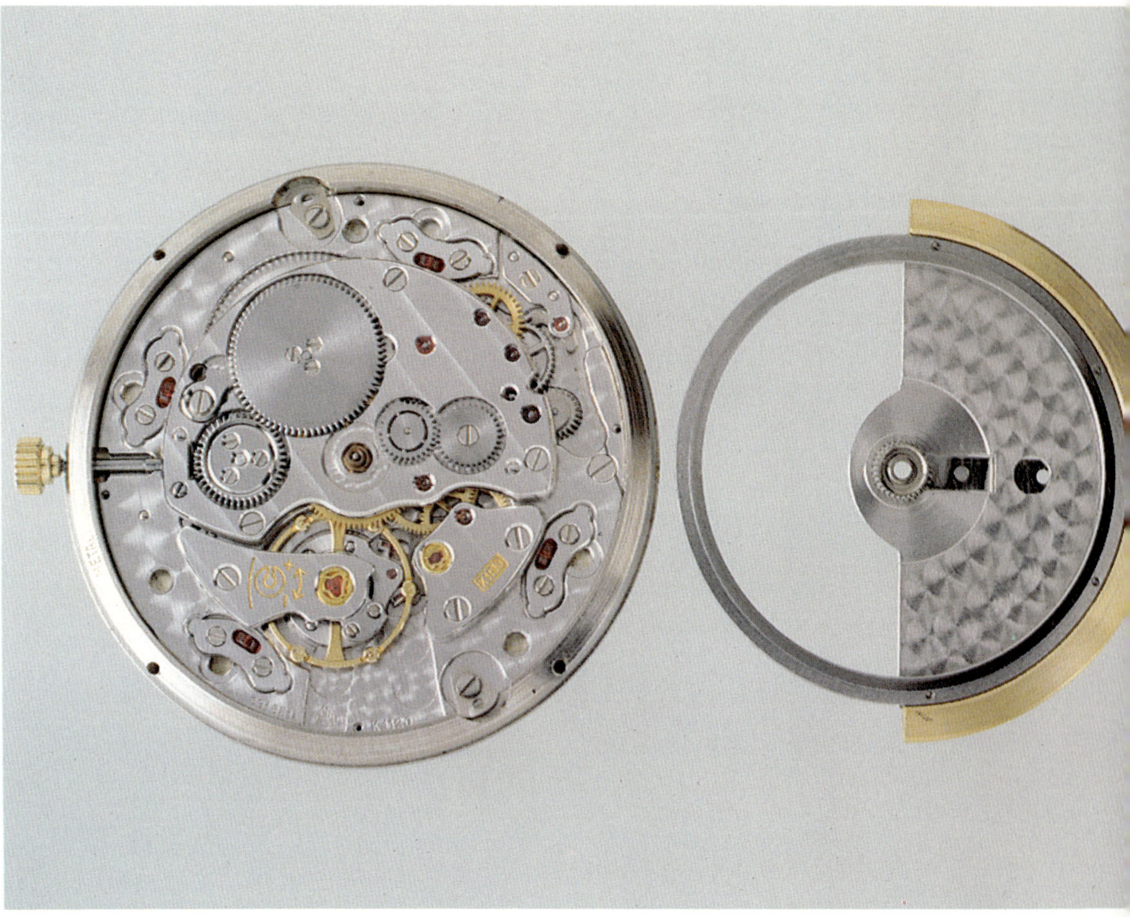

Caliber Development

The brothers John and Charles Reymond began to produce chronograph components in Les Bijoux in 1901 under the name of "J. & C. Reymond Frères". Only in 1910 did they begin to produce raw movements for chronographs, marked with the firm's "R" trademark. In 1929 the firm's name was changed to "Valjoux S.A., named after the Vallée de Joux, the valley in which Les Bijoux is located. The firm joined the Ebauches S.A. relatively late, in 1944.

The only Valjoux automatic caliber, numbered 7750, was introduced in 1974. It was a fast beat, at 21,600 or 28,800 BPH, featuring an automatic module with rotor winding in one direction as well as a chronograph mechanism.

This sturdy movement with its efficient winding has been used, and is still being used, by several well-known firms, including IWC, in their watch models.

Caliber Notes

Caliber 7750
D = 30.0 mm (13.25 lignes), H = 7.9 mm, 17 jewels
AUTOMATIC MODULE mounted on the basic movement, central rotor with ball bearings and heavy metal ring, winding in one direction via combined click and locking wheel (with leaf-shaped locking spring), two reduction wheels on the locking wheel of the barrel. Chronograph mechanism set on the basic movement beside the automatic drive, coulisse-lever switching.
BASIC MOVEMENT: three-spoke screwless nickel balance, Nivarox hairspring, 21,600 BPH, Incabloc shock resistance, fine regulator adjustment via eccentric lever, moveable hairspring block carrier, eccentric seconds hand, date and weekday indication with quick set.

Valjoux Caliber 7750
Automatic chronograph made by Helvetia circa 1975, with Valjoux Caliber 7750.
Barrel-shaped stainless steel case with screw bottom.
Rotor with heavy metal ring and ball bearings, winding in one direction via click wheel (at 6:30). Chronograph mechanism with coulisse-lever switching.
Screwless nickel balance, Nivarox hairspring, Incabloc shock resistance, fine regulator adjustment by eccentric level.

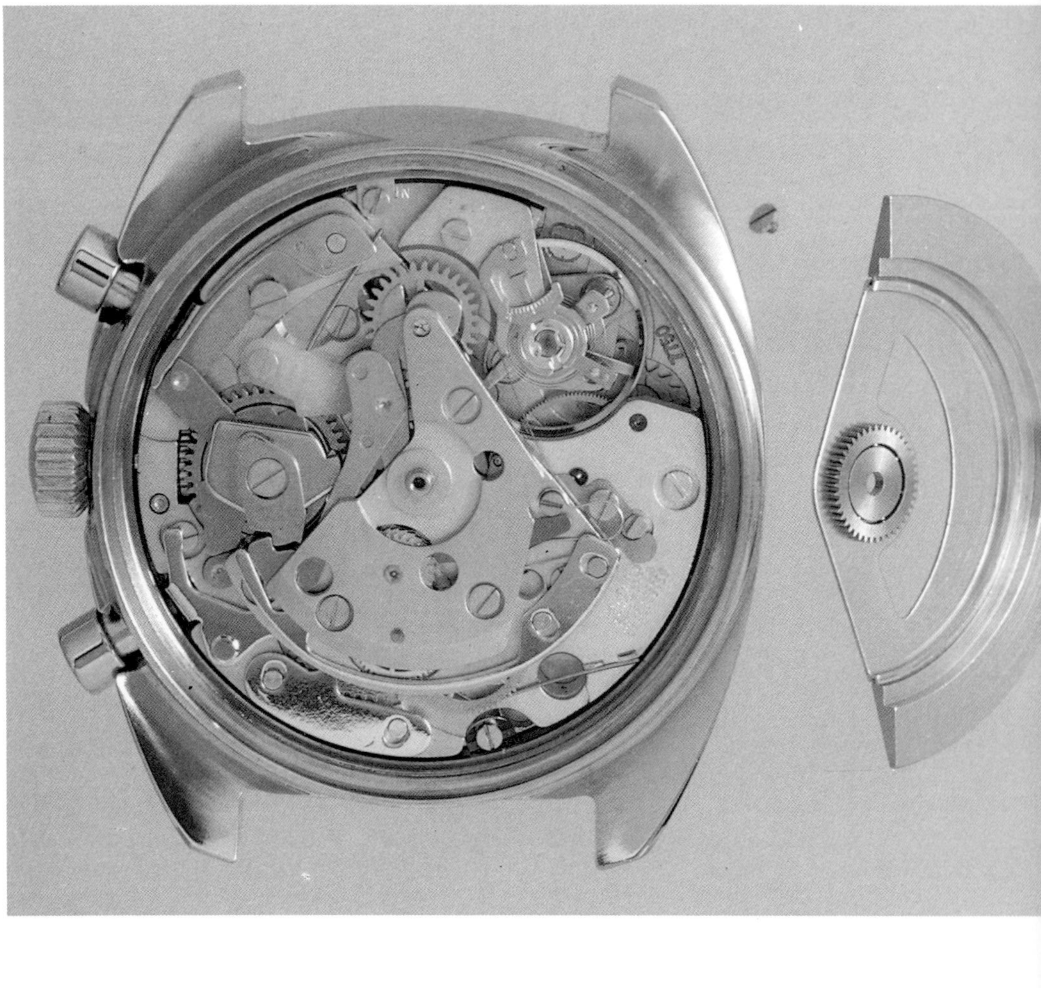

Caliber Development

The firm was founded by George Favre-Jacot at Le Locle in 1865. In Geneva, where the firm's headquarters was relocated later, it was registered under the name of "Zenith".

The firm's first automatic, Caliber 133, appeared in 1948 and had hammer winding. Its movement height was kept remarkably low for its time by special new design features, such as the minute wheel mounted out of the center, the direct center second, and the location of the large barrel far to the side in the "dead" angle of the hammer. Zenith was very successful with this movement, which ranked for a long time as the flattest automatic design with center second, and with its beat speed raised to 21,600 BPH (Caliber 133.8), it could also be sold with a chrometer certificate.

Only in 1954 did Zenith introduce its first rotor design, Caliber 2522 P, also known as the "Zenith-Rotor" and obviously much thicker. On the basis of the "Zenith-Rotor", there followed the improved Caliber 2542 PC in 1959, with a fast beat escapement and date indication, and the fast beat Caliber 2562 PC around 1968, using two miniature ball bearings and attaining chronometer quality.

The Zenith engineers developed a very unusual design for the ladies' Caliber 1724, introduced in 1967 as the "Zenitissimo". It was based on the concept that the long drive train, that was needed first to reduce the turning speed from rotor to barrel and then increase it from barrel to escapement, could be simplified by the intermediate switching of a differential drive. Therefore the number of wheels could be decreased, being especially advantageous in small movements. The differential, in the form of a planetary equalization drive, took the place of the intermediate wheel. With powerful rotor motion, the turning moment was transmitted directly to the second wheel via the differential and thus to the escapement, but at the same time it turned the barrel in opposition to its unwinding direction via the minute wheel.

With little or no rotor motion, the differential acted like a normal intermediate wheel; the greater part of the turning moment for the escapement was then supplied by the barrel. Hand-winding could take place completely separately via the locking wheel, thus dispensing with an additional uncoupling device for the automatic drive.

The chronograph Caliber 3019 PHC, introduced in 1969, was also a novel design and was not called "El Primero"—The First—without justification. Zenith was able to combine the fast beat (36,000 BPH) basic movement with the chronograph mechanism and automatic winding in a structure that still looks relatively flat. With it, times down to 1/10 of a second can be clocked accurately. The large barrel requires very efficient self-winding, which is guaranteed by a large heavy-metal rotor with ball bearings and the direction-changing mechanism.

Caliber Notes

Caliber 133.8
D = 29.5 mm (13 lignes), H = 4.9 mm, 20 jewels
AUTOMATIC MODULE partly integrated in the basic movement, hammer with ruby bearings, winding in one direction via click and central winding wheel on the hammer, two reduction wheels (the first with a spring click) on the locking wheel, uncoupling of the hand-winding mechanism by a Breguet coupling between the crown and locking wheels.
BASIC MOVEMENT: two-spoke Glucydur screw balance, Nivarox hairspring, 21,600 BPH, Incabloc shock resistance, direct center second.

Caliber 2522 P
D = 25.6 mm (11.5 lignes), H = 5.63 mm, 28 jewels
AUTOMATIC MODULE partly integrated into the basic movement as a complete unit, central rotor with ruby bearings and heavy metal ring, winding in both directions via springless twin click wheels on the crown wheel.
BASIC MOVEMENT: two-spoke Glucydur screw balance, Nivarox hairspring, 18,000 BPH, Incabloc shock resistance, direct center second.

Caliber 2542 PC
D = 25.6 mm (11.5 lignes), H = 5.63 mm, 25 jewels.
AUTOMATIC MODULE: see Caliber 2522 P.
BASIC MOVEMENT: three-spoke screwless Glucydur balance, Nivarox hairspring, 21,600 BPH, Incabloc shock resistance, moveable hairspring block carrier, fine regulator adjustment with eccentric screw, direct center second (all wheels of beryllium bronze), date indication with quick change.

Caliber 1725
D = 17.5 mm (7.75 lignes), H = 4.35 mm, 25 jewels.
AUTOMATIC MODULE screwed to the basic module, heavy metal rotor with ball bearings, winding in both directions via gliding pinion, one reduction wheel, planetary equalization drive retrograde via the eccentrically mounted minute wheel to the barrel, complete uncoupling between automatic and hand-winding drives.
BASIC MOVEMENT: three-spoke screwless Glucydur balance, Nivarox hairspring, 19,800 BPH, Incabloc shock resistance, moveable hairspring block carrier, fine regulator adjustment via toothed disc, center second wheel without hand arbor.

Caliber 2562 PC
D = 25.6 mm (11.5 lignes), H = 5.63 mm, 23 jewels
AUTOMATIC MODULE: see Caliber 2522 P, but with ball bearings for the rotor and crown wheel.
BASIC MOVEMENT: see Caliber 2542 PC, but with 28,800 BPH and date indication.

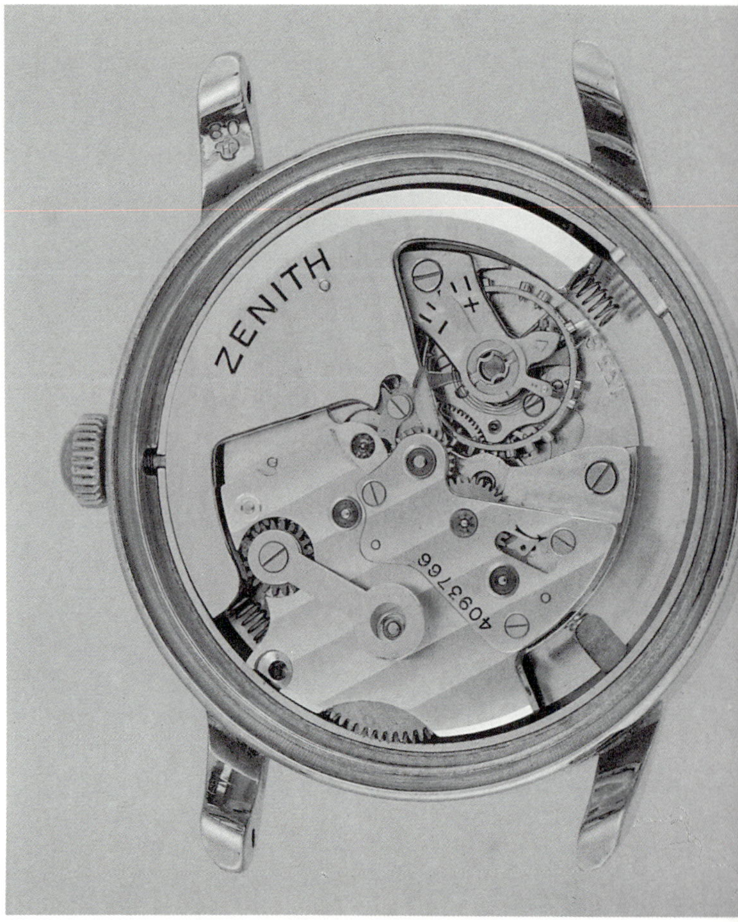

Zenith Caliber 133.8

Men's automatic made by Zenith circa 1952, with Caliber 133.8. Gilded case (80 u) with stainless steel screw bottom.

Hammer with ruby bearings, winding in one direction via Harwood-type drive, uncoupling of the hand-winding drive. Glucydur screw balance, Nivarox hairspring, Incabloc shock resistance, leaf spring for elastic support of the movement on the screw bottom.

Zenith Caliber 2522 P

Men's automatic made by Zenith in 1954, with Caliber 2522 P.
Eighteen karat gold case with screw bottom, winding crown with star symbol. Rotor with heavy metal ring and ruby bearings, winding in both directions via click-wheel switcher (at 6 and 7:00). Glucydur screw balance, Nivarox hairspring, Incabloc shock resistance, regulator with double hand.

Zenith Caliber 2542 PC

Men's automatic made by Zenith in 1959, with Caliber 2542 PC. Stainless steel case with screw bottom, winding crown with star symbol. Movement largely identical to Caliber 2522 PC (see page 319), but with screwless Glucydur balance and fine regulator adjustment by eccentric screw.

Zenith Caliber 1725

Ladies' automatic, "Zenitissimo" model, made by Zenith in 1967, with Caliber 1725.

Rectangular stainless steel case with bowl-shaped press-in bottom, winding crown with star symbol.

Heavy metal rotor with ball bearings, winding in both directions via gliding-pinion switcher (at 3:30), direct energy transmission with the help of a planetary equalizing drive to the second wheel.

Screwless Glucydur balance, Nivarox hairspring, Incabloc shock resistance, fine regulator adjustment by toothed disc.

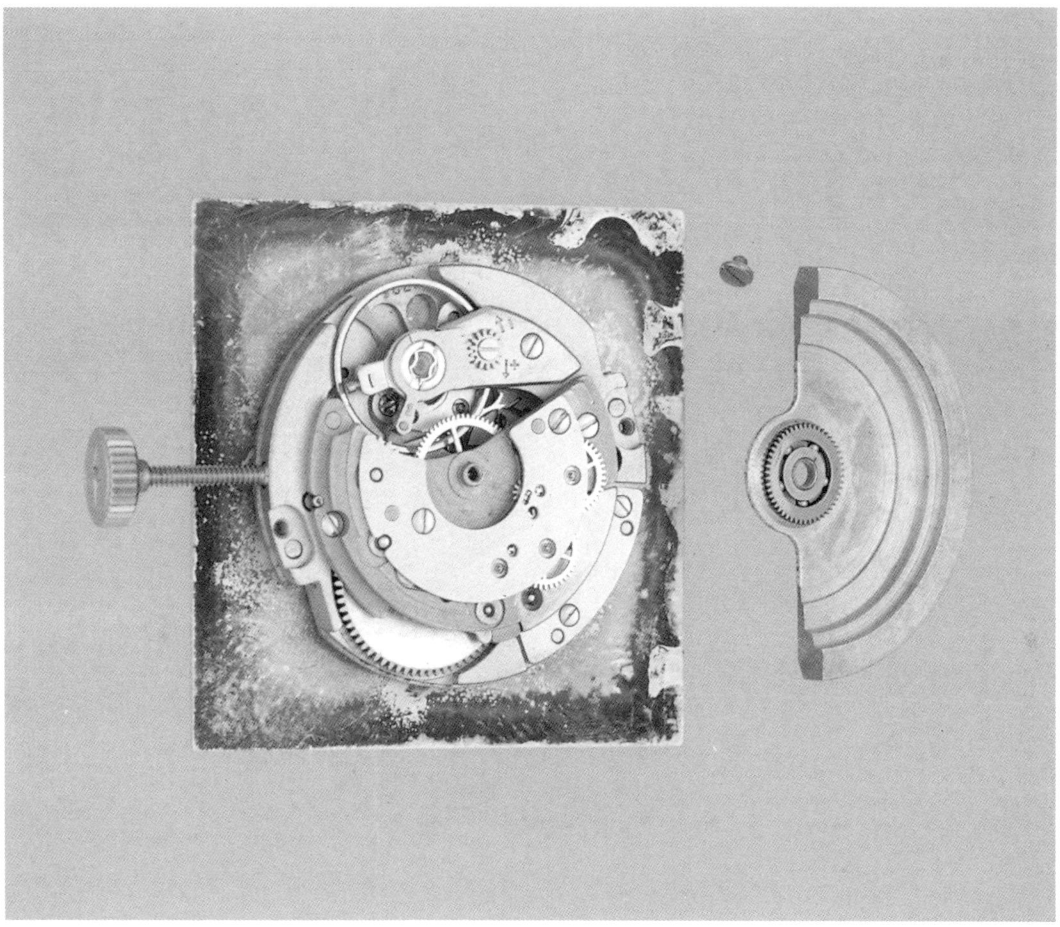

Zenith Caliber 2562 PC

Men's automatic, "Captain de luxe" chronometer model, made by Zenith circa 1970, with fast beat escapement Caliber 2562 PC.

Heavy angled 18-karat gold case (40 grams) with screw bottom, gold winding crown with star symbol, dial and hands.

Rotor with heavy metal ring and ball bearings, winding in both directions via click-wheel switcher on the crown wheel (at 9:00), also on ball bearings.

Movement otherwise identical to Caliber 2522 P or 2542 PC (see pp. 319, 320).

Zenith Caliber 3019 PHC
Automatic chronograph, "El Primero" model, as a diver's watch, made by Zenith circa 1975, with Caliber 3019 PHC, the world's first automatic chronograph caliber (fast beat escapement), made since 1969. Barrel-shaped stainless steel case with glassed screw bottom, screwed winding crown with star symbol. Rotor with heavy metal ring and ball bearings, winding in both directions via click-wheel switcher (at 11:00).
Chronograph mechanism with switching wheel (at 6:30).
Screwless Glucydur balance, Nivarox hairspring, Incabloc shock resistance, fine regulator adjustment by eccentric screw.

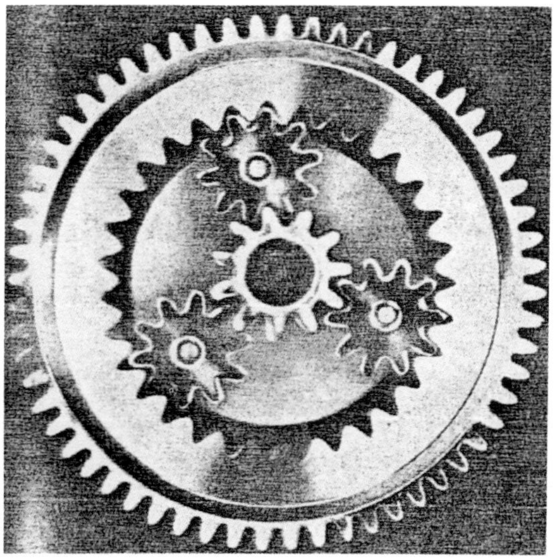

View inside the planetary equalizing drive of the "Zenitissimo" (Caliber 1725).

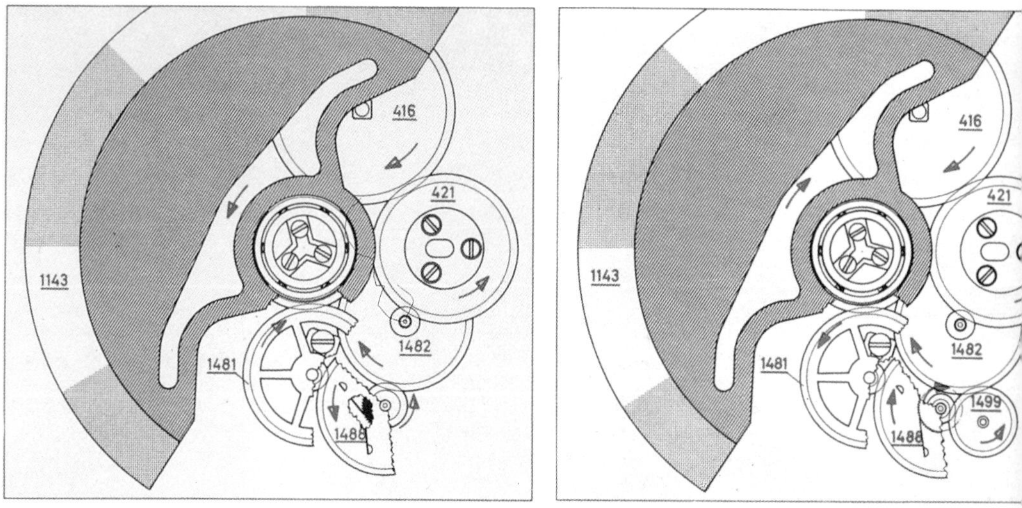

*Functioning of the automatic drive in Zenith Caliber 3019 PHC:
Rotor (1143) turns counter-clockwise (above): energy transmission via first reduction wheel (1481) to lower double click wheel (1488) and from its click pinion via second reduction wheel (1482) and crown wheel (421) to the locking wheel (416) of the barrel (the reversing wheel (1499), not shown, turns passively.*

Rotor (1143) turns clockwise (below): energy transmission via first reduction wheel (1481) to upper double click wheel (1488) and from its pinion to the reversing wheel (1499) that has been switched in, from there via the second reduction wheel (1482) via the crown wheel (421) to the locking wheel (416) of the barrel.

Caliber 3019
D = 30.0 mm, H = 6.5 mm, 31 jewels
AUTOMATIC MODULE mounted on the basic movement, central rotor with ball bearings and heavy metal ring, winding in both directions via double click wheel with click springs and reversing wheel, two reduction wheels in all on the crown wheel.
Chronograph mechanism with switching wheel set on the basic movement beside the automatic drive.
BASIC MOVEMENT: three-spoke screwless Glucydur balance, Nivarox hairspring, 36,000 BPH, Incabloc shock resistance, moveable hairspring block carrier, fine regulator adjustment by eccentric screw, eccentric seconds hand, date indication with instant change and quick correction.

Caliber List: Zenith

Caliber	Diameter Lignes	mm	Height mm	Other	BPH	Jewels	Running Reserve	Notes	Year	Photo
71	13	29,5		SCD, QG	21.600				ca. 52	
133	"	26,5	4,9	SCD	18.000				1948	
133.8	"	"	"	SCD	21.600	20	40		ca. 52	xx
1724	7 3/4	17,5		SCD	19.800	25		Zenitissimo	1967	
1725	"	"	4,35		"	"		"	1967	xx
2522 P	11 1/2	25,6	5,63	SCD	18.000	28	40		1954	xx
2522 PC	"	"	"	SCD, QG	"	"	"			
2532 P	"	"	4,35	SCD	21.600	"	45			
2532 PC	"	"	5,63	SCD, QG	"	"	"		1959	
2542 PC	"	"	"	SCD, QG	"	25	"		1959	xx
2562 PC	"	"	"	SCD, QG	28.800	23	48	Chronometer	ca. 68	xx
2572 PC	"	"	"	SCD, GG, RDR	"	17	"			
3019 PHC		30,0	6,5	QG, RDR, CHR	36.000	31	50	El Primero/Espada	1969	xx
3019 PHF				QG, JG, PH, RDR, CHR	"	"	"	"		

Magazine advertisement (from SUZ, 1949).

Caliber Development

The firm was founded by Ariste Calame at Le Locle in 1882. Zodiac did not develop any automatic movements itself, using raw movements made by Eta in its watch models for ladies and AS calibers in its men's watches.

In 1948, almost simultaneously with Jaeger LeCoultre, the firm introduced an automatic with up and down indicators, the "Astrographic". Here too, a basic movement by AS is used for the hammer-wound Caliber 1298, which was completed by Zodiac with a differential drive. The next "Astrographic", of 1953, used the rotor-driven AS Caliber 1424 and indicated the running reserve by a fourth central hand.

From 1961 to the mid-1960s, one basic design formed the standard equipment of all automatic Zenith models for men. This caliber was shared with Doxa, Eberhard, Favre-Leuba and Girard-Perregaux and produced by AS within the CHP (Communautée Horlogère de Précision).

The "Olympos" men's models introduced in 1961 used this Caliber 70 in cases with a striking shape based on a Greek amphora. They were waterproof up to a pressure of five atmospheres, had a protected crown by the 2, and indicated the hours on a calibrated disc.

The "Astrographic" model series followed at the beginning of the 1970s, and featured very sloped stainless steel cases in oval or rounded rectangular form, taking their name from the dot-like second marker that circled the dial like a satellite. All three hands were replaced by appropriately marked transparent discs placed one above the other, with their power source appearing to be "mysterious".

The "Astrographic SST" of 1973 was a fast beat escapement based on the CHP caliber, equipped with Clinergic 21 escapement by FAR (Fabriques d'Assortiments Réunies), and attaining chronometer quality.

Zodiac Caliber 1225
Men's automatic, "Autographic" model, made by Zodiac circa 1949, with Caliber 1225 (AS raw movement), one of the first models with up and down indicator (on the auxiliary dial over the 6). Stainless steel case with screw bottom.
Hammer winding in one direction via click-carrier disc. Automatic module removable as a complete unit. Glucydur screw balance, Nivarox hairspring, Incabloc shock resistance.

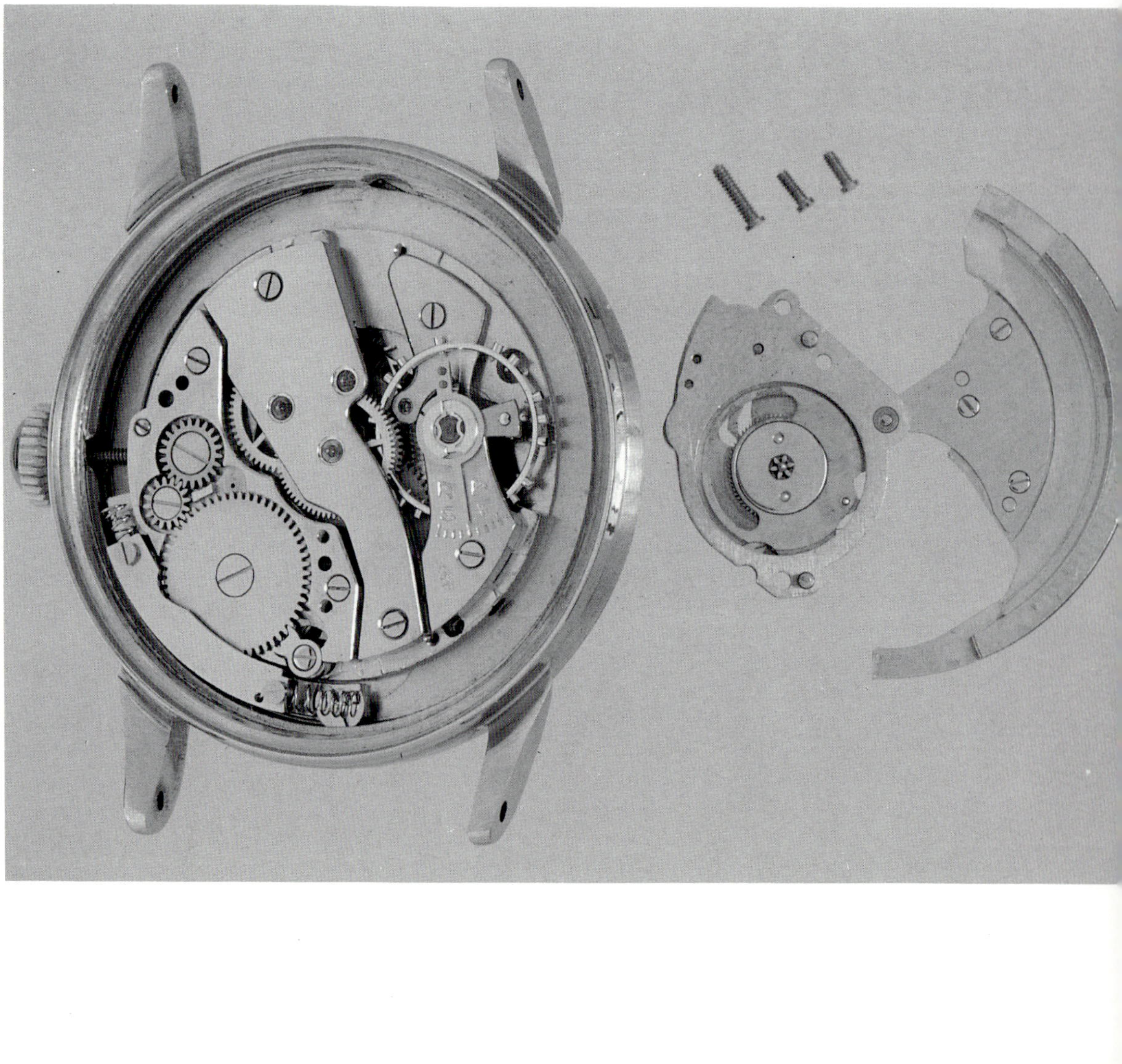

Zodiac Caliber 1424 N

Men's automatic, "Autographic" model, made by Zodiac in 1954, with Caliber 1424 N (AS raw movement).

Gilded case with stainless steel screw bottom, winding crown with Zodiac emblem.

Up and down indicator by a fourth central hand.

Movement largely identical to AS Caliber 1382 N (see page 33), but with Glucydur screw balance.

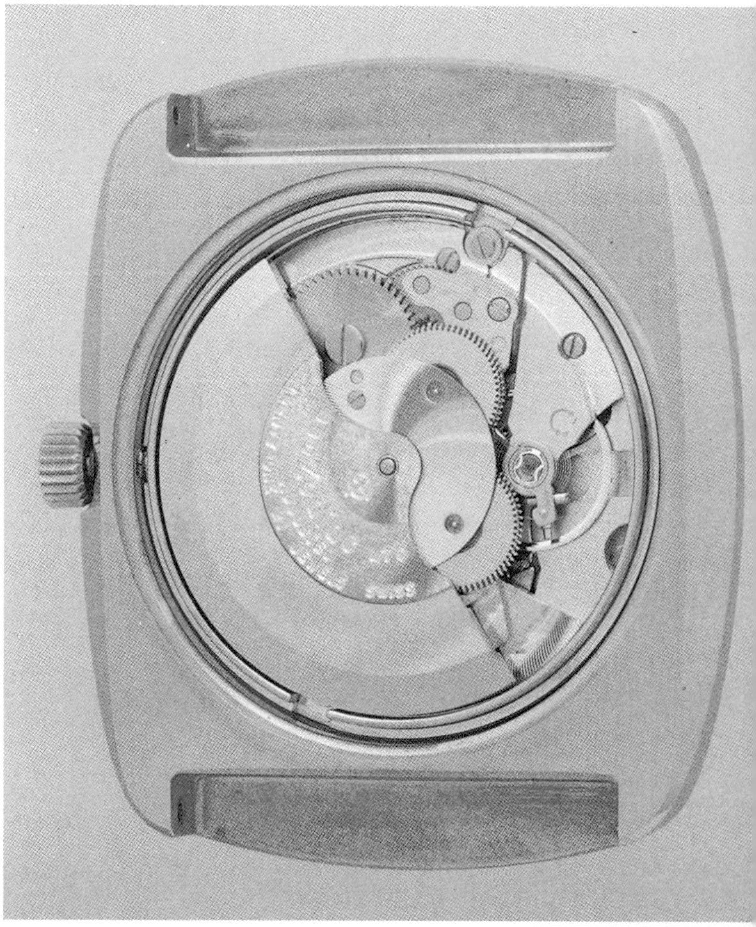

Zodiac Caliber 88 D

Men's automatic, "Autographic SST" model, made by Zodiac in 1973, with fast beat escapement Caliber 88 D (CHP raw movement by AS).

Rounded and sloped rectangular stainless steel case with screw bottom, winding crown with Zodiac symbol.

Mineral glass with printed-on index ring, so-called mysterious indication by transparent panels with rotating markers.

Rotor with heavy metal ring, winding in both directions via click-wheel switcher (at 1 and 4:00).

Screwless Glucydur balance, Nivarox hairspring, Incabloc shock resistance, Triovis fine adjustment.

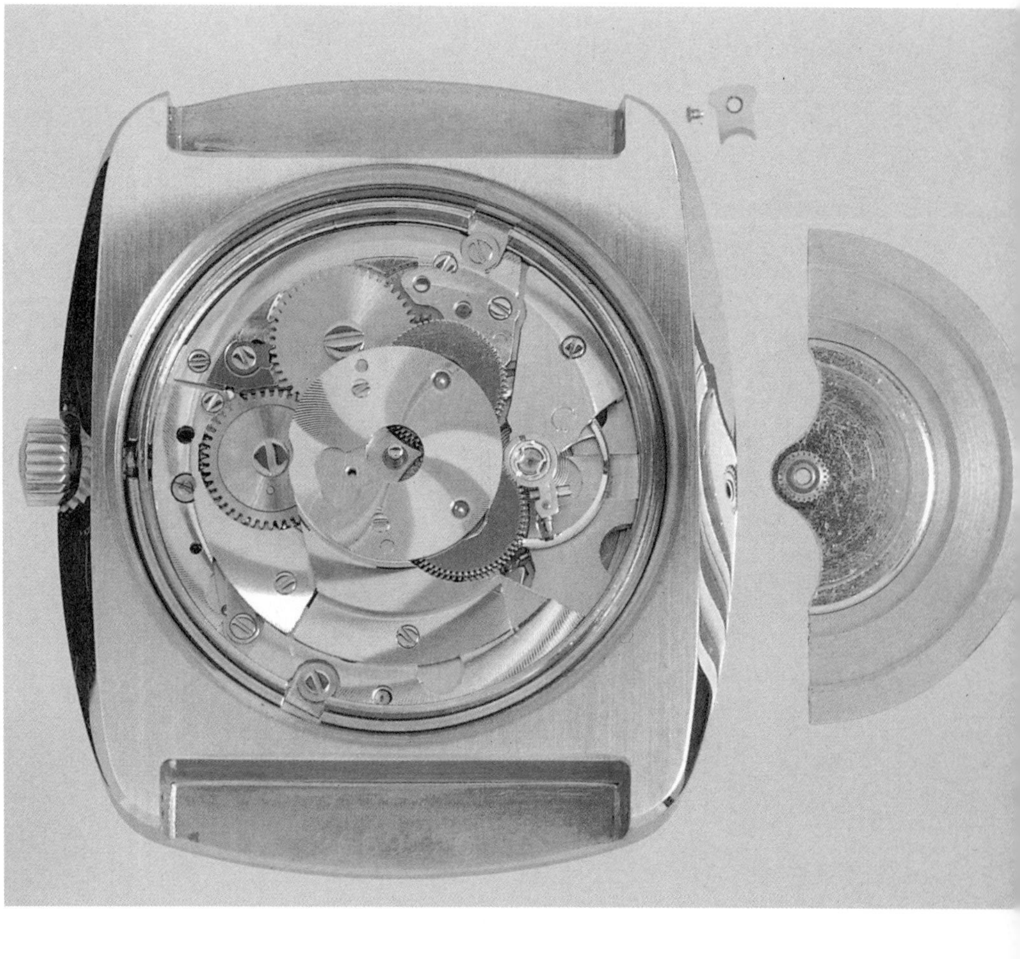

Caliber Notes

Caliber 1225
AS 1298 Ebauche
D = 28.0 mm (12.25 lignes), 17 jewels
AUTOMATIC MODULE screwed to the basic movement as a complete unit, hammer winding in one direction via partly toothed click-carrier disc, combined winding and reduction wheel on the crown wheel.
BASIC MOVEMENT: two-spoke Glucydur screw balance, Nivarox hairspring, 18,000 BPH, Incabloc shock resistance, direct center second, running reserve indication by differential drive with spur gears.

Caliber 1424 N
AS 1424 N Ebauche
D = 26.0 mm (11.5 lignes), H = 6.0 mm, 17 jewels
AUTOMATIC MODULE screwed to the basic movement, central rotor with heavy metal ring, winding in both directions via switching rocker, one tension wheel with locking click under a separate plate, two reduction wheels, the second with Breguet coupling for uncoupling during handwinding, on the crown wheel.
BASIC MOVEMENT: two-spoke Glucydur screw balance, Nivarox hairspring, 18,000 BPH, Incabloc shock resistance, direct center second, running reserve indication by a ball-wheel differential drive.

Caliber 88 D
AS Ebauche (CHP)
D = 25.6 mm (11.5 lignes), H = 5.05 mm, 21 jewels
AUTOMATIC MODULE partly integrated into the basic movement, central rotor with heavy metal ring, winding in both directions via springless twin click wheels, one reduction wheel on the locking wheel.
BASIC MOVEMENT: three-spoke screwless Glucydur balance, Nivarox hairspring, 36,000 BPH, Incabloc shock resistance, Triovis fine regulation with moveable hairspring block carrier, Clinergic 21 escapement, direct center second, date indication with quick change and correction, second stop.

Caliber List: Zodiac

Caliber	Diameter Lignes	mm	Height mm	Other	BPH	Jewels	Running Reserve	Notes	Year	Photo
34	7 3/4	17,2	4,8	SCD, QG	28.000			Eta-Ebauche	ca. 71	
35	"	"	5,35	SCD, QG, JG	"			"	ca. 71	
46	7 3/4	17,2	4,5	SCD	21.600			Eta-Ebauche	ca. 66	
47	"	"		SCD, QG	"			"		
70	11 1/2	25,6	4,75	SCD, ST	21.600	17	40	AS-Ebauche	1961	
72	"	"	5,05	SCD, QG, ST	"	"	"	(CHP- Movement)		
72 b	"	"	"	SCD, QG, RDR, ST	"	"	"	"		
74	"	"	6,5	SCD, QG, PH, ST	"	"	"	"		
75	"	"	5,05	SCD, QG, ST, GMT	"	"	"	"		
75 b	"	"	"	SCD, QG, RDR, ST	"	"	"	"		
76	"	"	5,8	SCD, QG, JG, ST	"	"	"	"		
78	"	"	5,05	SCD, QG, ST	36.000		34	"	ca. 71	
79	"	"	"	SCD, QG, ST	"		"	"		
86	"	"	5,8	SCD, QG, JG, RDR, ST	"	21	"	"	1973	
88 D	"	"	5,05	SCD, QG, RDR, ST	"	"	"	" „Astrographic SST"	1973	xx
90	13 3/4		7,7	QG, CHR	19.800	17		Buren/Breitling-Eb.		(xx)
1225	12 1/4	28,0		SCD, RMA	18.000 "	17	36	AS-Ebauche (1298) „Autographic"	1948	xx
1424 N	11 1/2	26,0	6,0	SCD, RMA	18.000	17	36	AS-Ebauche (1424)	1953	xx
1501	"	"		SCD, RMA	"	"	"	" „Powergraphic"	1956	

Technology of the Automatic Wristwatch

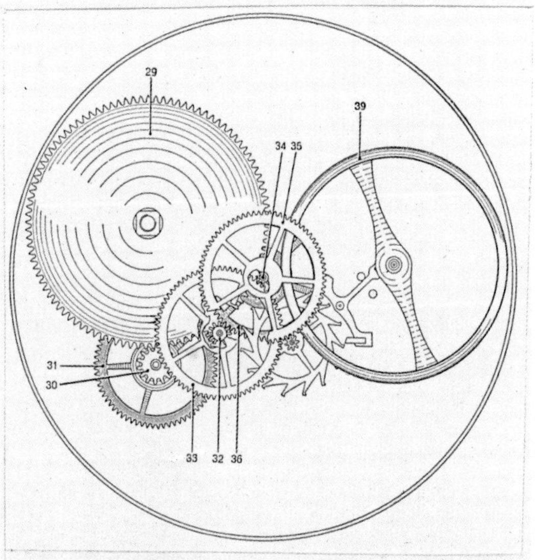

T1: Modern Basic Movement
Transmission of turning moment from barrel 29 to pinion 30 of the minute wheel 31, mounted out of the center, the pinion 32 of the intermediate wheel 33, to the pinion 34 of the center seconds wheel 35 (direct center seconds), from there to the pinion of the anchor wheel, which lies under the balance 39 with the anchor. Indirectly driven minute hand from pinion 32 of the intermediate wheel via catch wheel 36 (on the underside of the movement).

The automatic wristwatch differs from the non-automatic type essentially only in having an additional self-winding apparatus. This is usually mounted on the basic movement with two or three screws as a complete functioning unit, or recessed and thus integrated in the movement. In the 1950s, the special mounting screws were blued for easier recognition. The *basic movement* thus corresponds to the movement of a hand-wound watch and consists of the barrel with the mainspring, the wheel train, the escapement and the balance assembly which regulates the timing. The wheel train transmits the slow but strong rotation of the barrel into a quick but appropriately weak turning of the escapement or anchor wheel. It generally consists of four wheel-and-pinion combinations, also known as toothed-wheel organs: the *minute or "Great" wheel,* the *intermediate or "small" wheel,* the *second wheel,* and the *escape wheel* (see T1). In all quality watches one finds a ruby-anchor lever escapement of the Swiss type, while watches in the lower price ranges generally use a pin lever escapement.

The *escapement* includes the escape wheel and the anchor or lever, which intermittently stops and starts via its ruby pallets or pins, the turning motion of the anchor wheel and at the same time gives power to the swinging balance by way of the roller jewel which intersects the fork of the lever. The balance assembly, consisting of a ring-shaped balance wheel and its attached retrieving hairspring, determines the rhythm with which the anchor releases the wheel train so the second wheel makes one revolution in one minute. The minute wheel, therefore, will be driven at one revolution per hour because of the gear ratios. The hour hand is driven by the minute wheel via a separate 1:12 reduction drive under the dial.

Additional drives allow the hand-winding of the mainspring via the crown while the winding stem is in its inner resting position, and while in its outer position it sets the hands (see T 2). Seconds indication was done in pocket watches and early wristwatches on an subsidiary dial over the 6:00.

With this arrangement, the small seconds hand is fixed directly on the lengthened pivot of the seconds wheel and thus is "directly"driven by the seconds wheel. Later the larger center seconds hand became the standard because it is easier to read and gives a quick indication of the operating condition of the watch. For this purpose, while the previous order of functioning has been retained, an additional toothed wheel is attached to the lengthened pivot of the intermediate, or small, wheel and makes contact with the pinion of the center seconds arbor. Here, then, the seconds hand is not in the direct power flow of the drive train, but is driven without power via a subsidiary drive, and therefore called "indirect center seconds".

Unlike this, in the "direct center seconds" the power flows undivided via the intermediate wheel to the center second arbor and from there to the anchor wheel. The earlier eccentrically located second wheel can thus be eliminated. The mounting of the long center second arbor inside the bored-out minute wheel is not without peril and can lead to problems with the even transmission of the turning moment.

The disadvantage of the "indirect center second" is that of irritating hand fluttering between its step-by-step motions, caused by the so-called "tooth air" of the powerlessly working subsidiary drive. Help is provided by a leaf-shaped friction spring that presses on the upper pivot of the second arbor and is generally mounted very visibly.

An old watchmaker's saying goes, "The fine adjustment of a watch begins at the *barrel*", indicating the importance of an even transmission of power during the six turns between full winding and full unwinding of the mainspring.

If a barrel of larger diameter is now installed in the movement, then a longer mainspring is installed in it, and not only is the running reserve increased from the previous 36 to 45 or more hours, but the realm of constant power is expanded, offering better running regularity.

The minute wheel, formerly mounted in the center, must be moved outward to make room for the large barrel. The minute hand is then driven indirectly by an additional drive (Zenith 1948).

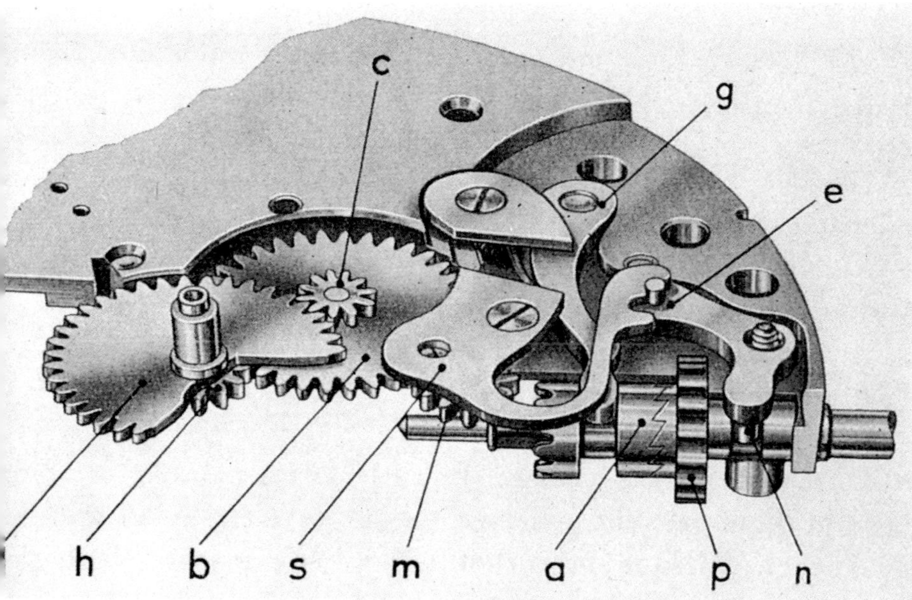

T2:
Winding and hand-setting drive on the underside of a movement with centrally mounted minute wheel. On the winding stem is the pusher pinion a, linked by a Breguet coupling with the transmission wheel p, which in turn contacts the crown wheel (not shown) situated on the other side (upper side!) of the movement. When the winding stem is pulled out, the angled or setting lever n/e moves out of the notch of the angled or setting-lever spring s into the next one and, at the same time, pushes, via the rocker g (rocker or Breguet spring concealed), the pusher pinion a into contact with the hand-setting wheel m, which, via the switching wheel b and the hour wheel d, on which the hour hand is mounted, while the minute tube h turns with the minute hand.

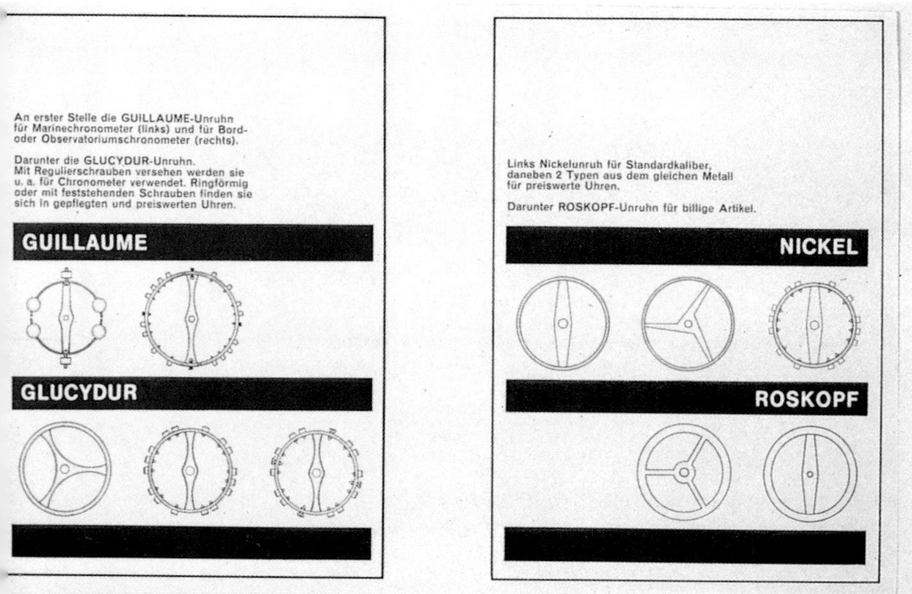

T3: Balance Wheel Types
The Glucydur balance, with which all of the higher-quality watches are equipped, is easy to recognize by the shape of its spokes.

The combination of the minute wheel mounted out of the center and the directly driven center second thus allows an optimal utilization of space for the benefit of the barrel and balance, and is the standard for modern watches (see T 1).

The *balance,* originally fitted with weight screws, was replaced by a smooth ring balance in the 1950s. It can be ordered from special factories complete with hairspring as a complete, adjusted unit. In addition to the balance made of Glucydur alloy, watches of the standard class also use those made of nickel, that differ by their particular shape (see T 3).

At the same weight and diameter, the smooth ring balance has a greater moment of inertia and thus swings more evenly. The increased air resistance caused by the screws, and the danger of braking by bits of dust hanging from them, are eliminated.

A further improvement in running performance can be attained by a faster *oscillation* of the balance (see page 23).

The so-called "fast beats" do not need such a great moment of inertia in the balance, so that the balance becomes lighter, and thus the danger of breaking the sensitive balance arbor becomes less.

To protect it, various *shock-resistance systems* were originally developed, but few of them actually proved themselves. The ones seen most often are the Incabloc system made by Le-Porte-Echappement Universel S.A. of La Chaux-de-Fonds and the various Kif systems made by Parechoc S.A. of Le Sentier.

Almost all calibers are fitted with a *self-compensating flat hairspring,* such as those made by Nivarox S.A. of St. Imier, which provides them at temperature compensation levels from I to V. The overcoiled or Breguet hairspring is found in only a few wristwatches today; it improves running reliability slightly, but also requires more height (see under IWC). The *moveable hairspring block carrier* was introduced in the 1950s and makes it possible to even out the swinging of the balance at the anchor-fork point of contact. The formerly required turning of the hairspring roller on

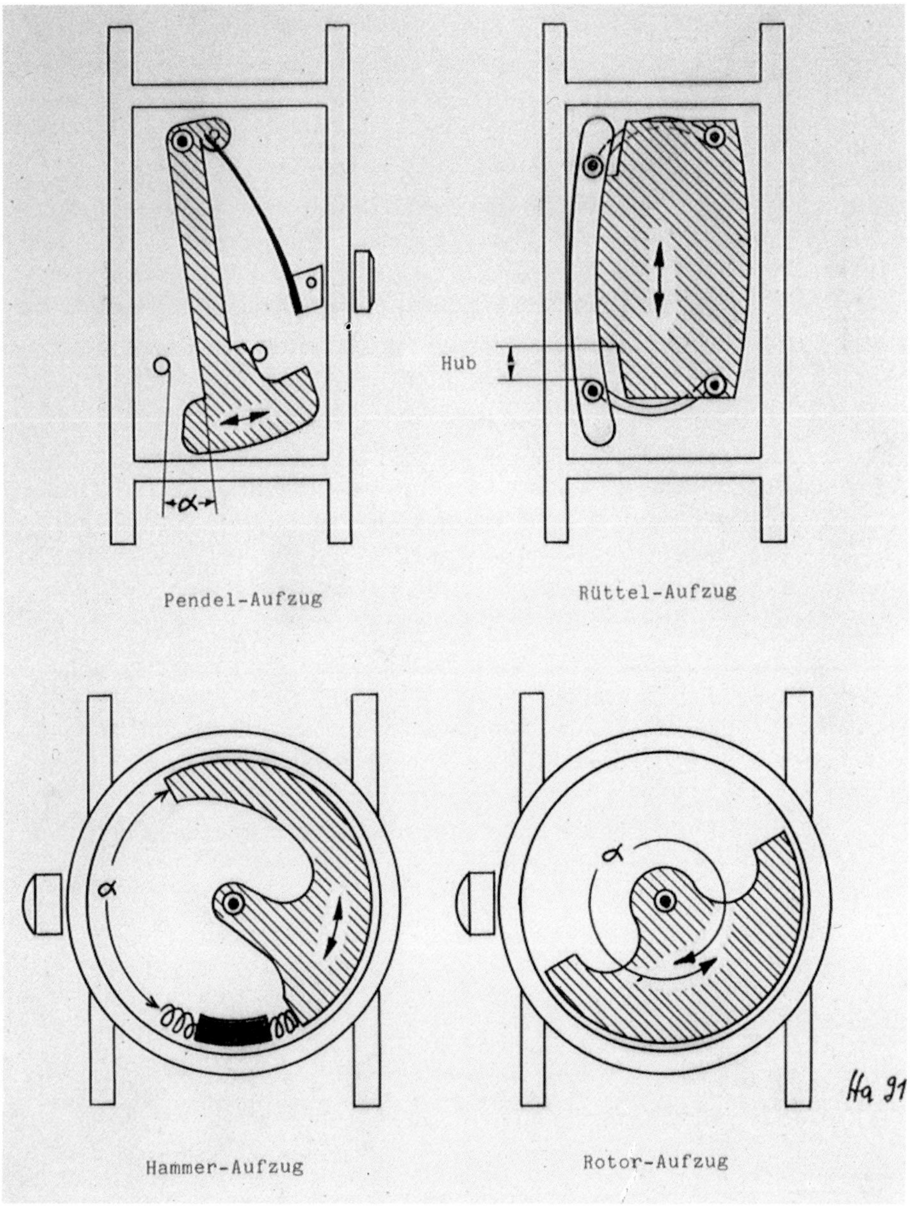

T4: Types of Winding
Pendulum winding: the turning point of its long lever is at the edge of the movement, with its angle of arc a measuring about 15 degrees.
Roller winding: the rolling weight moves in a straight line between the 6:00 and 12:00 positions, with a maximum movement of 3 millimeters.
Hammer winding: the centrally mounted hammer covers an arc a of at most 130 degrees and is stopped by buffer springs at both sides.
Rotor winding: the rotor turns without limit.

the balance arbor is no longer necessary. (The hairspring roller is the inner, the hairspring block the outer attachment point of the hairspring.)

The technological standards of quality for individual watch components that still apply today were established for the most part by Dr. Richard Straumann at his Waldenburg Research Institute between 1933 and 1958: they concern the mainspring, which consists of an unbreakable, rust-free beryllium steel alloy (nivaflex) and is coated with a lasting dry lubricant (Nivaglide); the very thin toothed wheels of beryllium bronze, which can be stamped out without a spine and, with their microgearing, transmit the turning moment much more evenly; and the self-compensating and therefore not temperature-dependent and also antimagnetic combination of hairspring (Nivarox, Isoval) and Glucydur bearing, both also made of special beryllium alloys.

All of these improvements have improved the reliability and performance of the wristwatch in such an impressive manner that even in professional circles there was talk of "miracle alloys".

Parallel to this progress, intensive work was done to improve the quality of the lubricants. The special requirements for lubrication in the construction of small watches can scarcely be understood by a layman. They depend, among other things, on the material combination of the mounting place (such as steel and ruby), the turning speed (a slow-turning barrel or fast-turning balance), the mounting forces that occur (high on the minute pinion, low on the anchor wheel), the acceleration forces (the constant stopping of the anchor wheel, especially in a fast-swinging watch, makes special demands on the lubricant's ability to adhere to the teeth). In addition, the oil should possess not only a consistency unaffected by temperature but also be insensitive to the influence of light, should not liquify and naturally not affect the surfaces of the mount.

It is thus understandable that an ideal universal watch oil cannot be created. A watch should therefore be cleaned thoroughly and lubricated properly after every three to at most five years if it is to remain reliable, precise and undamaged in continued use.

The automatic winding mechanism of a wristwatch complicates its construction to a great degree and requires not only sufficient space but also a stable mount for the swinging mass. The center of gravity of this mass lies outside the pivot point, as in the case of the rotor, and results in uneven pressure on the mount. The forces affecting the swinging mass when the wearer's arm moves are a combination of weight and impulse components. When the watch is turned evenly with the dial in a vertical position, the swinging mass always turns to its deepest point—doing its work only with the help of gravity. But if the watch is moved vigorously back and forth with the dial in a horizontal position, for example facing upward, only the impulse forces affect it.

The motion of the swinging mass relative to the watch movement is transmitted optimally by a reduction drive to the mainspring, winding it. In the early automatic watches, winding took place in only one direction, but later, with the installation of a direction changer, also known as a switcher, the second direction was also used. Additional uncoupling elements simultaneously make sure that the hand-winding drive is uncoupled and reduce friction losses. To avoid overwinding the mainspring by automatic winding, a special safety device, such as a sliding clutch or rotor-blocking mechanism, is provided.

Thus the complete automatic module comprises these moving parts: Swinging mass, Switcher, Reduction drive, Overwinding protection, and in addition, the mutual uncoupling devices between the automatic and hand-winding mechanisms, as well as an auxiliary device to indicate the running reserve.

The rotor, though it was the first type of swinging mass used in a self-winding pocket watch (Perrelet Sr., 1770), had to prevail over other types in the wristwatch, such as the pendulum, hammer and rolling mass (see T 4). The characteristic design feature of the *pendulum* is a long lever, at the end of which a weight is located which provides a relatively large turning moment for winding. Its swinging from a turning point at the other end of the lever is limited to a small arc and always winds from the 3:00 position when the watch is worn on the left arm. With the help of a recoil spring, the pendulum always has to be brought back to its original position. This optimal position of the pendulum has a great effect on the layout of the rest of the movement. During arm movements in the direction of the lever, this being between the 6 and the 12, no winding effect can be expected, so that this type of winding turns out not to be very effective.

In addition, the heavy strikes of the pendulum against the limiting pins have a disturbing effect on the uniformity of the balance swinging, make a loud noise, and can even loosen screws in the movement, with all the resulting troubles imaginable. These disadvantages occur with all non-continually swinging masses, including the hammer and the rolling weight.

The pendulum is found in all self-winding pocket watches with the exception of that of Perrelet, in the pedometer, and in some of the early automatic wristwatches (Leroy, Perpetual, Bulova, Buren). Even in the automatic tourbillon made by Audemars Piguet in 1986, this type of winding was revived for reasons of space.

Unlike it, the *hammer swinging mass* is centrally mounted and its motion is limited to a maximum swinging arc of 130 degrees by buffer springs. To increase its winding power, the hammer is weighted along its extent by lead inlays. To protect the hammer mount from shock damage, the shank of the hammer is generally made in the form of a spring, such as the S-shaped type used in a Movado caliber.

This type of winding was invented by Harwood and utilized in most automatic watches from 1935 to 1950. The reason for its success has less to do with its winding efficiency than with the problem-free mounting of the hammer between two plates and the simple drive layout, which allows the movement to be relatively flat. The buffer springs, on the other hand, are a disadvantage, as they often break and jam either the hammer or the whole movement.

The hammer automatic usually winds in one direction. A few calibers developed later, such as those by Revue, Lanco, Jaeger LeCoultre (Futurematic) and Baumgartner, are exceptions.

The *rolling weight* is a swinging mass that moves back and forth in a straight line between the 6:00 and 12:00 positions and whose contact with the case is absorbed in part by leaf springs. Various parts of the watch have served as producers of energy, sometimes the entire movement and the case, moving on a parallelogram frame over a plate attached to the band (Bucherer automatic to a Bulova patent); the entire movement with an additional weight that glides on rails in the case (Rolls) or moves on two lever arms in a case frame (Wig-Wag); or a weight recessed in the movement and led by two rails (Pierce). The rolling distance is no more than three millimeters and is not very efficient when winding in only one direction. And the disadvantages listed for the pendulum also apply here.

The *rotor* was able to prevail as the optimal source of energy only toward the end of the 1940s, after its mounting problems could be overcome thoroughly with the introduction of the miniature ball bearing (Eterna). The ball bearing simultaneously improved the central mounting of the rotor in several ways: In addition to the decreased bearing friction, the lessened play of the bearing also provided greater safety against the rim of the rotor rubbing on the plate or the case bottom, and the danger of breakage could be eliminated by doing away with the sensitive bearing arbor. The height of the movement could also be reduced.

On the other hand, the peripheral guiding of the rotor, supported by balls or rollers, also brought certain disadvantages: The loud rolling sounds were often regarded as disturbing, and the peripherally greater rolling friction of the weight diminished the degree of winding effect (FHF, Vacheron Constantin, Patek Philippe).

The general progress in technology, though, led via the decrease of required winding power with improved effectiveness of the automatic winding to a definite saving of weight with the rotor. This also diminished the problems of mounting it, so that numerous manufacturers chose the quieter type of mounting with ruby or sintered metal roll-

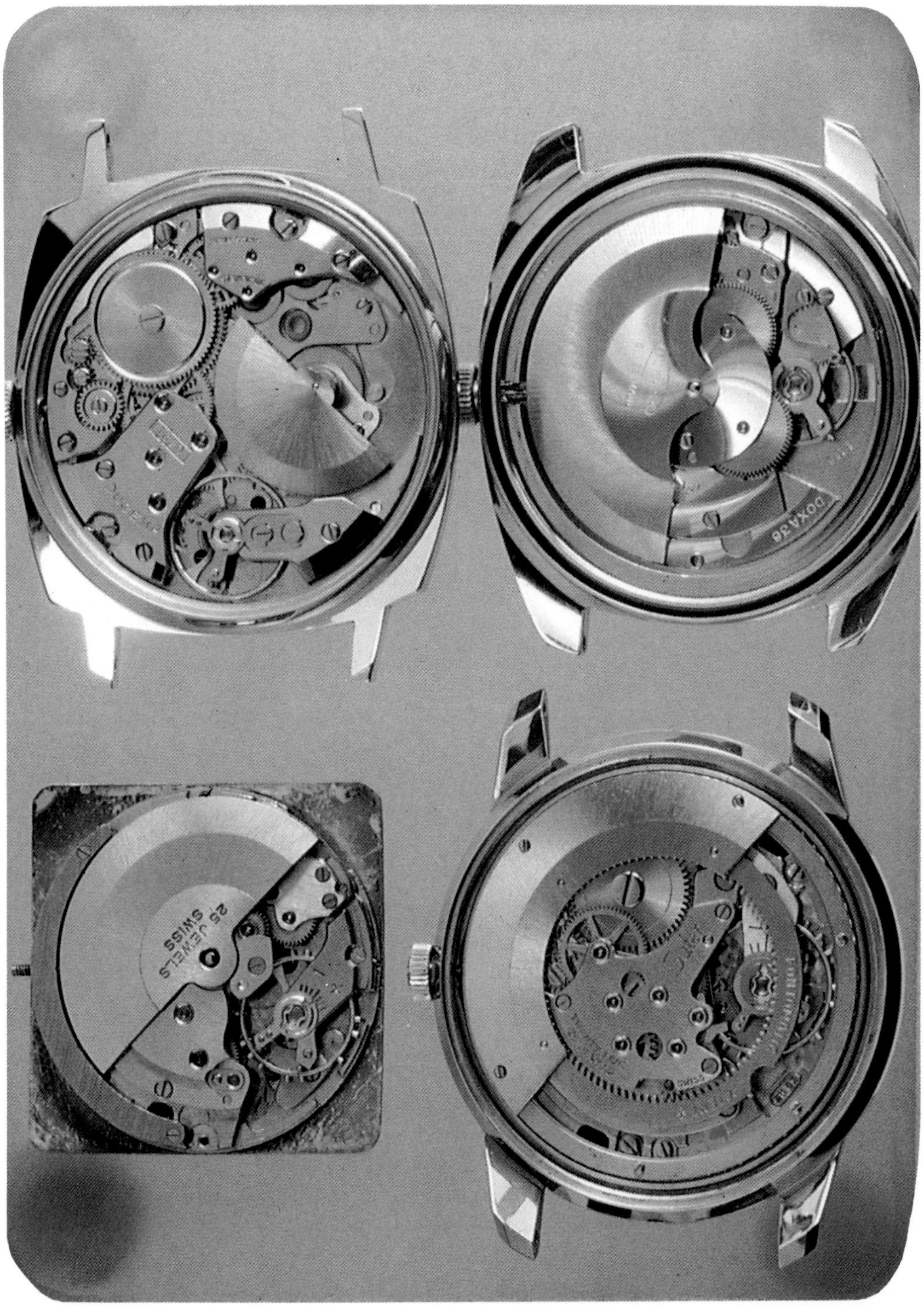

T5: Rotor Systems
Central rotor with gliding mount (upper right);
Ring rotor with no arbor, recessed in the movement and running on ball bearings, with peripheral gear coupling (lower right).
Microrotor completely recessed in the movement, described by Buren as "Planetary Rotor" and by Universal as "Microtor" (upper left).
Large rotor with central gear coupling, mounted out of the center—also made by Longines with peripheral coupling (lower left).

ers (see under Rolex). To prevent pivot breakage under stress, various solutions were invented, such as the elastic S-shaped rotor shank used by Movado or the sprung rotor axles used by IWC and Cyma.

Increasing the effective weight of the rotor by weighting its periphery improved winding performance. At first lead was poured into cutouts near the edge; later the entire rotor or peripheral segment was made of heavy metal, such as high-karat gold (Eterna, Audemars Piguet, Piaget, Patek, Vacheron Constantin). To diminish the movement height further, the rotor was mounted in numerous parts of the movement (see T 5), with many resulting variations in the manner of drive coupling. Peripheral coupling via a toothed ring is often found in especially flat calibers but requires an additional reduction wheel (see under FHF, Longines).

In the end, the central roller mounted on balls, with a heavy metal ring and drive coupling using a rotor pinion, proved to be the optimal swinging-mass combination.

While only the central screw has to be unscrewed to remove the rotor and its ball bearing, the rotor with a gliding mount requires several more or less simple detaching procedures.

For the axle mount, after a screw is removed from above, a bolt has to be pushed back or lifted out (see AS, Tissot) or a sprung lever pushed (see Felsa, Certina). On the other hand, the rotor running on one arbor (with a ruby hole jewel or bearing box in the plate) can be separated only

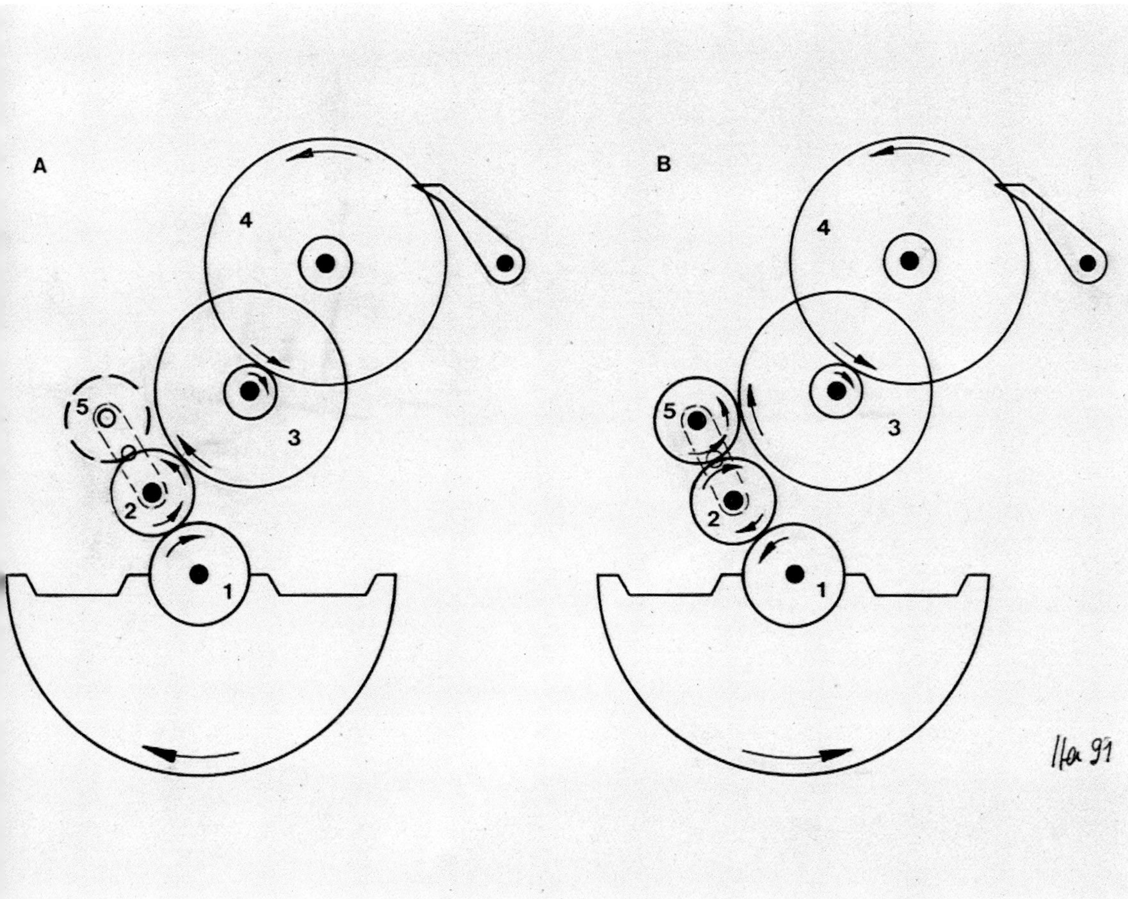

T6: Switching Rocker
A (Rotor turns clockwise): Turning moment transmitted via rotor pinion 1, direct rocker wheel 2 (indirect 5 turns freely with it), first reduction wheel 3 and second reduction and locking wheel 4.
B (Rotor turns counterclockwise): Turning moment transmitted via rotor pinion 1, direct rocker wheel 2, indirect rocker wheel 5 forced into contact by rotor turning, first reduction wheel 3 and second reduction and locking wheel 4.

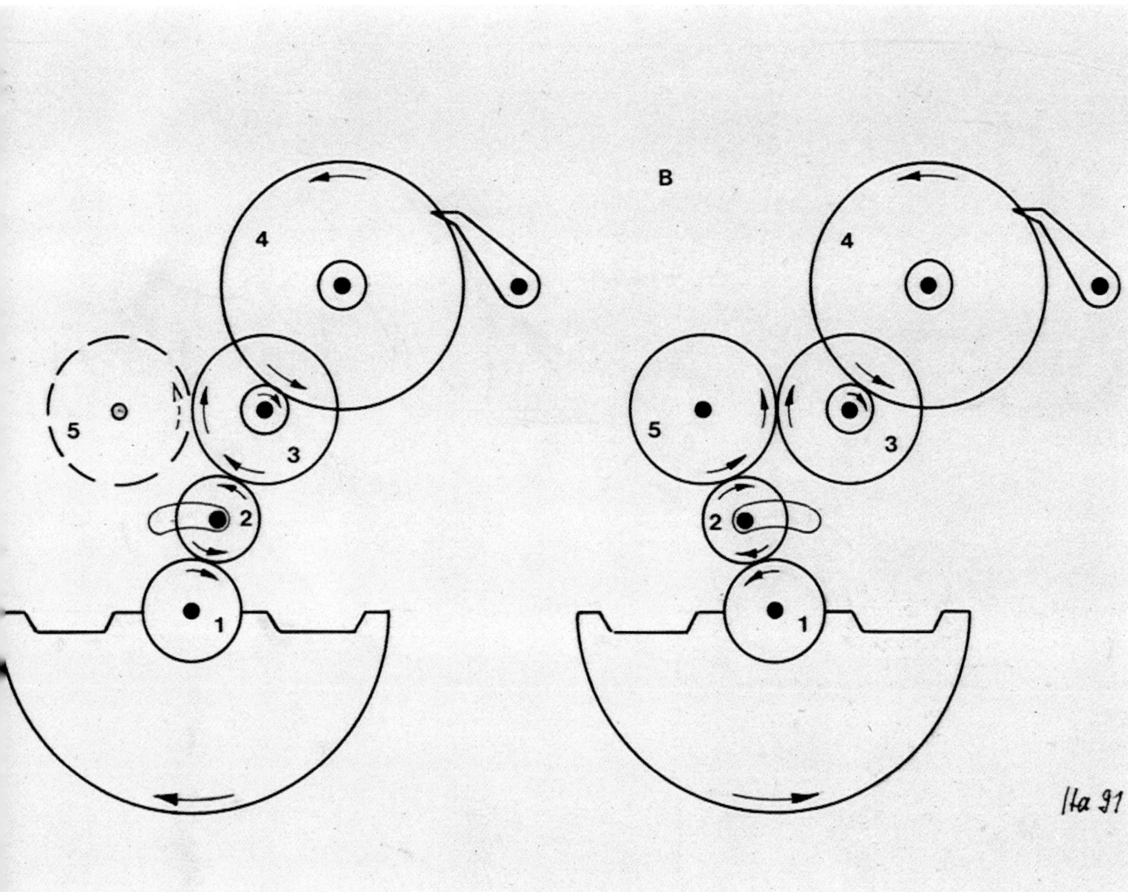

T7: Gliding-Pinion Switcher
A (Rotor turns clockwise): Turning moment transmission via rotor pinion 1, gliding pinion 2, first reduction wheel 3, second reduction and locking wheel 4.
B (Rotor turns counterclockwise): Turning moment transmitted via rotor pinion 1, gliding pinion 2, which is forced into position by rotor turning, additional tension wheel 5, first reduction wheel 3, second reduction and locking wheel 4.

T8: Click-Wheel Switcher with Identical Wheels

A (rotor turns counterclockwise): Turning moment transmission via rotor pinion 1, direct click wheel 2/3 (click on the drive wheel 2 takes the click pinion on the reverse wheel 3 along) to the first reduction wheel 4. The indirect click wheel 5/6/6' turns freely with it, whereby the click pinion 6 is driven backward via its reverse wheel 6' by the reduction wheel 4 and thus turns the opposite way to its drive wheel 5, whose click slides through. The locking function to prevent unwanted unwinding of the mainspring during a change of direction is now taken over by this click 5.
B (Rotor turns clockwise): Turning moment transmission via rotor pinion 1, drive wheel 2 of the direct click wheel, indirect click wheel 5/6/6' as additional wheel to reverse the turning direction (click on drive wheel 5 takes click pinion 6 on reverse wheel 6' along) to the first reduction wheel 4. The click pinion 3 is driven backward via its reverse wheel by the reduction wheel 4 and thus turns in the opposite direction to its drive wheel 2, whose click slides through, but at the same time now takes over the locking function to prevent unwanted unwinding of the mainspring.

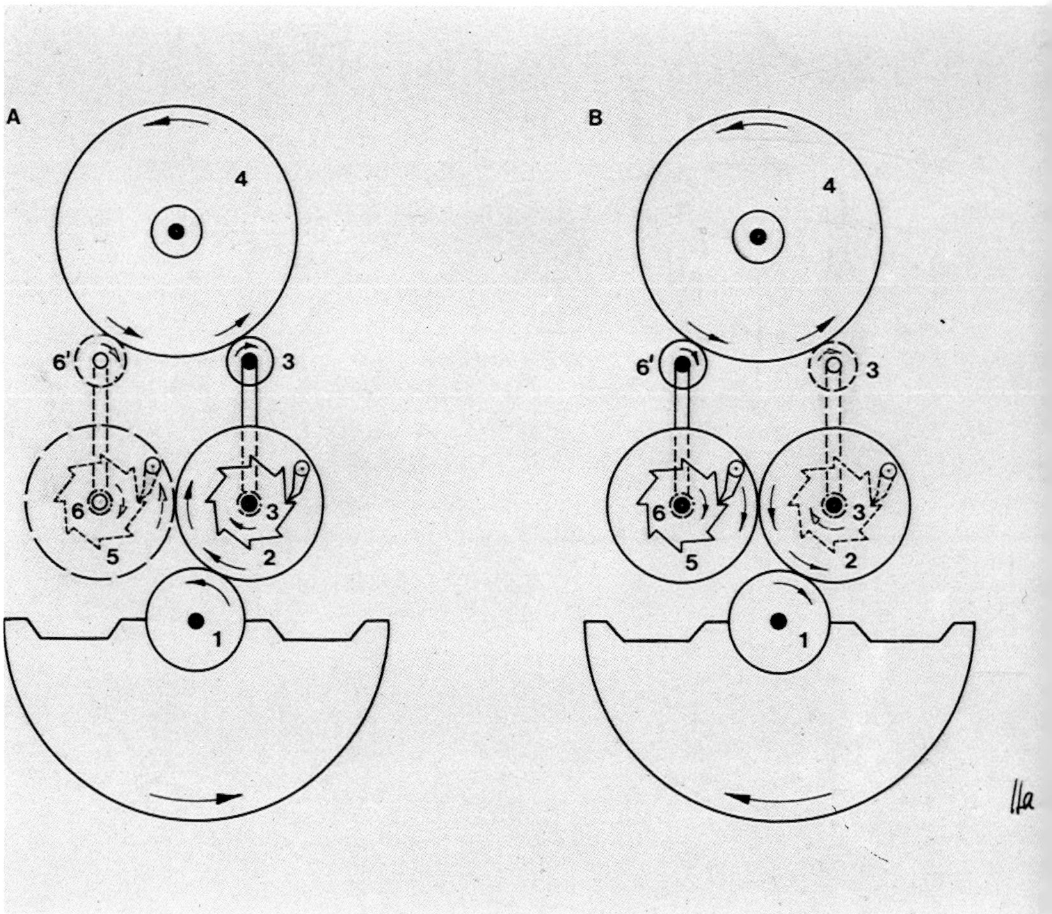

T9: Click-Wheel Switcher with Unequal Wheels

A (Rotor turns counterclockwise): Turning moment transmission via rotor pinion 1, direct click wheel 2/3 (click on drive wheel 2 takes click pinion 3 on reverse wheel 3 along) to first reduction wheel 4. The indirect click wheel 5/6/6' turns freely with it here, whereby the click pinion 6 is driven backward via its reverse wheel 6' by the direct click wheel 2/3 and thus turns in the opposite direction to its drive wheel 5, whose click slides through. The locking function to prevent unwanted unwinding of the mainspring during a change of direction is now taken over by this click 5.
B (Rotor turns clockwise): Turning moment transmission via rotor pinion 1, indirect click wheel 5/6/6' as additional wheel to reverse the turning direction (click on drive wheel 5 takes click pinion 6 on reverse wheel 6' along), reverse wheel 3 of the direct click wheel to the first reduction wheel 4. The click pinion 3 is driven backward over its reverse wheel by the indirect click wheel 5/6/6' and turns freely, but opposite to its drive wheel 2, whose click slides through. This time, this click 2 takes over the locking function to prevent unwanted unwinding of the mainspring.

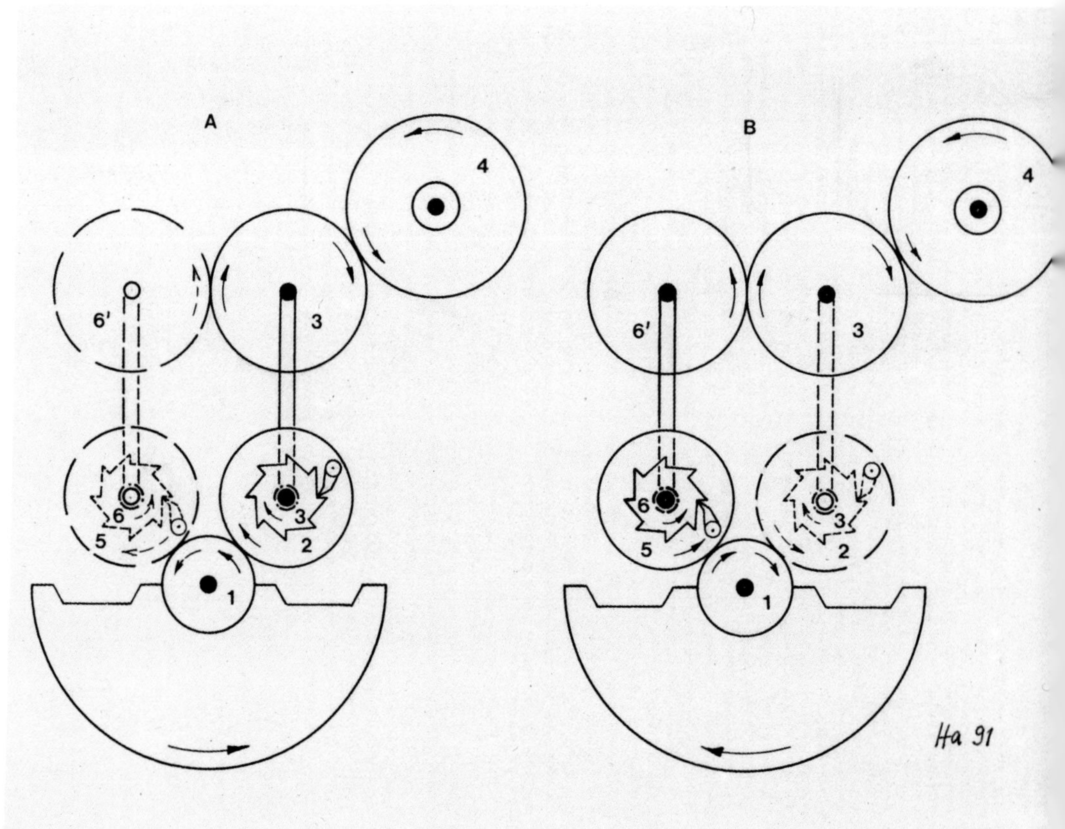

T10: Eccentric Switcher
The rotor turns clockwise: Turning moment transmission via rotor eccentric 1, click-carrier rocker 2, which moves downward on its turning point M and thus with its click lever 2", turns the click wheel 3 clockwise. When the rotor turns farther in this direction, the click-carrier rocker 2 is forced to turn upward by the changed eccentric position, and now moves the click wheel 3 farther in the same direction via the click lever 2'. While the one click lever moves, the other takes on the locking function to prevent unwanted unwinding of the mainspring when the rotor changes direction.
The rotor turns counterclockwise: Winding takes place similarly, but the click-carrier rocker 2, with the rotor eccentric in this position, is first moved upward, and thus the click wheel 3 is moved farther by the click wheel 2'.

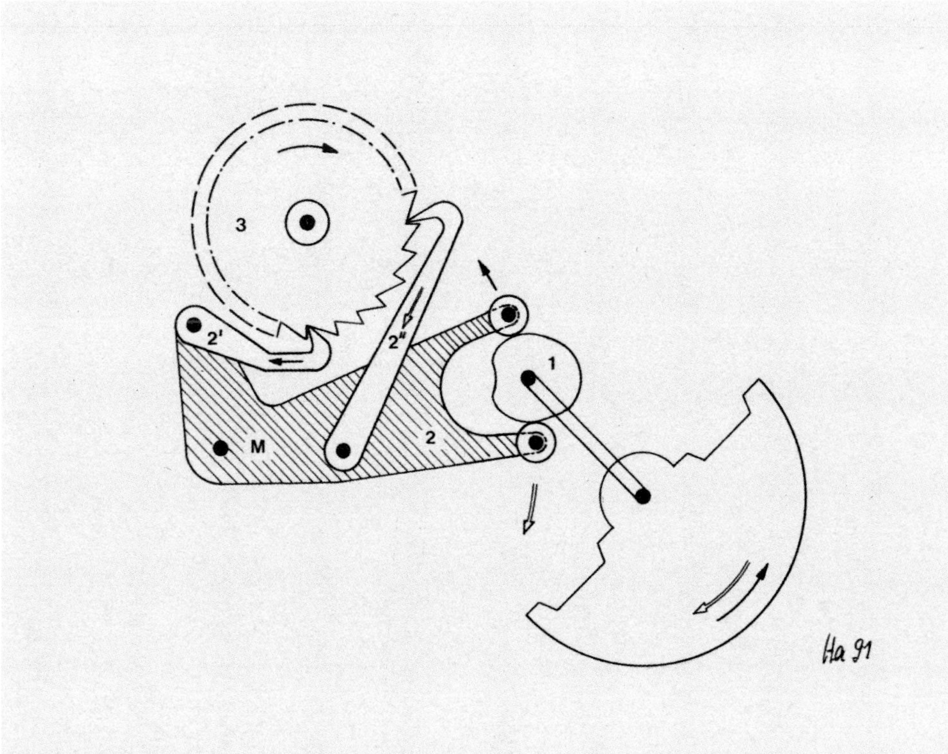

after removing the entire automatic module (see Rolex, Eta).

Until the "Bidynator" was first introduced by the Felsa firm in 1942, this swinging mass could wind only in one direction. To convert the two rotor directions into one winding direction to wind the mainspring, two essentially different solutions were developed: either continual equalization with toothed-wheel elements such as the switching rocker (AS) or click wheels (Eta), or the non-continual type using an *eccentric switcher,* as done by Longines (see T 6 to T 10).

In the majority of rotor automatics one finds the continually working **switcher,** which are based without exception on the principle of forced intermediate switching by an additional wheel when the rotor's turning direction changes. (An additional toothed wheel in a drive train reverses its turning direction at the end. See T 6 to T 9).

In no other area of automatic development does one find as much variety as in the click wheels (see T 11). The reasons for this are rooted in the efforts of the individual manufacturers to circumvent the many existing patents with more and more refined designs.

Among the constantly working switchers there is another difference between the designs with moveable toothed wheels, such as the *switching rocker* (see T 6) or the *pusher pinion* (see T 7), and those with fixed mounts, such as the *click wheels* or *free-running connections* (see T 8 and T 9).

Although the technology of rotor winding in two directions could be developed to a high level of perfection and efficiency, discussion of the most effective type of winding—in one or two directions—never ended. The raw-movement producer AS made an interesting experiment on this subject.

A rotor caliber originally winding in both directions was rebuilt by removing the indirect click wheel, so that winding could take place in only one direction. This modified automatic was compared with another unchanged model of the same caliber, with both of them being worn side by side on the arm of a test person.

The result favored the one-direction winding, which wound up the watch in just a short time. At AS they spoke of a rotor with "dynamic free running", which supported winding in the opposite direction disproportionately. For that reason, one finds one-direction winding in almost all modern caliber developments (see under Lassale, Longines, Patek Philippe, Fr. Piguet, Valjoux).

The individual wheels of the automatic drive are arranged so that the turning moment is transmitted from a small wheel, such as the central rotor pinion, to a large wheel, which turns more slowly because of the ratio of the numbers of teeth. This is the **reduction drive,** which reduces the fast movements of the swinging mass, depending on the moment of inertia, into 110 to 180 times slower but correspondingly more powerful turns of the locking wheel on the barrel (see T 12 and T 13). A small reduction ratio thus means fast winding but requires a correspondingly heavier or larger swinging mass.

A variant of the toothed-wheel reduction drive is the *click-carrier rocker* used with hammer winding and eccentrically guided rotor winding. The majority of the hammer automatics developed in the 1940s used it, while the simple reduction drive with the winding click on the hammer was found previously (i.e., Harwood, AS, Certina, Cyma; see T 13).

T11: Click Wheel Types (selection)
(A in powered, B in free-running function)
Type 1 with spring click (as used by Buser, Roamer, etc.)
Type 2 with springless anchor click (AS, Record, etc.)
Type 3 with springless double anchor (AS patent)
Type 4 with springless click (Eta patent)
Type 5 with toothed-wheel directing lock (Longines patent)
Type 6 with roller- or ball-clamp lock, also called free-running ("Gyrotron" patent by Girard-Perregaux, or Roamer patent)

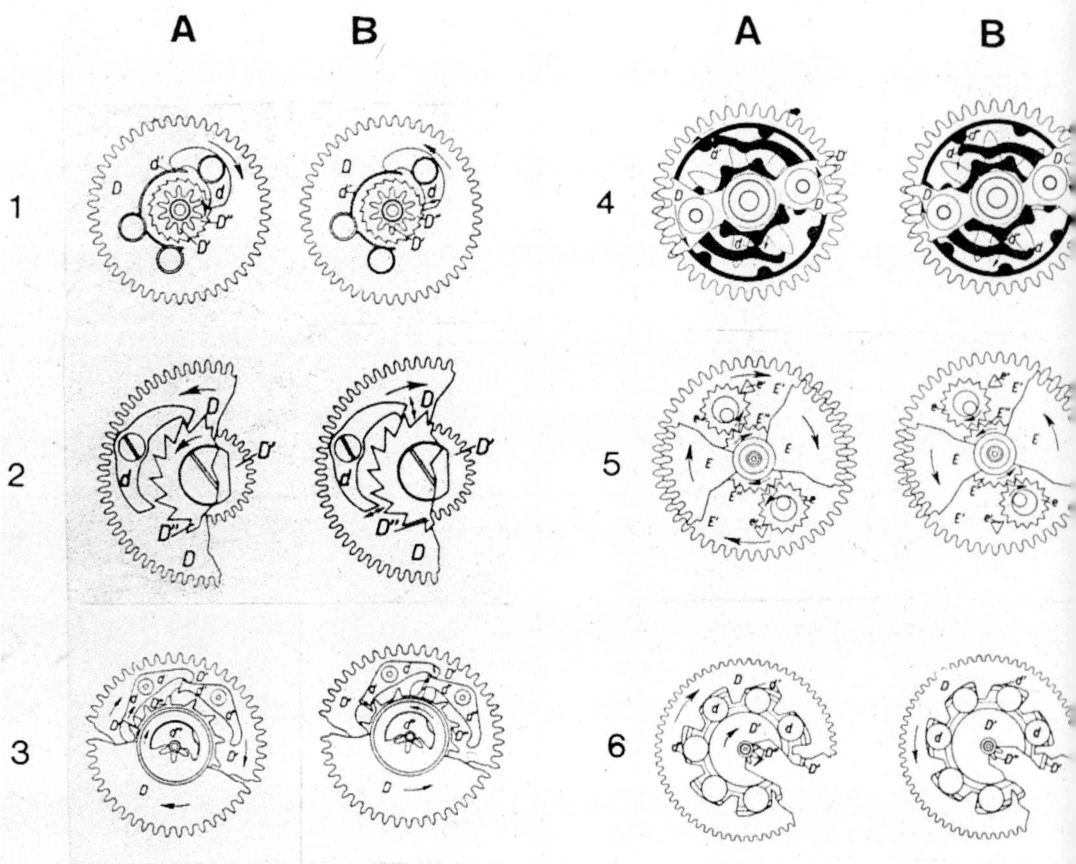

The device that protects against overwinding the mainspring during automatic winding prevents overly extreme swinging of the balance (more than 270 degrees), and thus rules out the danger of its over banking, which would result in considerable running irregularity. The sliding coupling in the barrel has proved to be a simple and reliable design, which goes back to Adrien Philippe and was installed by Roskopf in his reasonably priced pocket watches from 1860 on to avoid the danger of breaking the mainspring in case of overwinding by hand. A *sliding stud* is connected with the end of the mainspring which is set against the barrel walls and the tension is friction set. One can wind to a certain point before the spring slips (see T 15), which is set by the tension of the sliding stud. Therefore if winding continues, the mainspring with the sliding stud slips along the barrel wall and, no matter how much you wind, the maximum driving power is limited.

Systematic research work by Dr. Richard Straumann led around 1960 to the development of the GL barrel, made of aluminum with a smooth surface of Eloxal that, in combination with the Nivaflex spring, guaranteed unchanging functioning over many years of use. Until that time, incorrect functioning had to be expected after just one or two years, depending on the combination of materials and lubricants. The overwinding protection often ceased to work completely because the sliding stud ate its way into the barrel and jammed, which led to the danger of damage to the balance overbanking and made the watch run fast. But failure of the other kind through premature sliding of the stud was also a frequent reason why a watch stopped because of a too-small running reserve. The maximum running reserve is the time, in hours, that passes between full winding, when the overwinding protection is activated, and the stopping of the watch when winding by hand or automatically is lacking. As has been mentioned already, a large running reserve is an essential prerequisite for good running performance, and it naturally increases the functioning reliability of the watch as well. The lowest running reserve is 36 hours, modern automatics reach 50 hours, and the record is 100 hours, which is quoted for the two-barrel Caliber 11.50 made by Fr. Piguet.

Alternatives to the sliding coupling in the barrel are the blocking of the swinging mass at full winding and the friction coupling outside the barrel.

The most elegant, but also the most expensive solution to this problem is the stopping of the hammer regulated by a spring-arbor screw- or differential drive that indicates the running reserve, as used in the "Futurematic" of Jaeger LeCoultre or with the rotor of the "Inversator" caliber of Felsa. This basic idea also goes back to A. L. Perrelet, who used the height of the chain on the fusee, dependent on the degree of winding, to block the rotor in the first automatic pocket watch, where a moveable pin moved into a notch on the rotor at full winding.

In the friction couplings, along with that located on the hammer of the Harwood automatic, the "self-regulat-

T12: Reduction Drive in Rotor Winding

The automatic module of the Omega Caliber 501, seen from below, with unscrewed lower plate: Turning moment transmission via rotor pinion (projecting from under the rotor-arbor plate), direct wheel, indirect wheel of the switching rocker, first reduction and locking wheel (with sprung locking click), second reduction wheel, whose pinion contacts the locking wheel of the barrel in the basic movement, not shown here.

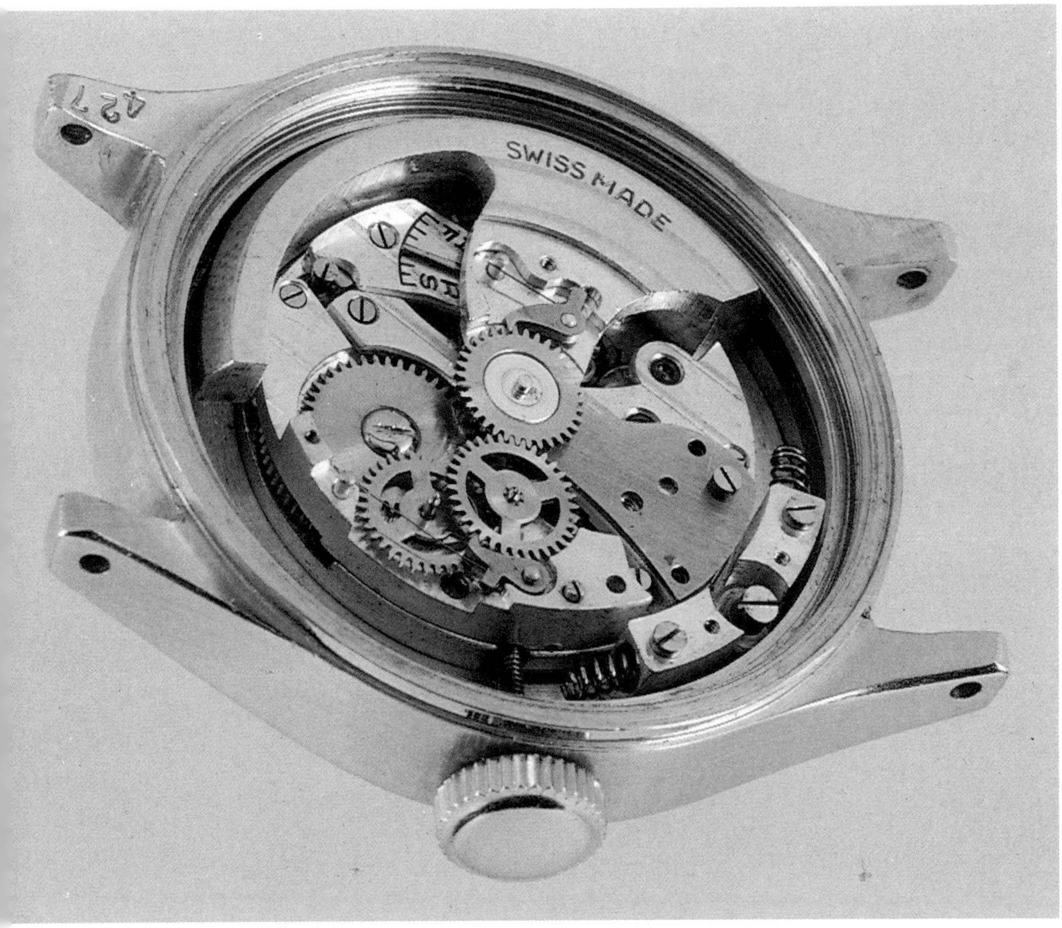

T13: Reduction Drive in one-direction hammer winding I (Harwood type)

AS Caliber 1171 with three plates removed:
Turning moment transmission via winding click on the hammer, centrally mounted winding wheel, first reduction and locking wheel (with sprung locking click in the foreground), second reduction wheel on the barrel locking wheel (hand-winding not provided).

T14: Reduction Drive for one-direction hammer winding
Omega caliber 342 with removed upper automatic module plate:
Turning moment transmission via toothed segment of the hammer shank Z1, click-carrier rocker Z2, which reduces like the sector of a twice-as-big wheel and moves the combined click and reduction wheel Z3, with the winding click C1, clockwise (the turning points of Z2 and Z3 are together) to the crown wheel Z4 and from there to the barrel locking wheel Z5. The locking click C2, screwed to the upper plate of the basic movement, is intended to prevent, with its wire spring, backward turning of the click wheel Z3 while the hammer is turning freely (clockwise).

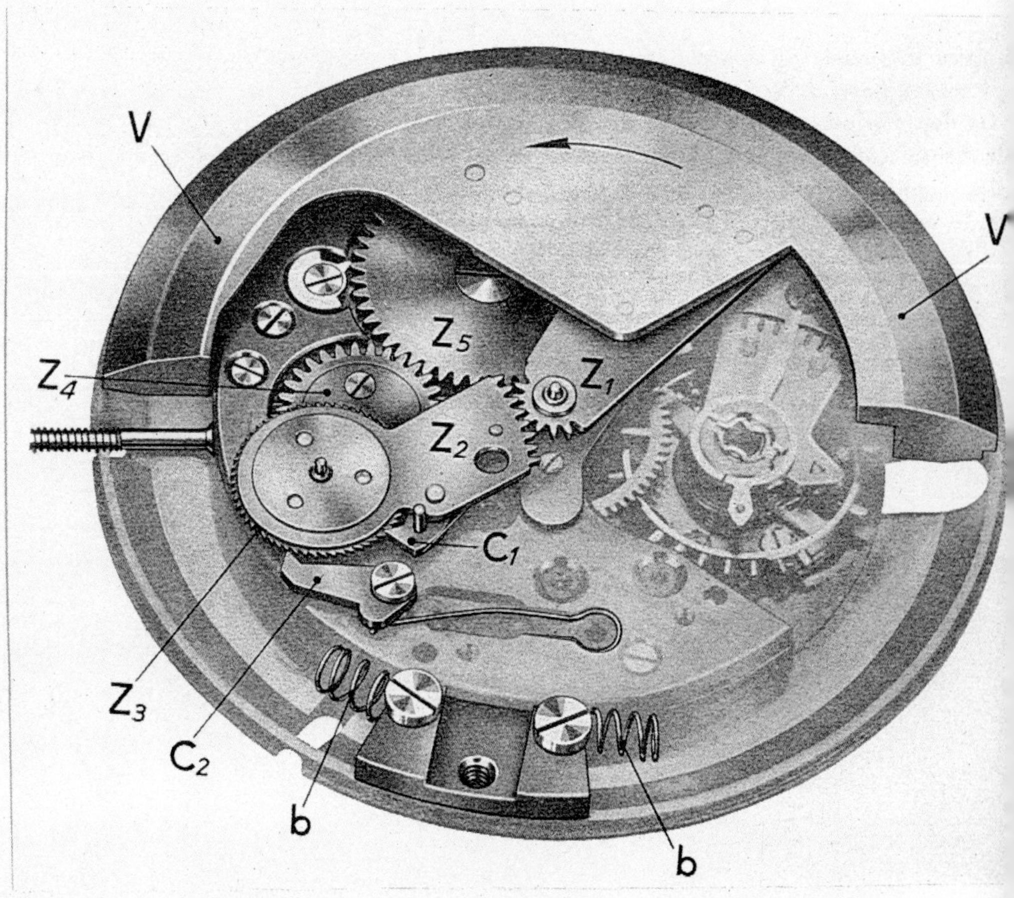

T15: Sliding Stud
In modern barrels, a sliding stud is located between the outermost and the next winding of the mainspring and welded to it. It presses the outer spring winding against the barrel wall and increases its contact friction. When the mainspring is fully wound, overwinding occurs, which overwhelms this friction—the spring slides directly on the inner surface of the barrel wall.

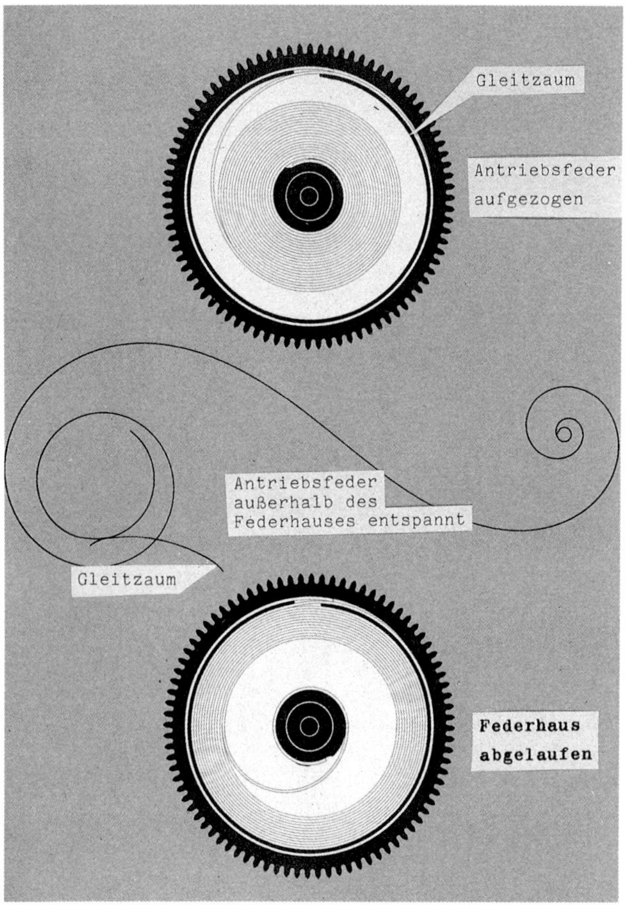

ing friction coupling" of the Gruen "Power-Glide" automatic deserves special attention. A moveable spring-boss nut, the position of which depends on the degree of winding, regulates a coupling in the reduction drive so that at full winding an interruption in the power flow between the rotor and the barrel takes place.

In several automatics of the 1930s a *separating coupling* between the reduction drive and the locking wheel was seen for the first time (i.e., Rolex, AS). Without this device, the wheels of the automatic drive would be accelerated and run prematurely, very fast and prone to damage. Automatic wristwatches should therefore, for safety's sake, be wound only slowly by the crown.

During automatic winding, on the other hand, the hand-winding mechanism is separated only at the Breguet coupling on the winding stem (between the transmission wheel and the pusher pinion, see T 2), which requires a finely tuned positioning-lever spring. The turning of the winding crown, which uses up a great deal of energy, is thus avoided. In the early automatics this problem did not exist at all, since hand-winding was not provided. The uncoupling of the whole hand-winding drive was first undertaken by Bulova in the mid-1930s. The crown wheel, mounted on a sprung lever, swings out during automatic winding and disconnects the transmission chain.

This principle of separation is also found in calibers with a large barrel, since the greater distance to the crown wheel can be bridged by an intermediate drive that swings out. A further uncoupling device on the barrel of

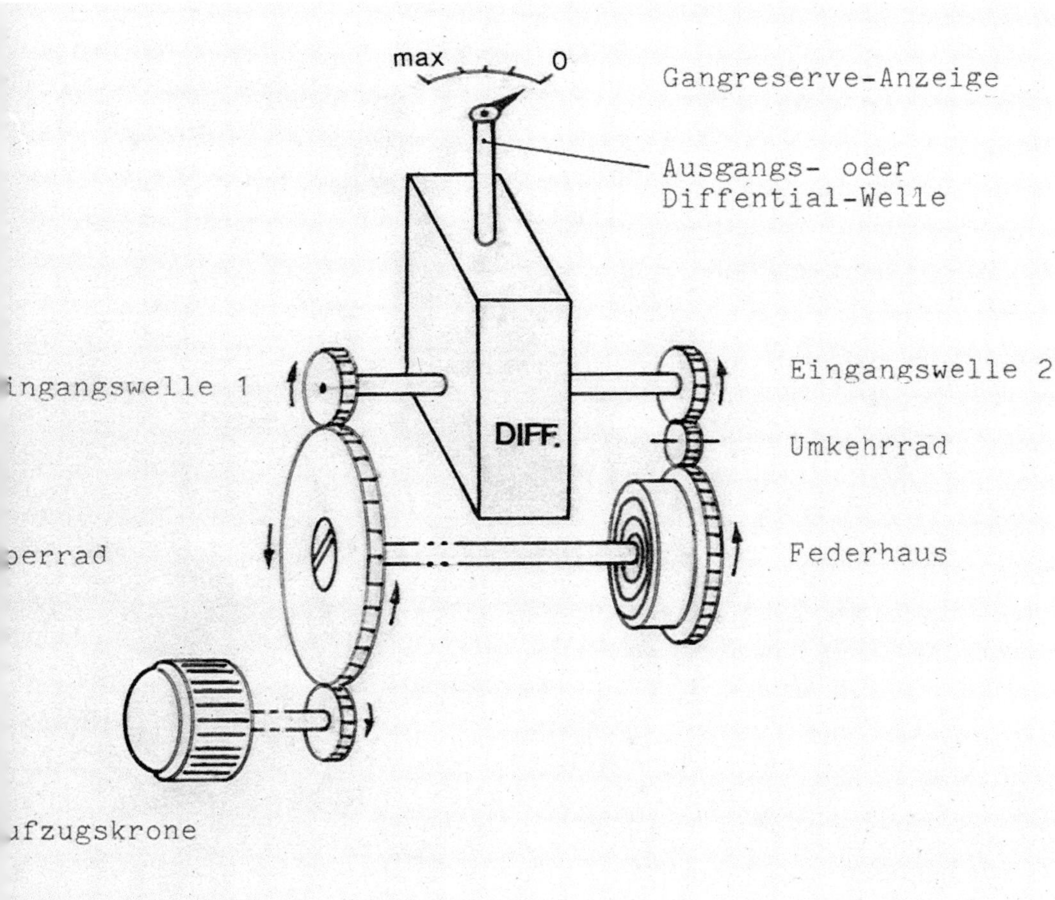

T16: Running Reserve Indication with a differential drive

Comparison of the positions of the locking wheel and barrel to each other via a differential drive, whereby the difference is indicated as the remaining spring tension to drive the movement (running reserve in hours).

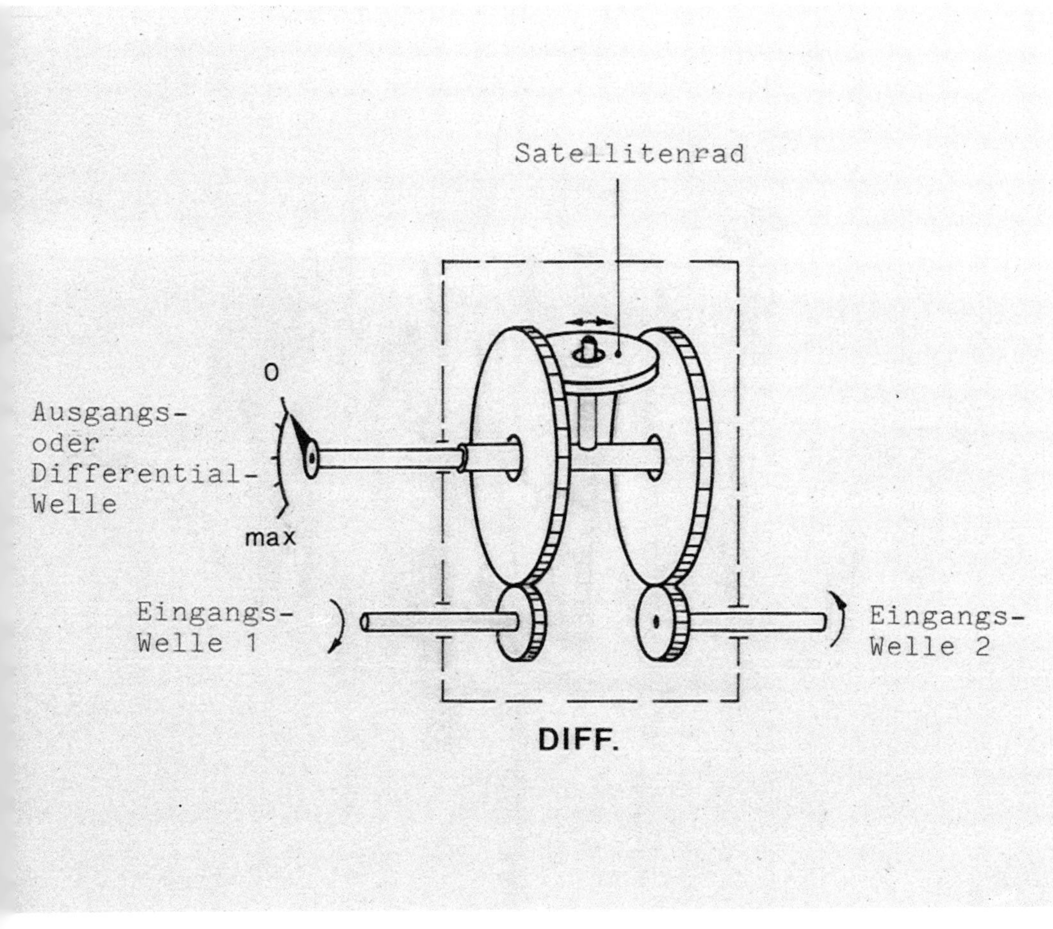

T17: Differential Drive

The exit or differential arbor is driven by a so-called planet or satellite wheel that is simultaneously in contact with the two large so-called sun wheels, each of which is connected to an entry arbor. The planet wheel is mounted on its axle so it can turn; the axle is vertical and firmly linked to the differential arbor. When both sun wheels turn equally fast but in opposite directions, then the planet wheel turns without having to change its axle position, which means the differential arbor is in the same position (winding of the mainspring and power transmission by the barrel take place at the same speed—the running reserve remains constant). During fast winding (entry arbor 1), the appropriate sun wheel forces the planet wheel to turn, which is only possible when it releases from the standing sun wheel 2. But the planet wheel can only be turned when it simultaneously changes its axle position—the differential arbor is thus turned in the direction of increasing running reserve.

T18: Running Reserve Indication with rotating scale

The position of the barrel to the locking wheel is indicated directly by a hand, under which a scale ring, linked to the locking wheel, turns to indicate the hours.

When winding on the mainspring and power transmission from the barrel to the movement take place at the same speed, the hand and scale ring turn clockwise at the same speed—the running reserve remains the same. During fast winding, the scale ring moves under the unmoving hand toward maximum running reserve. A sliding coupling between the hand arbor and the hand drive wheel makes possible the motion of the hand by a scale pin in case of further winding beyond the maximum, which amounts to automatic locking of the scale.

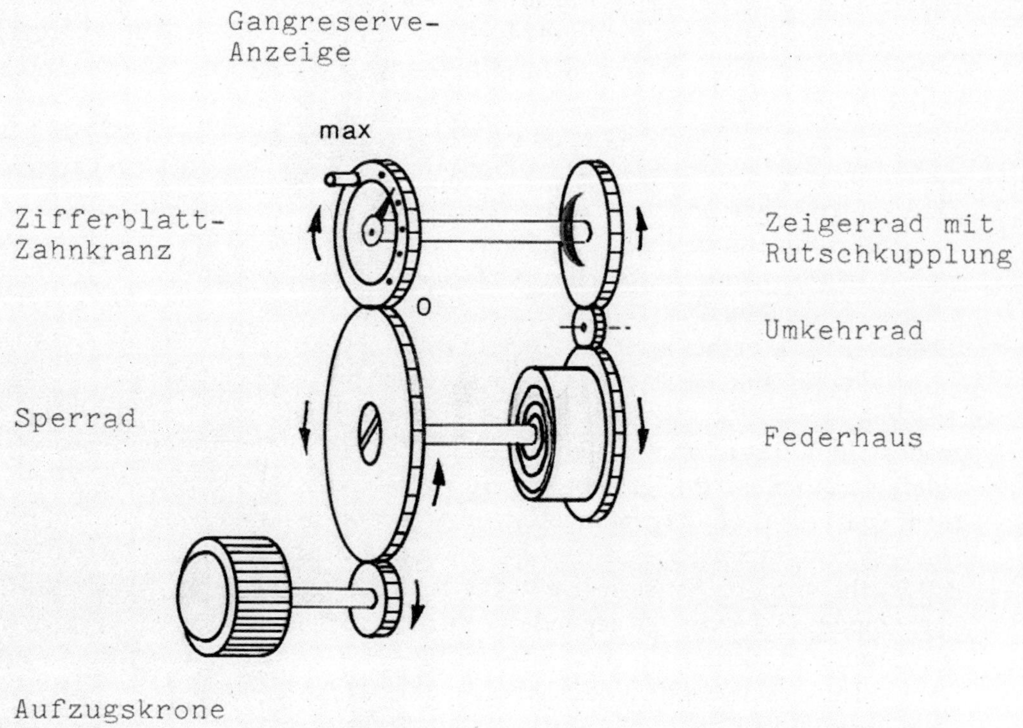

some automatics, the so-called double locking wheel, is constructed so that the opposite uncoupling, that of the reduction drive during hand-winding, can also take place (already used by Rolex in 1931).

The *up and down indicator* can be classified as special equipment for the automatic watch. It can be read from either a small hand on an auxiliary dial or a disc in a dial cutout. The indication is achieved by two different methods. The more elaborate utilizes a *differential drive* that compares the position of the spring boss of the locking wheel with that of the barrel and shows the existing difference.

The differential drive looks externally like a "black box" with two entry arbors and one exit or differential arbor. Entrance 1 is connected to the locking wheel, entrance 2 with the barrel by an additional reversing wheel, and the hand is coupled to the exit arbor (see T 16 and T 17). When winding is lacking, the entry arbor 1 does not move, while entry arbor 2 turns slowly as the barrel dispenses energy to the movement—and thus the exit arbor with the hand moves in the direction of decreasing running reserve.

Another possible way of indicating the running reserve is the direct indication of the locking-wheel and barrel positions on the dial. During winding, a toothed scale ring with printed running-reserve numerals (in hours) turns concentrically with the locking wheel under the pointer, which is linked with the barrel by a reversing wheel (see T 18). The pointer and scale ring can turn only clockwise, thus the indication results from their relative movement to each other: the zero point rotates.

While this simple system is found in several automatics of the 1950s (such as Longines, Record, and the Buren Rotowind), the system using the differential drive was already known in 1933 from an individually built square platinum watch made by the Breguet firm for Sir Percival David, but as a series-production product it was first seen in 1948 in the Jaeger LeCoultre "Powerwind" and the Zodiac "Autographic".

The information on the running reserve still present is actually much more important to the wearer of a hand-wound watch, for with the increasing perfection of self-winding and the simultaneously extended running reserve, the automatic watch will carry out its task dependably for years without being hand-wound.

So how, by the present-day state of knowledge, could a modern series-production automatic watch be described?

—*Automatic module* integrated into the basic movement, central rotor with ball bearings and heavy metal ring, winding in one direction via springless winding and locking wheel, two reduction wheels on the locking wheel of the barrel, dividing coupling between the crown and locking wheels.

—*Basic movement:* three-spoke screwless Glucydur balance, self-compensating flat hairspring, 28,800 BPH, shock resis-

tance, moveable hairspring block carrier, fine regulator adjustment, direct center second, Eloxal barrel with dry-lubricated unbreakable mainspring and 48-hour running reserve.
—*Special feature:* Date indication with quick change and correction.
—*Case:* Round waterproof stainless steel case with screw bottom and scratchproof glass.
The many complications above and beyond these which are available today can be regarded, other than an alarm, as more or less baggage (calendar, chronograph, repetition) and are usually combined with the automatic caliber as an additional module in the "sandwich" construction type.

Ladies' automatics, about which not much has been said here, are in principle smaller versions of men's automatics. The miniaturization of the components requires considerable sacrifices, though, not only in running performance but also in winding, since the inherent friction losses in the system do not decrease correspondingly, and in addition, the winding performance of the rotor decreases overproportionally through dependence to the third power of the radius of inertia.

This also resulted in the original concept of the ball-bearing rotor in a ladies' automatic made by Eterna, where the winding performance in the "Golden Heart" was enhanced by a rotor made of gold. But only the first "Ladymatic" by Omega and the "Zenitissimo" by Zenith can be classified as individual special designs.

References

1. Allgemeine Schweizerische Uhrenindustrie (Ed.): Das Herz der Uhr blossgelegt, ASUAG Biel, no date.
2. Brunner, G. L., Armbanduhren, Munich, Heyna-Verlag, 1990.
3. De Carle, Donald, Complicated Watches and their Repair, London, NAG Press Ltd., 1957 and 1977.
4. Château des Monts, Musée d'Horlogerie (Ed.), Horamatic-Montres à remontage automatique de 1770 à 1978, Le Locle, no date.
5. Ebauches S.A. (Ed.), Les Ebauches—Zwei Jahrhunderte Uhrenindustrie, Neuchâtel, 1951.
6. Ebauches S.A. (Ed.), Chronorama Expo 75, Neuchâtel, 1963.
7. Ebauches S. A. (Ed.), Die schweizerische automatische Uhr, Technical Service, Neuchâtel, 1963.
8. Ebauches S.A. (Ed.), Technologisches Wörterbuch der Uhrenbestandteile, Neuchâtel, 2nd Edition, 1953.
9. Erhardt, Sherry & Roy, Wrist Watches Price Guide, Vol. 1 (1984) to Vol. 5 (1990), Kansas City, Heart of America Press.
10. Engelkemper, Carl (Ed.), Automatic-Spiegel, 1960.
11. Engelkemper, Carl (Ed.), Werk-Spiegel mit Kaliberverzeichnis, 2nd ed. 1967 and Illustrations 1966.
12. Flume, Rudolf (Ed.), Der Flume-Kleinuhrschlössel K 1 (1958), K 2 (1962), K 3 (1972), Berlin and Essen.
13. Flume, Rudolf (Ed.), Die Information (Fachliche Diskussion über die Uhr 1952-1958), Selbstverlag, 1958.
14. Huber, Martin, & Banbery, Alan, Patek Philippe Genève, Munich, Verlag Ineichen, 1982.
15. Huber, Martin, & Banbery, Alan, with G. L. Brunner, Patek Philippe—Die Armbanduhren, Geneva, Antiquorum Editions, 1988
16. Humbert, B., Die Schweizer Uhr mit automatischem Aufzug, Lausanne, Edition Scriptar, 1956.
17. Jendritzki, Hans, Die Reparatur der Armbanduhr, Halle/Saale, Verlag Wilhelm Knapp, 6-7th ed., 1949.
18. Jendritzki, Hans, Der moderne Uhrmacher, Lausanne, Verlag Scriptar, 3rd ed., 1978.
19. Jobin, A. F. (Ed.), Klassifikation der schweizerischen Uhrwerke und Uhrenfurnituren, 2 vols, Geneva, Reliure, 1949.
20. Kahlert, H., Mühe, R., & Brunner, G. L., Armbanduhren, Munich, Verlag Callwey, 2nd ed., 1984.
21. Kocher, Hans, Automatische Uhren, Verlag der Neuen Uhrmacher-Zeitung, Ulm, 1969.
22. Kocher, Hans, Die Geschichte der Uhrmacherei in Büren, Munich, Verlag Callwey, 1991.
23. Mercier, Francois, Mechanische Uhren mit automatischem Aufzug, in Alte Uhren, No. 1 & 2 (1985), Munich, Callwey.
24. Pipe, R. W., The Automatic Watch, London, Heywood Co., 1952.
25. Rolex S.A. (Ed.), Das Rolex Jubiläums Vademecum, Vol. 1-4, Geneva, 1946.
26. Schweizer Uhrenfabrikanten (Ed.), Die automatische Schweizer Uhr—Der Zeitmesser von morgen, Neuchâtel, Edition Ebauches & Biel, FH, 1954.
27. Schweizer Uhrenfabrikanten (Ed.), Offizieller Katalog der Ersatzteile der Schweizer Uhr, Vol. 1 & 2, with additions, Solothurn, Vogt-Schild S.A., 1949.
28. Various, Parts and Sales Catalogs and Service Manuals of various Swiss Manufacturers.
29. Periodicals:
29a. Alte Uhren und Moderne Zeitmessung, Munich, Callwey (AU).
29b. Die Uhr (DU).
29c. Journal Suisse d'Horlogerie, Lausanne (JSH).
29d. Neue Uhrmacher-Zeitung, Ulm (NUZ).
29e. Schweizerische Uhrmacher-Zeitung, Lausanne (SUZ).
29f. Uhrmacher und Goldschmied, Graz (UG).
29g. Uhren-Juwelen, Vienna & Graz (UJ).
29h. Uhren-Magazin, Bremen (UM).

Index

Price Guide *by Gordon S. Converse*

The interest in collecting the various technically interesting forms of self-winding wristwatches has not yet become a field of intense interest for collectors. For that reason, many of the values of these intricate watches are quite low. In many cases, a rare innovation of a self-winding watch can still be purchased for less than several hundred dollars.

The values of these wristwatches are based on a number of factors, including the brand name, the watch style, its usefulness as a timepiece as well as the materials from which the case is made. Other factors regarding the condition of the watch movement and case will also greatly affect the evaluation of the watch.

The values of these wristwatches are based upon their fair market value–a price that a willing buyer and seller can agree on–and assuming that the piece is in reasonably good mechanical condition. Often complete service to a complicated, self-winding wristwatch can be costly compared to its original value, and we have not included that as part of the evaluation of these watches.

The evaluation of the wrist watch is preceded by the page number in the book.

Cover	$300-$350
13	$400-$500
14	$700-$900
15	$250-$350
16	$300-$400
17	$300-$400
18	$350-$450
19	$300-$400
20	$300-$400
21	$350-$450
24	$350-$450
25	$350-$450
28	$200-$300
28	$200-$300
30	$200-$300
31	$700-$850
32	$350-$450
33	$200-$300
34	$500-$600
35	$350-$450
37	$200-$300
38	$350-$450
39	$300-$400
40	$350-$450
47	$1,200-$1,500
48	$1,000-$1,200
49	$1,000-$1,200
54	$150-$250
55	$150-$250
58	$100-$150
59	$150-$175
60	$150-$250
61	$200-$300
64	$200-$300
65	$225-$300
66	$250-$350
67	$400-$500
68	$300-$400
69	$150-$250
70	$800-$900
74	$300-$400
75	$150-$500
76	$200-$300
77	$350-$450
78	$200-$300
79	$175-$225
80	$125-$150
81	$300-$400
85	$250-$300
87	$300-$400
88	$200-$300
90	$200-$300
92	$100-$150
94	$150-$200
95	$150-$200
96	$100-$125
97	$100-$125
98	$175-$225
99	$350-$450
100	$100-$150
106	$200-$300
107	$200-$300
108	$400-$500
109	$150-$200
110	$175-$200
111	$1,000-$1,200
112	$150-$250
113	$200-$250
114	$300-$400
115	$350-$450
116	$225-$250
117	$500-$600
118	$175-$250
119	$450-$550
120	$150-$200
121	$150-$250
127	$250-$350
128	$300-$400
131	$150-$200
132	$150-$200
133	$200-$300
135	$150-$250
136	$250-$400
139	$250-$350
140	$200-$250
142	$150-$250
143	$500-$600
144	$350-$450
145	$150-$200
146	$150-$250
152	$400-$500
153	$150-$250
154	$1,000-$1,200
155	$1,200-$1,500
156	$400-$500
157	$300-$450
158	$1,000-$1,200
161	$350-$600
162	$400-$500
163	$700-$800
165	$200-$300
166	$250-$350
167	$450-$550
168	$300-$400
170	$350-$450
174	$300-$400
176	$150-$250
177	$175-$225
178	$175-$225
179	$150-$300
180	$1,200-$1,500
182	$500-$600
183	$250-$350
184	$600-$700
185	$600-$750
186	$400-$500
187	$400-$500
189	$250-$350
190	$600-$750

191	$200-$250
192	$200-$300
193	$350-$450
194	$175-$225
195	$175-$225
196	$400-$500
197	$250-$350
198	$250-$350
199	$400-$500
200	$275-$375
203	$200-$450
204	$200-$450
205	$350-$450
207	$200-$300
208	$325-$450
209	$200-$300
210	$200-$300
211	$300-$450
212	$350-$450
213	$250-$300
216	$300-$400
217	$325-$400
218	$600-$700
219	$300-$400
220	$150-$250
221	$200-$300
225	$350-$450
226	$400-$500
227	$600-$700
228	$400-$500
229	$650-$750
230	$500-$600
231	$300-$400
233	$600-$750
234	$375-$425
235	$200-$300
236	$300-$400
237	$300-$350
238	$300-$400
239	$650-$750
244	$200-$300
245	$200-$300
247	$4,000-$5,000
248	$4,000-$5,000
249	$5,000-$6,000
250	$6,500-$7,500
251	$6,000-$7,000
252	High Value
255	$150-$250
256	$275-$325
258	$2,500-$3,000
260	$250-$550
261	$450-$550
264	$300-$450
265	$300-$350
266	$200-$350
267	$300-$350
268	$350-$450
269	$300-$350
270	$200-$250
273	$175-$225
274	$200-$250
275	$250-$350
276	$200-$250
277	$300-$400
278	$200-$250
279	$175-$200
281	$350-$450
284	$450-$500
285	$350-$400
286	$900-$1,000
287	$350-$450
288	$500-$550
289	$400-$500
290	$500-$600
291	$175-$225
294	$175-$225
295	$300-$400
296	$400-$500
297	$175-$225
298	$200-$250
299	$200-$250
302	$200-$250
303	$750-$1,200
304	$300-$400
305	$300-$350
306	$300-$400
307	$700-$800
308	$250-$300
311	$750-$850
312	$750-$850
313	$600-$700
314	$750-$850
315	$400-$500
318	$200-$250
319	$450-$550
320	$300-$350
321	$300-$350
322	$500-$600
323	$1,500-$1,800
325	$250-$450
326	$250-$300
327	$275-$325
328	$175-$225

Gordon S. Converse is an antiquarian horologist who owns and operates his business near Philadelphia, Pennsylvania where they buy, sell and restore antique clocks and watches. His two shops in Strafford and Berwyn include a wide selection of unusual antique clocks and some watches. He publishes a catalog about once a year which is available by subscription by writing Gordon S. Converse & Co., 1029 Lancaster Ave, Berwyn, PA. 19312, or calling (610)-296-4932.